Springer Series in Biophysics 17

Series editor: Boris Martinac

For further volumes:
http://www.springer.com/series/835

Reinhard Krämer • Christine Ziegler
Editors

Membrane Transport Mechanism

3D Structure and Beyond

Springer

Editors
Reinhard Krämer
Department Chemistry
Institute of Biochemistry
University of Cologne
Köln
Germany

Christine Ziegler
Faculty for Biology and Preclinical Medicine
University of Regensburg
Regensburg
Germany

Series editor:
Boris Martinac
University of New South Wales
Victor Chang Cardiac Research Inst.
Darlinghurst
New South Wales
Australia

ISSN 0932-2353 ISSN 1868-2561 (electronic)
ISBN 978-3-662-52305-6 ISBN 978-3-642-53839-1 (eBook)
DOI 10.1007/978-3-642-53839-1
Springer Heidelberg New York Dordrecht London

Printed on acid-free paper

Springer is part of Springer Science+Business Media (www.springer.com)

This book is dedicated to Amy Davidson for her great devotion and contribution to the field of membrane transport; her competence and kindness will always be remembered.

Preface

The alternating-access mechanism is a fundamental and intuitively convincing concept for the translocation of a substrate across a membrane catalyzed by a membrane-embedded transporter. Although proposed already a long time ago (ref), the mechanistic understanding of alternating access on a molecular level is still one of the great challenges in biology, and the prerequisite to unraveling the general mechanism of energized membrane transport. Moreover, a number of transport proteins are not only catalytically active but also show distinct regulatory features, adding a further level of complexity. To grasp the basic principles of active transport, we require the molecular details of the sequence of conformational changes that are involved in the coordinated action of these fascinating molecular machines.

Due to the impressive progress in crystallization of membrane proteins, which only 10 years ago was beyond imagination, our knowledge of transporter folds has improved substantially. Crystallography has provided an increasing number of atomic structures for a large variety of active membrane transporters. For secondary transporters in particular, it turned out that individual protein folds that diverge considerably one from another nevertheless share a striking feature: the appearance of inverted topology repeats, which can be considered as the building blocks of a given transporter fold. The conservation of inverted repeats, or more precisely the conserved way in which they interact with their symmetry mates during the transition from outward to inward facing states, elegantly reflects the anticipated universality of the alternating access mechanism, as well as the symmetric nature of the process itself. This presence of repeated structural motifs has therefore emerged as the common element in membrane transport; a universal feature independent of sequence, fold, and coupling mechanism, opening the door to a unifying transport mechanism.

The central question of the transition between different states of transport proteins in the course of their action has in the past been addressed by myriad kinetic, biochemical, biophysical experiments, and more recently also by a rapidly growing collection of structural data on different transporters in different conformations. However, each new structure, while providing valuable information on overall fold, substrate coordination and other important aspects, is only a starting point towards a complete understanding of the transport mechanism. For example, it remains difficult to unambiguously link an experimentally determined structural state to a particular functional state during the alternating-access cycle of transport. Therefore, only the combination of structural data with kinetic, biochemical, biophysical, and computational studies will allow a conclusive picture to be drawn. More precisely, especially considering to the dynamic nature of transport, such a combination of efforts is essential to realize the vision of the molecular movie of the transport cycle.

With this edition we aim to compile particularly significant work on transporters that has contributed to the fundamental understanding of the alternating access mechanism. We cover four of the major secondary transporter families: the Glutamate transporter family (Chap. 1), the LeuT-like fold family (Chaps. 2, 3, 4, and 5), the Major Facilitator family (Chaps. 6 and 7), and the SMR (RND) family (Chaps. 9, 10, and 11), as well as the maltose transporter (Chap. 8), an important member of the ATP-driven family of ABC transporters. The selected examples impressively demonstrate how the combination of functional analysis, crystallography, and investigation of dynamics contributes to the overall mechanistic picture of solute transport.

To understand the inherent flexibility of carrier proteins requires techniques that access their conformational dynamics. Therefore, biochemical and biophysical techniques such as NMR (Chap. 11), EPR, and single-molecule FRET (Chap. 1) have been providing important new insights. Furthermore, bioinformatics approaches, e.g., MD simulations (Chaps. 2 and 5), have been extremely helpful as they access the dynamics of a given conformational state, and therefore its free energy, and under some conditions also give insights into transitions between states.

We remain far from a complete understanding of the mechanism of transport proteins. Most of our knowledge has been gleaned from prokaryotic proteins, which represent only a fraction of the large group of transporters. More structures of eukaryotic transporters, and of the same transporter in different conformational states, are required to progress along our exciting journey from static snapshots to the complete molecular mechanism of active membrane transport.

Köln, Germany — Reinhard Krämer
Regensburg, Germany — Christine Ziegler

Acknowledgement

We are indebted to Boris Martinac for his encouragement in starting this project. We also thank Lucy Forrest for editing the preface and cover text.

Contents

Chapter 1
Dance Lessons for Proteins: The Dynamics and Thermodynamics of a Sodium/Aspartate Symporter

Olga Boudker and Nurunisa Akyuz

Abstract Secondary active transporters harvest the energy of the ionic gradients to drive concentrative uptake of their substrates. This process entails a series of protein conformational transitions that are coupled to binding and unbinding of ions and substrates on the extracellular and intracellular sides of the membrane. Over the last decade, crystallography has provided a growing number of structural snapshots of the transport cycle for several ion-driven transporters. Already these structures, although intrinsically static, have revealed a remarkable plasticity encoded in the architecture of these proteins. Because internal dynamics is an essential feature of transporters, it is necessary to complement crystallographic studies with other techniques that provide information on the ensemble properties of these proteins as well as on the conformational dynamics of individual molecules. Here, we will discuss the emerging approaches to obtain thermodynamic and dynamic information on transporters using a sodium/aspartate symporter from *Pyrococcus horikoshii*, Glt_{Ph}, as a model system. Glt_{Ph} is a bacterial homologue of the mammalian glutamate transporters, for which crystal structures of several states have been determined, providing a framework for further mechanistic studies. We will discuss how within this system the equilibrium and kinetic studies based on the isothermal titration calorimetry, fluorescence, and electron paramagnetic resonance spectroscopy inform us on the energetic relationship between the key functional states, and mechanisms of coupling between transport cycle and ionic gradients. We will further describe how single molecule studies open doors to a detailed characterization of the timing and order of the conformational transitions underlying transport processes.

Keywords Glutamate transporters • Sodium symporters • Functional dynamics • Anion permeation • Substrate binding • Alternating access • Secondary transport • Conformational changes

O. Boudker (✉) • N. Akyuz
Weill Cornell Medical College, New York, NY, USA
e-mail: olb2003@med.cornell.edu

R. Krämer and C. Ziegler (eds.), *Membrane Transport Mechanism*, Springer Series in Biophysics 17, DOI 10.1007/978-3-642-53839-1_1,

Secondary active transporters harvest the energy of the ionic gradients to drive concentrative uptake of their substrates. This process entails a series of protein conformational transitions that are coupled to binding and unbinding of ions and substrates on the extracellular and intracellular sides of the membrane. Over the last decade, crystallography has provided a growing number of structural snapshots of the transport cycle for several ion-driven transporters. Already these structures, although intrinsically static, have revealed a remarkable plasticity encoded in the architecture of these proteins. Because internal dynamics is an essential feature of transporters, it is necessary to complement crystallographic studies with other techniques that provide information on the ensemble properties of these proteins as well as on the conformational dynamics of individual molecules. Here, we will discuss the emerging approaches to obtain thermodynamic and dynamic information on transporters using a sodium/aspartate symporter from *Pyrococcus horikoshii*, Glt_{Ph}, as a model system. Glt_{Ph} is a bacterial homologue of the mammalian glutamate transporters, for which crystal structures of several states have been determined, providing a framework for further mechanistic studies. We will discuss how within this system the equilibrium and kinetic studies based on the isothermal titration calorimetry, fluorescence, and electron paramagnetic resonance spectroscopy inform us on the energetic relationship between the key functional states, and mechanisms of coupling between transport cycle and ionic gradients. We will further describe how single molecule studies open doors to a detailed characterization of the timing and order of the conformational transitions underlying transport processes.

1.1 Introduction

Glutamate transporters, also termed excitatory amino acid transporters (EAATs), are secondary active transporters that reside at glutamatergic synapses and pump the neurotransmitter from the synaptic cleft into the cytoplasm of glial cells and neurons (Danbolt 2001). They maintain steep trans-membrane concentration gradients of glutamate, which may reach a million fold, by coupling the uptake of each glutamate to symport of three sodium (Na^+) ions and a proton and to countertransport of one potassium ion (K^+) (Kanner and Sharon 1978; Nelson et al. 1983; Barbour et al. 1988; Zerangue and Kavanaugh 1996). They ensure properly functioning excitatory synapses by maintaining low transmitter background, allowing for rapid rounds of signaling and preventing excitotoxicity (Rothstein et al. 1996), which arises from the sustained excitation of neurons and increased calcium influx followed by apoptosis (Choi 1994). In addition, they mediate bidirectional chloride fluxes that are triggered by glutamate and sodium binding but are not stoichiometrically coupled to substrate uptake and do not require completion of the transport cycle (Fairman et al. 1995; Wadiche et al. 1995a). These fluxes, which usually entail influx of chloride into glial and neuronal cells, are thought to offset membrane depolarization during rapid electrogenic Na^+/glutamate uptake (Eliasof and

Jahr 1996). In some cases, these fluxes modulate neuronal excitability and may represent the dominant role of EAATs, particularly in retina (Arriza et al. 1997; Veruki et al. 2006).

The complete transport cycle of EAATs involves binding of the substrate and ions from the extracellular solution, their translocation across the membrane, and release into the cytoplasm, followed by K^+ binding and return into the initial state (Barbour et al. 1988; Bouvier et al. 1992; Kavanaugh 1998). Therefore, to define a molecular mechanism of transport, it is essential to provide structural description of the conformational transitions that underlie substrate binding and release from the extracellular and intracellular solutions, as well as the translocation of the binding sites across the membrane (Krishnamurthy et al. 2009). Furthermore, it is important to describe the structural and energetic basis of coupling between the co- and counter-transported ions and the substrate and to identify which state or states in the transport cycle permit permeation of anions. Finally, in order to gain a dynamic view of the cycle, the sequence and the timing of the transitions between the conformational states need to be uncovered. In particular, the rate-limiting steps need to be identified, as their modulation, either by the environmental factors, such as membrane composition, or by pharmacological tools would have the greatest effect on the overall transport rate. The effective glutamate clearance by EAATs improves recovery after stroke and traumatic brain injury (Yi and Hazell 2006) and attenuates neuronal damage during neurodegenerative processes such as amyotrophic lateral sclerosis (Lau and Tymianski 2010). However, pharmacological tools that would stimulate EAATs are scarce and are limited to the upregulation of EAAT's expression levels (Rothstein et al. 2005; Lee et al. 2008). Detailed understanding of the dynamic properties of the transport cycle may reveal new strategies to directly increase the transport rates.

To date, the only member of the glutamate transporter family with known high-resolution crystal structures is an archaeal homologue from *Pyrococcus horikoshii*, Glt_{Ph} (Yernool et al. 2004). Glt_{Ph} recapitulates several functional properties of EAATs: it symports an acidic amino acid aspartate (Asp) together with 3 Na^+ ions (Boudker et al. 2007; Groeneveld and Slotboom 2010) and mediates anion fluxes (Ryan and Mindell 2007). However, it has a simpler transport stoichiometry and does not couple substrate uptake to movements of protons or K^+ (Ryan et al. 2009). Furthermore, it is orders of magnitude slower than EAATs with the turnover times in minutes (Ryan et al. 2009) rather than milliseconds at ambient temperatures. Amenability to crystallographic experiments, simpler functional properties, and slow overall kinetics make Glt_{Ph} a useful model within which to probe the structural and dynamic underpinnings of this transporter family.

Conformational transitions between the outward- and inward-facing states, in which the substrate-binding site is accessible from the extracellular solution and the cytoplasm, respectively, are the linchpin of the transporter mechanism (Jardetzky 1966). These global structural rearrangements are thought to involve alternating opening of the extracellular and intracellular "thick" gates, which represent the main barrier for the substrate permeation across the membrane (Fig. 1.1) (Krishnamurthy et al. 2009). In addition, in each the outward- and inward-facing

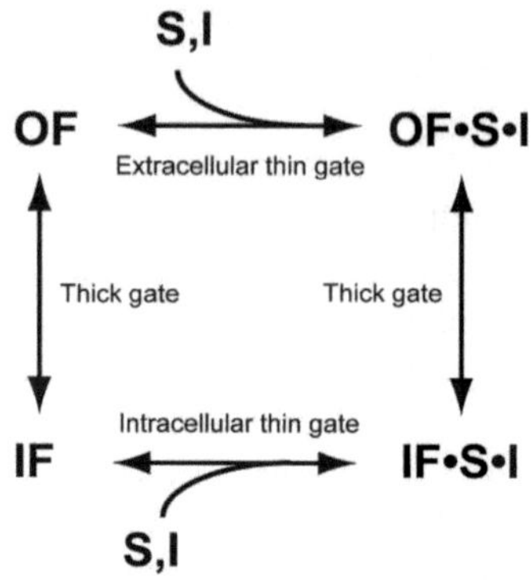

Fig. 1.1 *The schematic representation of the transport cycle.* A transporter in the outward-facing state (OF) binds substrate (S) and coupled ions (I). These events are associated with rearrangement of the extracellular thin gate. The fully bound transporter undergoes conformational transitions into the inward-facing state (IF), a process that is conceptualized as restructuring of the thick gate. In the IF, the substrate and ions equilibrate with the intracellular solution in a process coupled to the rearrangement of the intracellular thin gate. The unbound transporter completes the cycle by returning from the IF to the OF. Notably, every step of the cycle is reversible, and the overall direction is determined by the substrate and ion concentrations on the extracellular and intracellular sides of the membrane

states, the substrate and ions may remain occluded under the "thin gates," relatively dynamic structural elements that undergo opening and closing transition allowing for the equilibration of the solutes between the corresponding aqueous phase and their binding sites. Early on, it was thought that transporters were unlikely to yield well-diffracting crystals because of their dynamic nature, whereby they constantly sample outward- and inward-facing states. However, the growing number of crystal structures suggests that the crystallization conditions and crystal contacts within the crystals favor unique conformations. Remarkably, crystal structures of secondary transporters (see Chaps. 3 and 4) and ABC transporters (Chap. 8) have revealed a great variety of structural states, which include not only the outward- and inward-facing states, but also intermediates between them (Yin et al. 2006; Ressl et al. 2009; Oldham and Chen 2011; Perez et al. 2012; Verdon and Boudker 2012). Hence, the challenge is to identify conditions that favor distinct conformational states. Remarkably, in some cases the same crystallization conditions have yielded crystals of either the outward- or inward-facing states of transporters (Weyand et al. 2008; Shimamura et al. 2010). In others, the information about the range of conformational transitions has been deduced from the structures of distinct proteins sharing the same structural fold (Yamashita et al. 2005; Faham et al. 2008). In the case of Glt_{Ph}, an approach was developed to conformationally constrain the transporter using cross-linking of strategically placed cysteine residues (Reyes et al. 2009).

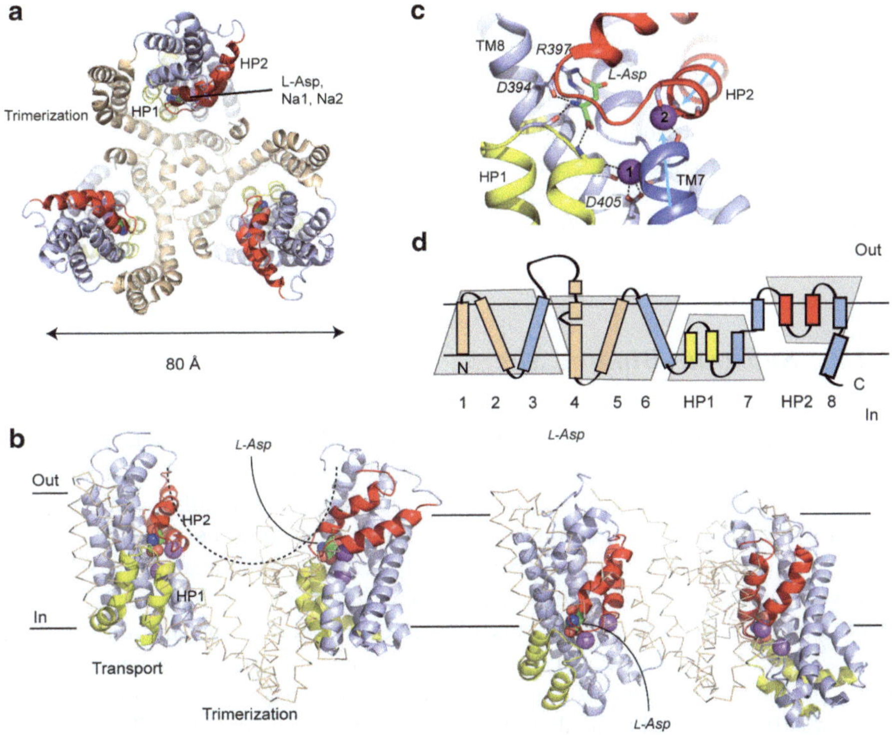

Fig. 1.2 *The overall structure of* Glt_{Ph}. (**a**) Glt_{Ph} trimer viewed from the extracellular space. The transport and trimerization domains are colored *blue* and *wheat*, respectively. HP1 is shown in *yellow* and HP2 is shown in *red*. The substrate and ions are highlighted as *spheres* and colored by *atom* type. (**b**) Two protomers of Glt_{Ph} are viewed from within the membrane plain, as captured in the outward-facing wild-type (*left*) and inward-facing Glt_{Ph}-K55C–A364C_{Hg} crystal structures (*right*). (**c**) Close-up view and the key residues in the substrate- and ion-binding sites. The helix dipoles that may contribute to Na^+ coordination are shown as *blue arrows*. (**d**) Topology cartoon of Glt_{Ph} colored as in (**a**). Highlighted in *gray*, are the regions of two inverted structural repeats

1.2 Outward-Facing State

1.2.1 Overall Structure

Glt_{Ph} was identified as a suitable crystallographic target in a process of screening of bacterial homologues of EAATs (Yernool et al. 2004). The crystal structure revealed a trimeric assembly (Fig. 1.2a), common to all characterized members of the family (Yernool et al. 2003; Gendreau et al. 2004), with a large extracellular bowl reaching approximately half way across the membrane (Fig. 1.2b). The structure of each protomer can be divided into a centrally located trimerization domain, mediating inter-subunit interactions in the trimer, and a peripheral so-called transport domain (Reyes et al. 2009). The central bowl is lined with polar residues and provides aqueous access to the substrate-binding sites located at its bottom within each individual transport domain. The existence of three sets of substrate- and ion-binding binding sites distant from each other is consistent with each protomer providing independent permeation pathway, as has been postulated based on functional studies (Grewer et al. 2005; Koch and Larsson 2005; Groeneveld and Slotboom 2007).

1.2.2 Substrate-Binding Site

The modest resolution of Glt_{Ph} crystals (~3.5 Å) precluded unambiguous placement of the substrate in its binding site and the identification of Na^+-binding sites. To circumvent this difficulty, the ligands were replaced in crystallographic experiments with analogues containing heavy atoms, which could be detected directly using anomalous scattering data (Boudker et al. 2007). Specifically, the substrate Asp was replaced with L-cysteine sulfinic acid and Na^+ ions were replaced with thallium (I) (Tl^+). These experiments allowed localization of the substrate-binding site, where a highly conserved arginine in the trans-membrane segment (TM) 8, R397, formed an ionic bond with the side chain carboxylate of Asp (Fig. 1.2c). Consistently, the residue analogous to R397 in mammalian EAATs (R446 in EAAT3) is known to be responsible for their selectivity for the acidic amino acids (Bendahan et al. 2000). Further, the substrate coordinating D394 residue (Fig. 1.2c) corresponds to D443 in EAAC1. Modification of this residue was also shown to impair the uptake (Teichman and Kanner 2007).

Crystallographic studies further revealed that the substrate-binding site was largely defined by two reentrant helical hairpin motifs (HP1 and 2) related by a pseudo-twofold structural symmetry (Fig. 1.2c). Only the tip of HP2 occludes the substrate-binding site from the extracellular solution, defining it as a likely extracellular "thin" gate. In contrast, the site is distant from the cytoplasmic solution (Fig. 1.2b). Hence, this state was identified as outward facing. The hypothesis that HP2 may serve as an extracellular gate was further supported by the structure of the transporter bound to a blocker L-threo-β-benzyloxyaspartate (TBOA), in which the benzyl group of the inhibitor props HP2 in an open conformation (Boudker et al. 2007). Molecular dynamic studies have also revealed an intrinsic flexibility of HP2 that allows it to sample an open conformation providing substrate access to the binding sites from the extracellular solution (Shrivastava et al. 2008). The structural plasticity of the HP2 tip is attributed to the presence of four conserved glycines (Huang and Tajkhorshid 2008).

1.2.3 Sodium-Binding Sites

Two Tl^+-binding sites were identified crystallographically. Only Na^+ ions could outcompete Tl^+ in these sites, suggesting that they defined *bona fide* Na^+-binding sites (termed Na1 and Na2) (Fig. 1.2d). Consequent computational and functional studies supported these crystallographic results and suggested that these Na^+-binding sites were conserved between Glt_{Ph} and EAATs (Shrivastava et al. 2008; Tao et al. 2008; DeChancie et al. 2010; Huang and Tajkhorshid 2010). The location of the expected third binding site is still debated (Holley and Kavanaugh 2009; Huang and Tajkhorshid 2010; Larsson et al. 2010; Tao et al. 2010; Rosental et al. 2011; Bastug et al. 2012; Teichman et al. 2012).

Of the two crystallographically identified sites, neither is coordinated by the substrate Asp. Na1 is located deep at the core of the transport domain, coordinated by D405 in TM8 and main chain carbonyl oxygen of a non-helical highly conserved motif in the middle of TM7 (Fig. 1.2d). Na2 site is located between the carboxyl-termini of two helices: the amino-terminal half of TM7, and the amino-terminal arm of HP2. It is coordinated exclusively by the main chain carbonyl oxygen decorating the ends of the helices, with helical dipole moments pointing toward the site.

1.3 Inward-Facing State

1.3.1 Constraining Glt_{Ph} in the Inward-Facing State

While the structure of the outward-facing state of Glt_{Ph} explained how the substrate and ions accessed their binding sites from the extracellular solution, the mechanism of the release into the cytoplasm remained unknown. Some clues were provided by the earlier biochemical studies. For example, it was shown that cross-linking protomers had no effect on the transport activity, suggesting that the trimerization domain remained rigid during transport process (Groeneveld and Slotboom 2007). On the other hand, regions in the transport domain, which are buried in the outward-facing structure, were shown to be accessible to the intracellular solvent (Grunewald et al. 1998; Seal et al. 1998; Slotboom et al. 1999; Shlaifer and Kanner 2007), suggesting that the transport domain restructured during transport cycle. Interestingly, one study in particular was inconsistent with the structure of the outward-facing Glt_{Ph}: Vandenberg and colleagues showed that an EAAT double cysteine mutant, in which one cysteine was in the trimerization domain (residue 55 in Glt_{Ph}) and the other was in the transport domain HP2 region (residue 364), formed a spontaneous disulfide bond even though these residues were over 25 Å apart (Ryan et al. 2004). These results suggested that at some point during the transport cycle the protein underwent large-scale conformation changes, most likely associated with the transition from the outward- to the inward-facing state. When reproduced in Glt_{Ph}, the two cysteines could be cross-linked under oxidizing conditions, suggesting that such cross-linking can be used to trap the protein in a conformation profoundly different from that pictured by the available crystal structures (Reyes et al. 2009).

However, quantitative cross-linking could not be achieved, probably because of the competing oxidation of the cysteine thiols to sulfo-acids. To circumvent this problem, we employed mercury (II) (Hg^{2+}), which forms a highly polarized ionic bond with thiols and can be used to bridge two suitably positioned thiol moieties. These bonds are not covalent and can be reversed by reducing or metal-chelating agents. To test whether mercury forms a bridge or simply decorates individual cysteines, it is necessary to determine the stoichiometry of Hg^{2+} binding to the double cysteine mutant, by, for example, isothermal titration calorimetry (ITC). Binding of only one Hg^{2+} ion per protomer indicates the formation of the bridge (Verdon and Boudker 2012; Reyes et al. 2013).

1.3.2 Structure of the Inward-Facing State and the Mechanism of Transport

The 55C/364C double cysteine mutant cross-linked with Hg^{2+} yielded diffracting crystals, phases for which were determined by molecular replacement using the trimerization domain (Reyes et al. 2009). The transport domains were then built into the excess density manually. The structure of the cross-linked mutant revealed that the transport domains were also structurally unchanged, but rather moved as rigid bodies approximately 15 Å across the lipid bilayer. As a result of this movement, the substrate and ion binding were translocated from being near the extracellular solution to being near the cytoplasm (Fig. 1.2f), suggesting that the captured structure was, indeed, inward facing. Comparison of the interactions between the trimerization domain and the transport domain showed that the same region of the trimerization domain was involved in both the outward- and inward-facing states. In contrast, essentially non-overlapping regions of the transport domain participated in the packing of the interface in the two states. In the outward-facing state, HP1 comprised the largest part of the interface, while HP2 lined the surface of the extracellular bowl. In the inward-facing state, HP2 provided most of the interfacial interactions, while HP1 was now lining the cytoplasm-exposed surface of the transport domain. Based on these structures, a model of substrate transport was proposed, which is sometimes termed "elevator model" (Fig. 1.3). According to the model, HP2 serves as an extracellular gate of the transporter, permitting ions and substrate access to their binding sites. Upon inward movement of the transport domain, HP2 becomes locked on the domain interface while HP1 is released from the interface and is poised to serve as an intracellular gate (Figs. 1.2f and 1.3). This mechanism explains how the alternating access to the binding sites is achieved and why the extracellular and intracellular gates cannot open at the same time, a condition that would turn the transporter into channel and short-circuit the ionic gradient. However, additional structures would be required to establish unambiguously the gating roles of the two helical hairpins.

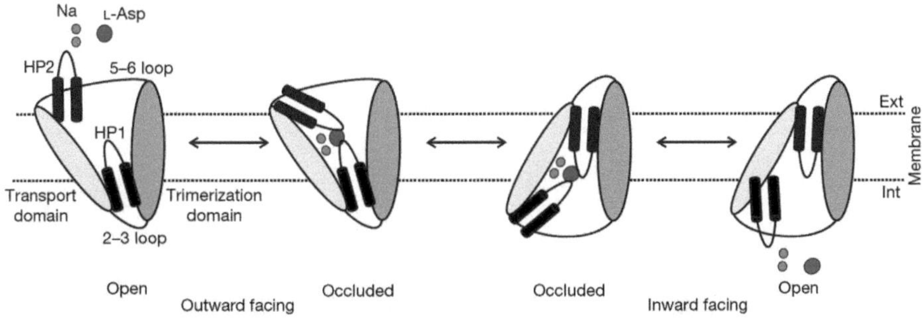

Fig. 1.3 *"Elevator" mechanism*. Shown is a schematic representation of a single Glt_{Ph} protomer with the trimerization and transport domains colored dark and light *gray*, respectively. HP1 and HP2 (*black*) are proposed to serve as the intracellular and the extracellular gates. During concentrative uptake, HP2 serves as a gate to allow binding of the substrate and ions to the outward-facing apo transporter. Trans-membrane movements of the transport domain underlie translocation of the substrate and ions. It has been proposed that the release of the substrate and ions into the cytoplasm is mediated by the structurally symmetric HP1. The cycle is completed when the apo transport domain transitions back into the outward-facing orientation (adapted from Reyes et al. 2009)

1.3.3 Inverted Structural Repeats

Many secondary transporters unrelated in sequence and in structure contain inverted structural repeats, whereby two regions of the transporter are related to each other via a pseudo-twofold structural symmetry, and are inserted into membrane in opposite orientations (Forrest and Rudnick 2009; Boudker and Verdon 2010). Glt_{Ph} contains two such repeats: one comprising TM1–TM3 and TM4–TM6 and the other comprising HP1–TM7 and HP2–TM8 (Fig. 1.2e). Of these, TM1–TM2 and TM4–TM5 make up the rigid scaffold of the trimerization domain, while TM3, TM6, and the second set of repeats form the dynamic transport domain. Remarkably, similarly to other transporters, the transitions between the outward- and inward-facing states involve the conformational switch between the dynamic portions of the repeats in relation to the scaffold. Thus, TM3, HP1, and TM7 exchange their conformation with TM6, HP2, and TM8 (Crisman et al. 2009).

1.4 Transport Cycle Intermediates and Potential Mechanism of Channel Formation

One of the intriguing questions about the mechanism of EAATs and Glt_{Ph} is how these molecules are able to serve as both ion-coupled transporters and substrate/sodium-gated anion channels (Fairman et al. 1995; Picaud et al. 1995; Wadiche et al. 1995b; Eliasof and Jahr 1996). Neither the structure of the outward- nor that of the inward-facing states showed a potential anion permeation pathway. The early hypothesis that the channel may be located at the trimer symmetry axis was also

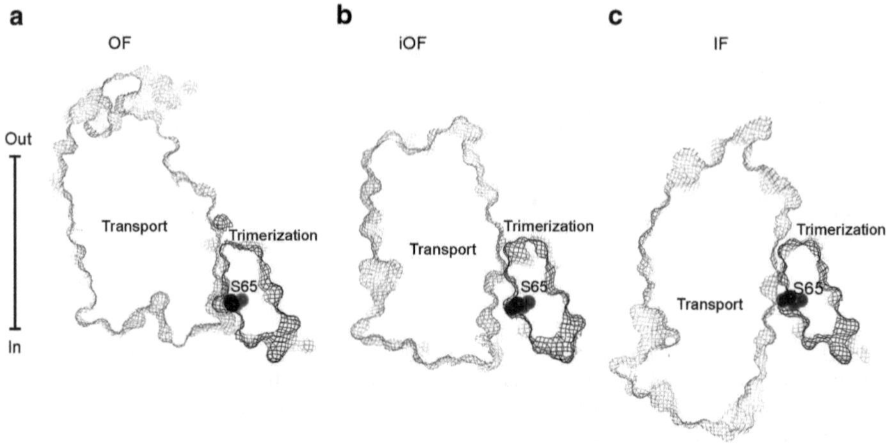

Fig. 1.4 *Transitions through intermediate states*. Shown are the thin slices through single protomers in the outward- (OF, **a**), inward-facing (IF, **b**) structures, and an intermediate close to the outward-facing sate (iOF, **c**). The transport and trimerization domains are contoured in light and dark *gray*, respectively. The *bar* represents an approximate membrane boundary. While the interfaces between the domains are well packed in the outward- and inward-facing states, the interface shows the formation of a large cavity in the intermediate state. The cavity is formed near residue S65 (colored *black*), which has been implicated in anion permeation. The disruption of the interface packing suggests that the translocation intermediates may provide permeation pathways for water, anions, and other small solutes

discarded because (1) the functional experiments showed that anion conductance occurs within the individual protomers independently (Leary et al. 2007); (2) the trimer interface is very hydrophobic; and (3) residues important for anion permeation map to both the trimerization and transport domains (Ryan and Vandenberg 2002; Ryan et al. 2004).

Although this question still awaits resolution, one possible clue comes from a structure of Glt_{Ph} in an intermediate state between the outward and inward facing (Verdon and Boudker 2012). This structure was a result of a study, in which a range of motions of the transport domain was investigated. A number of double cysteine mutants were generated, treated with Hg^{2+} and crystallized. One such mutant pictured a trimer in an asymmetric state with two protomers in the inward-facing state and one protomer in an intermediate state, close to the outward facing. Although no cross-links were observed in this mutant, with individual thiols decorated with Hg^{2+} ions, the structures of the inward-facing protomers were indistinguishable from those of the cross-linked 55C/364C mutant. The structure of the third protomer pictured the transport domain, which transitioned approximately one-quarter to one-third of the way from the outward- to the inward-facing orientation. A remarkable, albeit not unexpected, consequence of such partial transition is that the packing between the trimerization and transport domains is far less optimal than in either the outward- or inward-facing states (Fig. 1.4). In particular, a potentially hydrated cavity is formed at the thinnest region of the interface adjacent to a conserved serine S65. This is of significance because this residue has been implicated in anion conductance and selectivity, suggesting that it directly lines the anion permeation pathway (Ryan et al. 2004).

This structure led us to hypothesize that such transient intermediates during outward to inward transition of the transport domain may have functionalities of their own, serving as anion channels. This hypothesis is further supported by the mutational data, which suggest that modifications of cysteine residues placed in HP2 with polar thiol-specific reagents selectively abrogate substrate transport and not anion conductance in EAATs: (Seal et al. 2001; Borre et al. 2002; Ryan and Vandenberg 2002). Remarkably, these residues are exposed to the aqueous extracellular solution in the outward-facing state and in at least a subset of the potential transition intermediates, but not in the fully inward-facing protomers, where they become buried on the interface with the trimerization domain. The idea that the transition intermediates yield hydrated domain interfaces was further supported by consequent computational studies (Stolzenberg et al. 2012). Interestingly, ion and water fluxes have been observed not only in EAATs but also in a number of other diverse transporters (Fischbarg 1988; Zhang and Grewer 2007; Vandenberg et al. 2011), and computational studies suggest that water-conducting transition states might be a common feature (Li et al. 2013).

1.5 The Dynamic Equilibrium Between the Outward- and Inward-Facing States

1.5.1 Experimental Approaches

The transport cycle (Fig. 1.1) is sometimes described in deterministic terms, whereby the transporter progresses from the outward-facing apo state to the substrate- and ion-bound state, then to the inward-facing bound and apo states, and finally back to the outward-facing apo state. However, it is important to realize that each of the transport cycle steps is reversible and that only on average does the system move in the direction of the productive uptake in the presence of favorable ionic gradients. The relative energies of the key functional states and the height of the kinetic barriers between them determine which are the rate-limiting steps of the cycle. For example, if the equilibrium between the apo states is shifted toward the inward-facing orientation, then the outward-facing apo state becomes a transient high-energy intermediate of the cycle, and attaining this state in order to progress to the outward-facing substrate- and ion-bound state may become a rate-limiting step. It has been argued, indeed, that matching energies of the outward- and inward-facing states would assure the highest possible turnover rate of the cycle (Andersen and Koeppe 2007). To probe the relative energies of the key transport cycle states and the energy barriers between them, approaches have been developed to (1) probe the overall distribution of the outward- and inward-facing states; (2) establish how they are modulated by the environment; and (3) study the dynamics of the transitions between them.

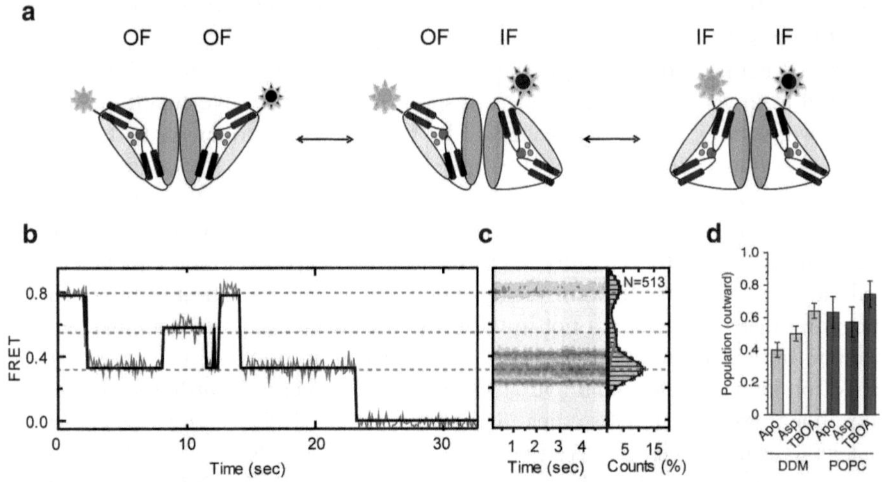

Fig. 1.5 *Conformational ensemble of* Glt_{Ph} *trimer.* (**a**) Labeling strategy for the PDS and sm-FRET experiments. Two protomers are labeled either with the spin label or with donor and acceptor fluorophores using a single cysteine mutant within the transport domain. In the case of sm-FRET experiments, the third cysteine in the trimer is used to attach the protein to the surface of the microscope slide. The distance between the labels depends on the relative orientation of the protomers. (**b**) sm-FRET recordings, reflecting movements of the transport domains within the individual trimers, revealed three FRET efficiency states corresponding to distinct protomer arrangements. Transitions between all states were observed, suggesting that the transport domains move independently of each other. (**c**) A contour plot of the state populations from hundreds of individual molecules as a function of time, as well as a 1D cumulative histogram (panels (**b**) and (**c**) adapted from Akyuz et al. 2013). (**d**) The *bar* graph of the populations of the outward-facing conformation in the apo, Asp-, and TBOA-bound Glt_{Ph} in detergent micelles (labeled as DDM) and in lipid bilayers (labeled as POPC), as estimated from PDS experiments (panel adapted from Georgieva et al. 2013)

Two techniques have been employed to take advantage of the fact that the inter-protomer distance between equivalent residues in the transport domains of Glt_{Ph} changes upon transitions between the outward- and inward-facing states (Fig. 1.5). These “spectroscopic ruler” techniques are pulsed dipolar electron spin resonance (PDS) spectroscopy (Schiemann et al. 2003) and single molecule fluorescence resonance energy transfer (sm-FRET) imaging (Weiss 1999). PDS measures inter-spin coupling between paramagnetic probes, which is inversely proportional to the cube of distance between them. PDS experiments are performed on frozen samples and provide information on the overall distance distribution in the molecular ensemble. FRET efficiency is inversely proportional to the sixth power of the distance between the donor and acceptor fluorophores, but may also be affected by environmental factors. In the context of the single molecule imaging using total internal reflection fluorescent (TIRF) microscope, this technique can inform on the dynamic changes of the distance between the fluorophores in real time at physiologically relevant temperatures. Both approaches rely on the site-directed labeling of the protein using cysteine mutants and are most sensitive in a similar range of distances, approximately between 20 and 60 Å. Hence, the two techniques complement each other and have been applied to Glt_{Ph} (Akyuz et al. 2013; Georgieva et al. 2013; Hanelt et al. 2013) and to other transporters (Majumdar et al. 2007; Claxton et al. 2010; Zhao et al. 2010, 2011) to gain useful equilibrium and kinetic information on the structural transitions.

1.5.2 Matching Energies of the Outward- and Inward-Facing States

In Glt_{Ph}, both PDS and sm-FRET showed broad distance distributions suggesting that the outward- and inward-facing states were both populated and nearly equienergetic (Akyuz et al. 2013; Georgieva et al. 2013; Hanelt et al. 2013) (Fig. 1.5). Binding of Na^+ ions and Asp modestly shifted the distributions toward the outward-facing state (Akyuz et al. 2013; Georgieva et al. 2013), favoring this state by about 1 kcal mol^{-1}. Furthermore, the observed distributions were consistent with the transporter sampling the symmetric states (with both observed protomers in either the outward- or inward-facing states) and the asymmetric states (with one protomer in the outward- and the other in the inward-facing states or vice versa) supporting previous functional and crystallographic results indicative of the subunit independence in the trimer (Grewer et al. 2005; Verdon and Boudker 2012). Similar distributions were observed in detergent-solubilized purified transporters and in transporters reconstituted into lipid bilayer for the majority of the tested mutants. However, some mutants revealed intriguing inconsistencies. For example, a spin-labeled Glt_{Ph} variant, in which the labeled residue was at the interface of the transport domain in the outward- but not inward-facing state, showed typically broad distance distributions in detergent, but a single narrow peak in lipid bilayers, indicative of an exclusive outward-facing state and conformational constrains on the label (Georgieva et al. 2013). The nitroxide spin label is both hydrophobic in nature and shows hydrogen bond acceptor potential. Hence, it may form stabilizing interactions on the domain interface. It appears that these interactions are energetically more favorable in the context of the bilayer compared to detergent micelles. The indirect effects of the bilayer on the strength of the protein:protein interactions that shift the state distributions of the transporter and likely affect its activity may represent an unexplored modality of the bilayer-mediated regulation of the transporters and are worth further exploration.

1.5.3 The Dynamics of the Transport Domain Movements

The equilibrium measurements suggest that the transport domain of Glt_{Ph} samples outward- and inward-facing states with nearly the same probability both when it is bound to its cargo and when it is empty. The dynamics of these processes were probed by sm-FRET approach, which is unique in its ability to detect transitions in real time. Glt_{Ph}, which originates from a hyper-thermophilic archaebacterium, is particularly well suited for these studies because it is very slow at room temperature with turnover times of nearly 200 s (Ryan et al. 2009). sm-FRET studies revealed clear transitions between the FRET-efficiency states, consistent with outward/outward, inward/inward, and mixed orientations of the subunit pairs in agreement with the equilibrium studies (Fig. 1.5b) (Akyuz et al. 2013). The transitions between the states occurred with average frequency of about 0.5 s^{-1} in the apo transporter and 0.01 s^{-1} in the Na^+/Asp-bound transporter. Hence, the transport domain movements are much faster in the unloaded transporter, and the movements of the substrate-loaded transporter may be rate limiting to the transport cycle (Fig. 1.6). Most remarkably, the transporter showed heterogeneous dynamics, alternating between periods or rapid transitions and periods of quiescence. This behavior is highly reminiscent of channels that alternate between periods of rapid opening and closing transition, interspersed by long closed periods due to inactivation or desensitization. The origin of this behavior in Glt_{Ph} remains to be investigated. However, one hypothesis suggests that in both the outward- and inward-facing states the transporter isomerizes between stable, long-lasting states and dynamic states. This isomerization may include breaking of specific interactions on the interface between the transport and trimerization domains, perhaps leading to the intermediate states with hydrated domain interface, reminiscent of that observed crystallographically (Verdon and Boudker 2012). The sm-FRET studies suggest that this dislodging of the transport domain, which is particularly slow in the Na^+/Asp-bound transporter, is the rate-limiting step of the transport domain movements and, hence, of the transport cycle overall.

1.6 The Mechanism of Gating, Binding, and Coupling Between Sodium and Substrate Potential Coupling Mechanisms

Secondary active transporters catalyze concentrative uptake of their substrates by utilizing the energy stored in the form of electrochemical potentials of the co-transported ions. In principle, the only requirement of the coupled symport is that neither the substrate nor the ions alone are efficiently transported, and only together the solutes are robustly translocated. Such coupling can be achieved by several related but distinct mechanisms. For example, the ions and the substrate may bind cooperatively to the transporter, i.e., neither the substrate nor ions alone bind with sufficient affinity, but together they bind well. It is also possible that either the substrate or the ions can bind alone, but binding of both is required to achieve translocation-competent protein conformation. Finally, functional coupling can be achieved if an ion is not required for substrate binding but stabilizes a

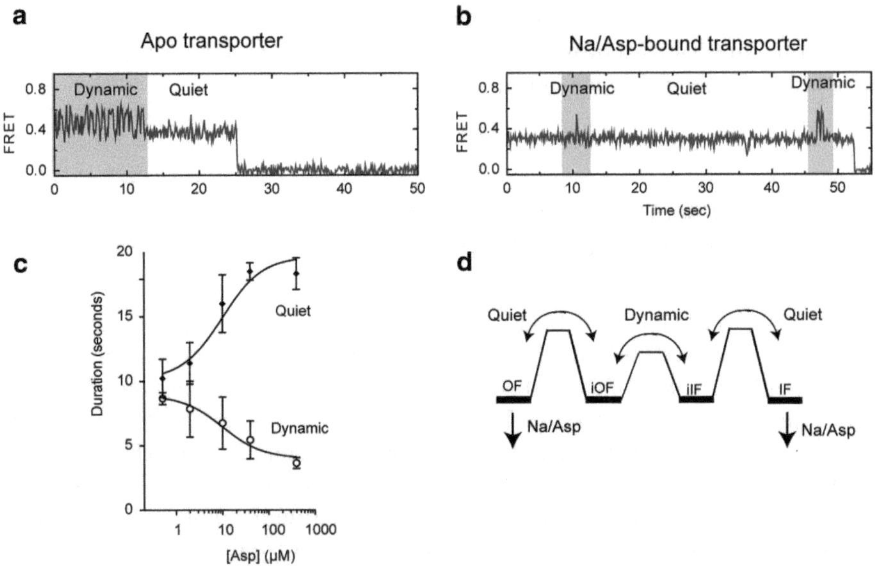

Fig. 1.6 *Single molecule dynamics of Glt_{Ph}.* (**a**) and (**b**) Single molecule trajectories revealed that both apo (substrate-free) (**a**) as well as Na^+/Asp-bound (**b**) transporters transition between dynamic and quiet periods. (**c**) Na^+/Asp binding shifts distributions in favor of longer quiescent periods and shorter periods of dynamics in a concentration-dependent manner (panel adapted from Akyuz et al. 2013). (**d**) A hypothetical model, in which the "dynamic" outward- and inward-facing states represent transient intermediates (iOF and iIF) on the reaction path between the stable outward- (OF) and inward-facing (IF) conformations

transition intermediate on the conformational pathway between the outward- and inward-facing states.

In Glt_{Ph}, a model was proposed based on distance measurements by a continuous wave EPR suggesting that the unloaded transport domain is compact with the substrate-binding site occluded in the outward-facing state by the tip of HP2 similarly to the fully loaded transporter (Focke et al. 2011). Binding of Na^+ ions to the unloaded outward-facing transporter induces opening of HP2, reminiscent of that observed in the transporter complex with TBOA. Such gate opening both provides access for Asp to its binding site and prevents transport domain from translocating across the membrane. This model is in line with functional studies on EAATs, suggesting that at least one Na^+ binds before the substrate (Wadiche et al. 1995b; Watzke et al. 2001; Koch et al. 2007; Zhang et al. 2007). On the other hand, binding studies using detergent-solubilized purified Glt_{Ph} showed that Asp binding to the transporter was thermodynamically coupled to binding of at least 2 Na^+ ions (Boudker et al. 2007), suggesting that neither Asp nor Na^+ ions could bind well to the transporter alone, but bound cooperatively. Indeed, only weak Na^+ binding was observed in the absence of Asp using single tryptophan mutants (Ewers et al. 2013; Hanelt et al. 2013). The shortcoming of these studies was that the transporter was sampling both the outward- and inward-facing states, and binding to a mixture of these states was detected. Hence, it was impossible to evaluate whether the mechanisms of binding and coupling differed on the extracellular and cytoplasmic sides of the membrane.

1.6.1 Na^+ and Asp Coupling in Isolated Outward- and Inward-Facing States

To study the mechanism of binding from the extracellular and the intracellular sides, Hg^{2+}-mediated cross-linking was employed to arrest the transporter in either the outward- or inward-facing states (Reyes et al. 2009). In addition, a binding assay using voltage-sensitive dye RH421 was developed to detect Na^+ as well as coupled Na^+ and Asp binding. In this assay, the amphipathic RH421 dye is incorporated into the protein/detergent particle and reports on the changes of the local electric field associated with binding of Na^+ ions to the protein. Combination of the fluorescent assay and ITC provided access to a broad range of binding constants (Fig. 1.7) (Reyes et al. 2013). These studies revealed that Asp did not bind to the transporter with an appreciable affinity in the absence of Na^+ ions. Moreover, binding in both states was coupled to binding of all three Na^+, at least in the range of Na^+ concentrations between 1 and 100 mM. Interestingly, similar dissociation constants were measured in both states, consistent with only a modest effect of Na^+ and Asp on the distribution between outward- and inward-facing states. Na^+ alone bound weakly with the K_D of ~100 mM in the outward-facing state and ~250 mM in the inward-facing state, but bound much tighter when Asp was present. These results suggest that the key, although possibly not the only, mechanism to ensure coupled Na^+ and Asp transport is cooperativity between their binding reactions. Specifically, Na^+ concentration in the cytoplasm is typically ~10 mM, far below the measured K_D, suggesting that ions alone would not bind to the transporter. The concentration of Na^+ in the extracellular solution is above 100 mM, and Na^+ will substantially occupy the binding sites in this state. Hence, opening of HP2, triggered by Na^+ binding in this state, may provide an important additional mechanism to ensure coupling.

1.6.2 Structural Basis of the Allosteric Coupling Between Substrate and Na^+ Ions

The structural basis of coupling remains to be elucidated. Notably, this coupling must be allosteric in nature because Asp does not directly coordinate the ions. A computational study proposed that a key residue coordinating substrate, N401, mediates coupling between Na1- and substrate-binding sites (Huang and Tajkhorshid 2008). Accordingly, in the absence of Na^+ ion in the Na1 site, N401

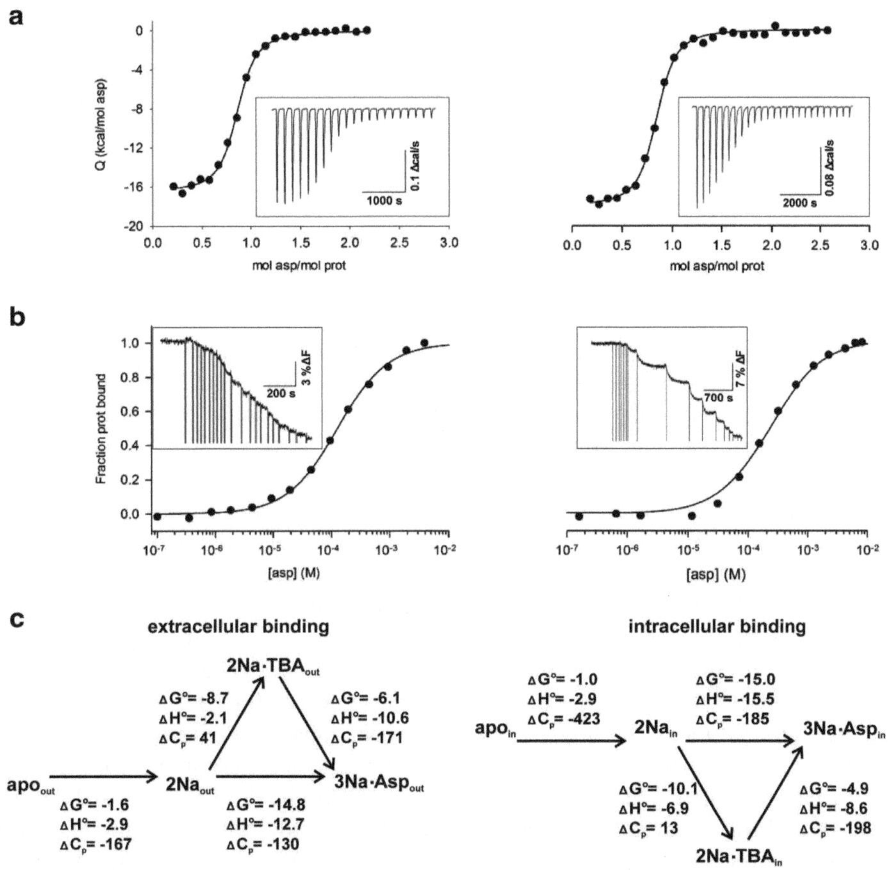

Fig. 1.7 *Binding thermodynamics of* Glt_{Ph}. Binding data for the outward- and inward-facing states shown on the *left* and *right*, respectively. (**a**) Asp binding measured by ITC in the presence of 10 mM Na^+. *Insets* show Asp-binding thermal powers. *Solid black lines* through the data are the fits to independent binding sites model. (**b**) Asp binding detected by the fluorescence-based bulk assays conducted in the presence of 1 mM Na^+. Insets show the percent change in signal as a function of time. *Vertical lines* mark the times of Asp addition. (**c**) A thermodynamic scheme of Na^+ and Asp binding with corresponding parameters, for the outward- and inward-facing states shown on the *left* and *right*, respectively (figure adapted from Reyes et al. 2013)

is oriented away from the binding pocket, while Na^+ binding to the site resulted in a correct positioning of the residue to coordinate the substrate. These results should be considered with caution, because they presuppose that the structures of the transporter in the apo and substrate-bound states are highly similar. Since the structure of the apo Glt_{Ph} has not yet been reported, this remains a conjecture. Moreover, as described below, this is likely not the case. Movements of HP2 could be responsible for the coupling between the Na2- and substrate-binding sites. The structure of Glt_{Ph} bound to TBOA revealed an open conformation of HP2 (Boudker et al. 2007). In this structure, not only is the substrate-binding site exposed to the solution, but also the Na2 site is destroyed. Hence, it was proposed that closure of HP2 is coupled to both binding of Na^+ in Na2 and binding of Asp. However, it is not clear whether HP2 opening observed in TBOA complex faithfully reflects its movements in the absence of the blocker.

1.6.3 The Heat Capacity Changes Associated with Na^+ and Substrate Binding

Using ITC, the enthalpy of coupled Na^+ and Asp binding was measured at different temperatures yielding the heat capacity changes (ΔCp) upon Na^+/Asp binding (Reyes et al. 2013). ΔCp is a unique experimental measure of the complexity of the underlying reaction. Binding of small solutes, such as Asp, to preformed binding sites is typically associated with small decrease of the heat capacity. Strikingly, Na^+/Asp binding in both the outward- and inward-facing states was associated with negative ΔCps that were an order of magnitude larger than expected. Also the ΔCp measured in the inward-facing state was much larger than that in the outward-facing states, suggesting that the binding mechanisms were different in these two states. Furthermore, the measured ΔCps were too large to be accounted for by the local conformational changes associated with opening and closure of the gates, such as those expected for HP2 in the outward-facing state. Although the origin of the large heat capacity changes remains to be identified, two possibilities could be considered. Large decrease of the heat capacity could be associated with the burial of a large hydrophobic area, exposed to the solvent in the apo, but not in the bound states. This could occur, for example, if the core of the transporter was largely hydrated in the apo state or if the interface between the domains was hydrated in the apo but not in the bound state. Alternatively, the large ΔCp may reflect the fact that the apo state is not a unique state but an ensemble of enthalpically distinct states, which is collapsed to a unique state upon Na^+/Asp binding. This would be in agreement with sm-FRET measurements, which suggest that in the apo state the dynamic and non-dynamic states of Glt_{Ph} are approximately equally populated, while in the Na^+/Asp-bound state the protein is mostly non-dynamic. However, it is unclear whether these states could be sampled in the conformationally constrained outward- and inward-facing variants.

1.7 Concluding Remarks

The ability of the secondary active transporters to catalyze coupled trans-membrane transport of solutes and ions is rooted in their structural plasticity, i.e., in their sampling of many conformationally distinct states. This structural flexibility presents unique challenges when we endeavor to understand the mechanisms of the transporters. From the structural perspective, it necessitates development of multiple approaches to stabilize unique states for determination of their crystal structures. From the mechanistic point of view, it requires establishment of the energetic relationships between these states and understanding of the nature of the energetic barriers between them. Hence, it is essential to complement the structural studies with approaches that probe the equilibrium and the dynamic properties of the conformational ensembles. These may include the spectroscopic, thermodynamics, and computational methods, as we have illustrated in this chapter (see also Chap. 11). Only the combination of multiple approaches is likely to yield a physically meaningful picture that would allow us to rationally modulate the activity of these transporters either by mutagenesis or by development of pharmacological tools.

References

Akyuz N, Altman R et al (2013) Transport dynamics of a glutamate transporter homologue. Nature 502(7469):114–118

Andersen OS, Koeppe RE 2nd (2007) Bilayer thickness and membrane protein function: an energetic perspective. Annu Rev Biophys Biomol Struct 36:107–130

Arriza JL, Eliasof S et al (1997) Excitatory amino acid transporter 5, a retinal glutamate transporter coupled to a chloride conductance. Proc Natl Acad Sci U S A 94(8):4155–4160

Barbour B, Brew H et al (1988) Electrogenic glutamate uptake in glial cells is activated by intracellular potassium. Nature 335(6189):433–435

Bastug T, Heinzelmann G et al (2012) Position of the third Na^+ site in the aspartate transporter Glt_{Ph} and the human glutamate transporter, EAAT1. PLoS One 7(3):e33058

Bendahan A, Armon A et al (2000) Arginine 447 plays a pivotal role in substrate interactions in a neuronal glutamate transporter. J Biol Chem 275(48):37436–37442

Borre L, Kavanaugh MP et al (2002) Dynamic equilibrium between coupled and uncoupled modes of a neuronal glutamate transporter. J Biol Chem 277(16):13501–13507

Boudker O, Verdon G (2010) Structural perspectives on secondary active transporters. Trends Pharmacol Sci 31(9):418–426

Boudker O, Ryan RM et al (2007) Coupling substrate and ion binding to extracellular gate of a sodium-dependent aspartate transporter. Nature 445(7126):387–393

Bouvier M, Szatkowski M et al (1992) The glial cell glutamate uptake carrier countertransports pH-changing anions. Nature 360(6403):471–474

Choi DW (1994) Calcium and excitotoxic neuronal injury. Ann N Y Acad Sci 747:162–171

Claxton DP, Quick M et al (2010) Ion/substrate-dependent conformational dynamics of a bacterial homolog of neurotransmitter:sodium symporters. Nat Struct Mol Biol 17(7):822–829

Crisman TJ, Qu S et al (2009) Inward-facing conformation of glutamate transporters as revealed by their inverted-topology structural repeats. Proc Natl Acad Sci U S A 106(49):20752–20757

Danbolt NC (2001) Glutamate uptake. Prog Neurobiol 65(1):1–105

DeChancie J, Shrivastava IH et al (2010) The mechanism of substrate release by the aspartate transporter Glt_{Ph}: insights from simulations. Mol Biosyst 7(3):832–842

Eliasof S, Jahr CE (1996) Retinal glial cell glutamate transporter is coupled to an anionic conductance. Proc Natl Acad Sci U S A 93(9):4153–4158

Ewers D, Becher T et al (2013) Induced fit substrate binding to an archeal glutamate transporter homologue. Proc Natl Acad Sci U S A 110(30):12486–12491

Faham S, Watanabe A et al (2008) The crystal structure of a sodium galactose transporter reveals mechanistic insights into Na^+/sugar symport. Science 321(5890):810–814
Fairman WA, Vandenberg RJ et al (1995) An excitatory amino-acid transporter with properties of a ligand-gated chloride channel. Nature 375(6532):599–603
Fischbarg J (1988) On the possible permeation of water across the glucose transporter. Mol Cell Biochem 82(1–2):107–111
Focke PJ, Moenne-Loccoz P et al (2011) Opposite movement of the external gate of a glutamate transporter homolog upon binding cotransported sodium compared with substrate. J Neurosci 31(16):6255–6262
Forrest LR, Rudnick G (2009) The rocking bundle: a mechanism for ion-coupled solute flux by symmetrical transporters. Physiology (Bethesda) 24:377–386
Gendreau S, Voswinkel S et al (2004) A trimeric quaternary structure is conserved in bacterial and human glutamate transporters. J Biol Chem 279(38):39505–39512
Georgieva ER, Borbat PP et al (2013) Conformational ensemble of the sodium-coupled aspartate transporter. Nat Struct Mol Biol 20(2):215–221
Grewer C, Balani P et al (2005) Individual subunits of the glutamate transporter EAAC1 homotrimer function independently of each other. Biochemistry 44(35):11913–11923
Groeneveld M, Slotboom DJ (2007) Rigidity of the subunit interfaces of the trimeric glutamate transporter GltT during translocation. J Mol Biol 372(3):565–570
Groeneveld M, Slotboom DJ (2010) Na(+):aspartate coupling stoichiometry in the glutamate transporter homologue Glt(Ph). Biochemistry 49(17):3511–3513
Grunewald M, Bendahan A et al (1998) Biotinylation of single cysteine mutants of the glutamate transporter GLT-1 from rat brain reveals its unusual topology. Neuron 21(3):623–632
Hanelt I, Wunnicke D et al (2013) Conformational heterogeneity of the aspartate transporter Glt (Ph). Nat Struct Mol Biol 20(2):210–214
Holley DC, Kavanaugh MP (2009) Interactions of alkali cations with glutamate transporters. Philos Trans R Soc Lond B Biol Sci 364(1514):155–161
Huang Z, Tajkhorshid E (2008) Dynamics of the extracellular gate and ion-substrate coupling in the glutamate transporter. Biophys J 95(5):2292–2300
Huang Z, Tajkhorshid E (2010) Identification of the third Na^+ site and the sequence of extracellular binding events in the glutamate transporter. Biophys J 99(5):1416–1425
Jardetzky O (1966) Simple allosteric model for membrane pumps. Nature 211(5052):969–70
Kanner BI, Sharon I (1978) Active transport of L-glutamate by membrane vesicles isolated from rat brain. Biochemistry 17(19):3949–3953
Kavanaugh MP (1998) Neurotransmitter transport: models in flux. Proc Natl Acad Sci U S A 95 (22):12737–12738
Koch HP, Larsson HP (2005) Small-scale molecular motions accomplish glutamate uptake in human glutamate transporters. J Neurosci 25(7):1730–1736
Koch HP, Hubbard JM et al (2007) Voltage-independent sodium-binding events reported by the 4B–4C loop in the human glutamate transporter excitatory amino acid transporter 3. J Biol Chem 282(34):24547–24553
Krishnamurthy H, Piscitelli CL et al (2009) Unlocking the molecular secrets of sodium-coupled transporters. Nature 459(7245):347–355
Larsson HP, Wang X et al (2010) Evidence for a third sodium-binding site in glutamate transporters suggests an ion/substrate coupling model. Proc Natl Acad Sci U S A 107(31):13912–13917
Lau A, Tymianski M (2010) Glutamate receptors, neurotoxicity and neurodegeneration. Pflugers Arch 460(2):525–542
Leary GP, Stone EF et al (2007) The glutamate and chloride permeation pathways are colocalized in individual neuronal glutamate transporter subunits. J Neurosci 27(11):2938–2942
Lee SG, Su ZZ et al (2008) Mechanism of ceftriaxone induction of excitatory amino acid transporter-2 expression and glutamate uptake in primary human astrocytes. J Biol Chem 283(19):13116–13123

Li J, Shaikh SA et al (2013) Transient formation of water-conducting states in membrane transporters. Proc Natl Acad Sci U S A 110:7696–7701
Majumdar DS, Smirnova I et al (2007) Single-molecule FRET reveals sugar-induced conformational dynamics in LacY. Proc Natl Acad Sci U S A 104(31):12640–12645
Nelson PJ, Dean GE et al (1983) Hydrogen ion cotransport by the renal brush border glutamate transporter. Biochemistry 22(23):5459–5463
Oldham ML, Chen J (2011) Crystal structure of the maltose transporter in a pretranslocation intermediate state. Science 332(6034):1202–1205
Perez C, Koshy C et al (2012) Alternating-access mechanism in conformationally asymmetric trimers of the betaine transporter BetP. Nature 490(7418):126–130
Picaud SA, Larsson HP et al (1995) Glutamate-gated chloride channel with glutamate-transporter-like properties in cone photoreceptors of the tiger salamander. J Neurophysiol 74(4):1760–1771
Ressl S, Terwisscha van Scheltinga AC et al (2009) Molecular basis of transport and regulation in the Na(+)/betaine symporter BetP. Nature 458(7234):47–52
Reyes N, Ginter C et al (2009) Transport mechanism of a bacterial homologue of glutamate transporters. Nature 462(7275):880–885
Reyes N, Oh S et al (2013) Binding thermodynamics of a glutamate transporter homologue. Nat Struct Mol Biol 20(5):634–640
Rosental N, Gameiro A et al (2011) A conserved aspartate residue located at the extracellular end of the binding pocket controls cation interactions in brain glutamate transporters. J Biol Chem 286(48):41381–41390
Rothstein JD, Dykes-Hoberg M et al (1996) Knockout of glutamate transporters reveals a major role for astroglial transport in excitotoxicity and clearance of glutamate. Neuron 16(3):675–686
Rothstein JD, Patel S et al (2005) Beta-lactam antibiotics offer neuroprotection by increasing glutamate transporter expression. Nature 433(7021):73–77
Ryan RM, Mindell JA (2007) The uncoupled chloride conductance of a bacterial glutamate transporter homolog. Nat Struct Mol Biol 14(5):365–371
Ryan RM, Vandenberg RJ (2002) Distinct conformational states mediate the transport and anion channel properties of the glutamate transporter EAAT-1. J Biol Chem 277(16):13494–13500
Ryan RM, Mitrovic AD et al (2004) The chloride permeation pathway of a glutamate transporter and its proximity to the glutamate translocation pathway. J Biol Chem 279(20):20742–20751
Ryan RM, Compton EL et al (2009) Functional characterization of a Na^+-dependent aspartate transporter from *Pyrococcus horikoshii*. J Biol Chem 284(26):17540–17548
Schiemann O, Weber A et al (2003) Nanometer distance measurements on RNA using PELDOR. J Am Chem Soc 125(12):3434–3435
Seal RP, Leighton BH et al (1998) Transmembrane topology mapping using biotin-containing sulfhydryl reagents. Methods Enzymol 296:318–331
Seal RP, Shigeri Y et al (2001) Sulfhydryl modification of V449C in the glutamate transporter EAAT1 abolishes substrate transport but not the substrate-gated anion conductance. Proc Natl Acad Sci U S A 98(26):15324–15329
Shimamura T, Weyand S et al (2010) Molecular basis of alternating access membrane transport by the sodium-hydantoin transporter Mhp1. Science 328(5977):470–473
Shlaifer I, Kanner BI (2007) Conformationally sensitive reactivity to permeant sulfhydryl reagents of cysteine residues engineered into helical hairpin 1 of the glutamate transporter GLT-1. Mol Pharmacol 71(5):1341–1348
Shrivastava IH, Jiang J et al (2008) Time-resolved mechanism of extracellular gate opening and substrate binding in a glutamate transporter. J Biol Chem 283(42):28680–28690
Slotboom DJ, Sobczak I et al (1999) A conserved serine-rich stretch in the glutamate transporter family forms a substrate-sensitive reentrant loop. Proc Natl Acad Sci U S A 96(25):14282–14287

Stolzenberg S, Khelashvili G et al (2012) Structural intermediates in a model of the substrate translocation path of the bacterial glutamate transporter homologue Glt_{Ph}. J Phys Chem B 116 (18):5372–5383
Tao Z, Gameiro A et al (2008) Thallium ions can replace both sodium and potassium ions in the glutamate transporter excitatory amino acid carrier 1. Biochemistry 47(48):12923–12930
Tao Z, Rosental N et al (2010) Mechanism of cation binding to the glutamate transporter EAAC1 probed with mutation of the conserved amino acid residue Thr101. J Biol Chem 285 (23):17725–17733
Teichman S, Kanner BI (2007) Aspartate-444 is essential for productive substrate interactions in a neuronal glutamate transporter. J Gen Physiol 129(6):527–539
Teichman S, Qu S et al (2012) Conserved asparagine residue located in binding pocket controls cation selectivity and substrate interactions in neuronal glutamate transporter. J Biol Chem 287 (21):17198–17205
Vandenberg RJ, Handford CA et al (2011) Water and urea permeation pathways of the human excitatory amino acid transporter EAAT1. Biochem J 439(2):333–340
Verdon G, Boudker O (2012) Crystal structure of an asymmetric trimer of a bacterial glutamate transporter homolog. Nat Struct Mol Biol 19(3):355–357
Veruki ML, Morkve SH et al (2006) Activation of a presynaptic glutamate transporter regulates synaptic transmission through electrical signaling. Nat Neurosci 9(11):1388–1396
Wadiche JI, Amara SG et al (1995a) Ion fluxes associated with excitatory amino acid transport. Neuron 15(3):721–728
Wadiche JI, Arriza JL et al (1995b) Kinetics of a human glutamate transporter. Neuron 14 (5):1019–1027
Watzke N, Bamberg E et al (2001) Early intermediates in the transport cycle of the neuronal excitatory amino acid carrier EAAC1. J Gen Physiol 117(6):547–562
Weiss S (1999) Fluorescence spectroscopy of single biomolecules. Science 283(5408):1676–1683
Weyand S, Shimamura T et al (2008) Structure and molecular mechanism of a nucleobase-cation-symport-1 family transporter. Science 322(5902):709–713
Yamashita A, Singh SK et al (2005) Crystal structure of a bacterial homologue of Na+/Cl– dependent neurotransmitter transporters. Nature 437(7056):215–223
Yernool D, Boudker O et al (2003) Trimeric subunit stoichiometry of the glutamate transporters from Bacillus caldotenax and Bacillus stearothermophilus. Biochemistry 42(44):12981–12988
Yernool D, Boudker O et al (2004) Structure of a glutamate transporter homologue from *Pyrococcus horikoshii*. Nature 431(7010):811–818
Yi JH, Hazell AS (2006) Excitotoxic mechanisms and the role of astrocytic glutamate transporters in traumatic brain injury. Neurochem Int 48(5):394–403
Yin Y, He X et al (2006) Structure of the multidrug transporter EmrD from *Escherichia coli*. Science 312(5774):741–744
Zerangue N, Kavanaugh MP (1996) Flux coupling in a neuronal glutamate transporter. Nature 383 (6601):634–637
Zhang Z, Grewer C (2007) The sodium-coupled neutral amino acid transporter SNAT2 mediates an anion leak conductance that is differentially inhibited by transported substrates. Biophys J 92(7):2621–2632
Zhang Z, Tao Z et al (2007) Transport direction determines the kinetics of substrate transport by the glutamate transporter EAAC1. Proc Natl Acad Sci U S A 104(46):18025–18030
Zhao Y, Terry D et al (2010) Single-molecule dynamics of gating in a neurotransmitter transporter homologue. Nature 465(7295):188–193
Zhao Y, Terry DS et al (2011) Substrate-modulated gating dynamics in a Na^+-coupled neurotransmitter transporter homologue. Nature 474(7349):109–113

Chapter 2
Characterizing the Structure, Function, and Evolution of Human Solute Carrier (SLC) Transporters Using Computational Approaches

Avner Schlessinger

Abstract Solute Carrier (SLC) transporters are membrane proteins that transport a broad range of solutes including metabolites, ions, toxins, and prescription drugs. In humans, there are about 400 SLC members, many of which are of medical importance. They can be drug target themselves (e.g., the serotonin transporter, SERT) or regulate the absorption, distribution, metabolism, and excretion (ADME) of drugs (e.g., the organic cation transporter 1, OCT-1). An important step toward describing the mechanisms of solute transport by SLC transporters includes computational or experimental characterization of their ligand-bound or unbound structures in different conformations. Due to a variety of technical issues, human SLC transporters are challenging targets for both experimental and computational characterizations. However, recent advances in computational approaches such as molecular docking and comparative modeling, coupled with the atomic structure determination of several membrane transporters, expanded our ability to characterize the human SLC families using in silico structure-based approaches. In this chapter, we first provide an overview of the structure, function, and pharmacology of the human SLC transporters. Second, we describe different computational methods, including sequence analysis, structural modeling, and ligand docking, that are commonly used, in combination with experimental testing, to characterize the human SLC transporters. Third, we demonstrate the utility of these approaches to characterize SLC members with three examples—the norepinephrine transporter (NET), the γ-aminobutyric acid (GABA) transporter 2 (GAT-2), and the L-type amino acid transporter (LAT-1). Finally, future directions in the field of computational structural biology of human SLC transporters are discussed.

A. Schlessinger (✉)
Department of Pharmacology and Systems Therapeutics, Icahn School of Medicine at Mount Sinai, One Gustave L. Levy Place, Box 1603, New York, NY 10029, USA
e-mail: avner.schlessinger@mssm.edu

R. Krämer and C. Ziegler (eds.), *Membrane Transport Mechanism*, Springer Series in Biophysics 17, DOI 10.1007/978-3-642-53839-1_2,

Keywords Comparative modeling • Membrane protein • Molecular docking • Protein function prediction • Structure-based ligand discovery • Virtual screening

2.1 The Human SLC Transporters

2.1.1 Sequences, Structures, and Functions of Human SLC Transporters

The solute carrier (SLC) transporters or the SLC superfamily include membrane transporters, ion channels, and other membrane proteins that mediate the movement of solutes across biological membranes via diverse mechanisms (Hediger et al. 2004; Povey et al. 2001; Saier 2000; Saier et al. 2009; Forrest and Rudnick 2009; Forrest et al. 2011; Jardetzky 1966). In humans, there are about 400 SLC members that are expressed ubiquitously and transport a broad spectrum of molecules, including metabolites, toxins, and drugs (Hediger et al. 2004). For example, the peptide transporter 1 (PepT1, SLC15A1) is a proton-dependent transporter of di and tripeptides that is primarily expressed in the brush border membrane of the intestinal epithelium, where it plays a key role in the uptake of digested proteins and absorption of various peptidomimetic drugs (Smith et al. 2013). Another instance is the γ-aminobutyric acid (GABA) transporter 1 (GAT-1, SLC6A1), an Na^+-dependent transporter localized in the brain's cerebral cortex, where it transports the neurotransmitter GABA at inhibitory synapses, terminating GABA's action and regulating inhibitory postsynaptic signaling (Guastella et al. 1990; Kanner and Zomot 2008). Therefore, genetic variation in SLC members can be associated with various diseases and disorders, and when occurs in transporters that are important for drug absorption, distribution, and excretion (ADME), it can also lead to differential drug response among individuals (pharmacogenetics) (Giacomini et al. 2010). For example, mutations in the organic cation transporter 1 (SLC22A1 or OCT1) affect the intracellular concentrations of the antidiabetic drug metformin (Shu et al. 2003, 2007, 2008; Dresser et al. 2000).

The Gene Nomenclature Committee (HGNC) of the Human Genome Organization (HUGO) defined the human SLC transporters as one "superfamily" (Povey et al. 2001; Hediger et al. 2004). The SLC superfamily is grouped into 52 families based on the number of predicted or observed transmembrane α-helices (TMH; usually 10–14) and sequence similarity, where members of each family typically share sequence identity of 20 % or more to at least one other family member (Hediger et al. 2004). Moreover, some SLC families were classified as such because they are functionally linked to other SLC members. For example, the SLC3 family includes single transmembrane helix proteins that physically interact with members of the SLC7 family of amino acids transporters to form active heterodimers (Verrey et al. 2004). Furthermore, SLC51A and SLC51B, which also function as a heterodimer (similarly to SLC3 and SLC7 families), are grouped into the same family (i.e., SLC51), despite not sharing detectable sequence relationship. Thus, the use of the term "SLC superfamily," which is commonly used among pharmacologists of membrane transporters (Hediger et al. 2004; Giacomini et al. 2010; Schlessinger et al. 2013b), can be confusing because it generally implies a common ancestor and/or detectable evolutionary relationship, which is not true for the SLC members (Murzin et al. 1995; Saier et al. 2009; Schlessinger et al. 2010, 2013b).

The only human SLC member with experimentally determined atomic resolution structure is the Rhesus glycoprotein ammonium transporter (RhCG, SLC42A3), whose X-ray structure has been determined at resolution of 2.3 Å (Gruswitz et al. 2010). Furthermore, the mouse mitochondrial uncoupling protein 2 UCP2/SLC25A8, which exhibits sequence identity of ~96 % to its human ortholog, has recently been determined via NMR (Berardi et al. 2011). There are several high-resolution structures of proteins from prokaryotic and other eukaryotic organisms that are sufficiently similar in sequence to the human SLC members to inform about their general topology and fold (Gether et al. 2006; Schlessinger et al. 2010, 2013a, b; Faraldo-Gomez and Forrest 2011; Pieper et al. 2013). These structures confirmed that the human SLC transporters are diverse in structure, as was expected from their divergent sequences. A recent analysis suggested that the two largest structural classes or folds in the human SLC members are the Major Facilitator Superfamily (MFS; e.g., GLUTs/SLC2) and the Neurotransmitter: Sodium Symporter (NSS) fold, which is also dubbed the APC (amino acid-polyamine-organoCation) superfamily (Lomize et al. 2006) or the LeuT-like fold (e.g., SLC6) (Schlessinger et al. 2010). Conversely, these atomic structures also revealed that some SLC families are related structurally, despite sharing weak sequence relationships (sequence identity lower than 10 %). For example, the structures of several prokaryotic homologs of the human families SLC5 (the Na^+-glucose co-transporter SGLT) (Faham et al. 2008), SLC6 (the leucine/alanine transporter LeuT) (Yamashita et al. 2005), and SLC7 [arginine/agmatine antiporter (AdiC) (Gao et al. 2009; Fang et al. 2009; Shaffer et al. 2009)], as well as other transporters, such as the uracil transporter UraA (Lu et al. 2011), the Na^+-hydantoin transporter Mhp1 (Weyand et al. 2008), the Na+/betaine symporter BetP (Ressl et al. 2009), the glutamate/gamma-aminobutyrate antiporter (Ma et al. 2012), and the L-carnitine/gamma-butyrobetaine antiporter CaiT (Schulze et al. 2010), all belong to the NSS fold (Schlessinger et al. 2010; Lomize et al. 2006).

The majority of the human SLC members are thought to transport substrates via the "alternating access" transport mechanism, in which the transporter interchangeably exposes its binding site at either side of the membrane (Forrest and Rudnick 2009; Kanner and Zomot 2008; Krishnamurthy et al. 2009; Forrest et al. 2008, 2011; Jardetzky 1966; Guan and Kaback 2006; Krishnamurthy and Gouaux 2012; Abramson and Wright 2009; Perez and Ziegler 2013). The alternating access is highly conserved among secondary active transporters, even among families that are unrelated in sequence and structure, such as proteins of the MFS (see Chaps. 6 and 7), NSS (see Chaps. 3, 4, and 5), and the glutamate transporter folds (see Chap. 1). Importantly, this mechanism is enabled by inverted structural repeats that are highly divergent in sequence (Saier 2000; Forrest et al. 2011; Krishnamurthy et al. 2009; Abramson and Wright 2009; Khafizov et al. 2010; Yamashita et al. 2005; Madej et al. 2013). For example, LeuT includes two five transmembrane helices repeats that are related to each other via a twofold pseudo-symmetry axis parallel to the membrane plane; the repeats make a channel and the first helix of each repeat coordinates ligand binding (Forrest et al. 2011; Krishnamurthy et al. 2009). The evolution of these pseudo-repeats in the SLC structures and how they contribute to the diverse energy couplings of the transporters are not completely understood and are likely to differ among different families. (Forrest et al. 2008; Forrest and Rudnick 2009; Perez and Ziegler 2013; Yamashita et al. 2005; Khafizov et al. 2010; Madej et al. 2013; Forrest 2013). Recent studies support the notion that the transport mechanism is conserved among members of the NSS fold (Abramson and Wright 2009; Krishnamurthy et al. 2009; Shaffer et al. 2009; Ressl et al. 2009; Shi and Weinstein 2010). For example, using comparative modeling and molecular dynamics simulations, Shi and Weinstein revealed that the positive charge of the a sodium ion (NA2) in LeuT is equivalent to a proton in ApcT (i.e., Lys158) in regulating specific conformational changes that are associated with transport (Shi and Weinstein 2010). Interestingly, a recent analysis of the MFS transporters LacY and FucP suggested that although these transporters share only weak sequence similarities, their structural repeats consist of triple helix units that are related to each other and have likely emerged from a common ancestor (Madej et al. 2013) (see Chap. 6). This suggests a simple, parsimonious chain of events that may have led to the sequence divergence among members of the MFS fold, which might be relevant to other SLC families including those in human.

2.1.2 Description of Substrate Specificity in Human SLC Transporters

The SLC transporter's function is determined by its structure and dynamics. In particular, the shape and physicochemical properties of the binding site(s) on the transporter surface (i.e., specificity determinants) determine the ligands (molecules or ions) that bind to the transporter ("binding specificity determinants"), which contribute to define the ligands that get transported by the transporter ("substrate specificity determinants"); the mechanism of transport describes how the specificity determinants of a transporter result in the binding specificity and substrate specificity. Description of specificity determinants in SLC families includes the characterization (computationally and/or experimentally) of different conformational states that the transporters undergoes during transport and inhibition, as well as the identification of structural relationships between the transporter and its ligands, including inhibitors and substrates.

Understanding these structure–function relationship in the SLC transporters can contribute to three key clinically important aspects of their functions (Schlessinger et al. 2013b). First, and perhaps obvious, identification of structural relationships in protein–ligand complexes of structurally related transporters can provide molecular basis for how ions and molecules get transported selectively across the membranes of human cells and organelles, which is essential for many of life's processes. For example, charge specificity within the organic ion transporter family SLC22, which is important for drug ADME in the kidney and liver, has been previously proposed based on biochemical data (Feng et al. 2001) and was recently rationalized using comparative models of representative members of the family (Pedersen et al. 2013); it was suggested that the organic anion transporter 3 (OAT3, SLC22A8) binding site residue Arg454, has its charge reversed in the organic cation transporter 1 (OCT1, SLC22A1) (Asp 474), in agreement with the role of the corresponding residue of the template structure in substrate charge specificity (Pedersen et al. 2013).

Second, the functional effect of clinically important genetic variants, including those variants that are associated with disease states or pharmacogenetics, can be rationalized based on their location on the protein structure. For example, the negative effect of variants in the proton-dependent Multidrug and toxin extrusion 2 (MATE2-K, SLC47A2), on the response to the antidiabetic drug metformin, was rationalized by their predicted impact on the MATE2-K comparative model (Choi et al. 2011); Gly393 mutation to arginine occurs in close proximity to the putative proton-binding site and likely affect interaction with cationic ligands and protons.

Third, unknown endogenous (e.g., metabolites) and exogenous (e.g., prescription drugs) ligands can be predicted by virtual screening of large small molecule libraries against the transporter structure. The experimentally validated ligands can then guide drug–drug interaction studies, provide novel chemical probes to characterize the transporter function, and serve as lead-like molecules for drug development. For example, virtual screening against the structural model of the norepinephrine transporter NET, followed experimental testing of top-scoring hits, revealed that several prescription drugs interact with NET. These results rationalize some of the positive and/or negative effects of some of these drugs (Schlessinger et al. 2011).

We begin this book chapter by describing computational tools, including sequence analysis, comparative modeling, model refinement and assessment, and virtual screening, that are commonly used to predict the structures and functions of SLC transporters. We then illustrate the applications of these tools in combination with experiments to define specificity determinants for selected SLC transporters. We conclude by discussing future directions in field.

2.2 Structure-Based Ligand Discovery for SLC Transporters

2.2.1 Overall Approach

In structure-based ligand discovery, a large virtual library of small organic molecules is computationally docked against an atomic structure of a protein target (Shoichet 2004) (Fig. 2.1). For a protein without an experimentally determined structure, virtual screening or molecular docking can be performed against structural model of the target protein (Jacobson and Sali 2004). In general, structural models for proteins can be constructed by using three main approaches. They include (1) comparative or homology modeling, which relies on detectable similarity between the target protein and least one known experimentally determined structure, (2) ab initio modeling, which uses sequence alone, without relying on similarity at the fold level between target protein and experimentally determined structures, and (3) hybrid methods modeling, which combines low-resolution experimental data such as that obtained from electron microscopy of 2D crystals using computational modeling with comparative modeling, ab initio modeling, or both (Baker and Sali 2001; Russel et al. 2012; Rohl et al. 2004). Considerable progress has been made recently in all three modeling categories, particularly for membrane proteins (Yarov-Yarovoy et al. 2006; Marks et al. 2011; Hopf et al. 2012; Nugent and Jones 2012). Notably, ligand discovery with molecular docking requires highly accurate and precise description of the target protein binding site. For example, incorrect configuration of even one side chain in the transporter binding site can yield results similar to that of random prediction (Schlessinger et al. 2011). Therefore, models calculated using comparative modeling are expected to be the most useful for virtual screening.

We take an iterative comparative modeling and ligand docking approach to predict ligands for human SLC transporters. In brief, we begin with a sequence of a transporter with unknown structure ("target") and identify homologous proteins with known structures, using sequence-based alignment and/or fold recognition methods (e.g., Promals3D (Pei et al. 2008)) (Fig. 2.1a). The alignment is corrected manually for different aspects of protein structure and function, including transmembrane helices (TMH), sequence motifs, and experimentally verified conserved residues (Fig. 2.1b). Initial structural models are then built with modeling programs such as MODELLER (Sali and Blundell 1993) and assessed using various statistical potentials that are derived from experimentally determined sequences (e.g., Z-DOPE (Shen and Sali 2006)) (Fig. 2.1c). The models are refined by modeling the loops, repacking the side chains on a fixed backbone, as well as by energy minimization using Molecular Dynamics (MD) simulations (Fig. 2.1d) (Hess et al. 2008; Schlessinger et al. 2012). Next, comparative models are evaluated based on their ability to discriminate known ligands from decoys by using docking against the binding site of the target (i.e., "enrichment calculations") (Fig. 2.1e)

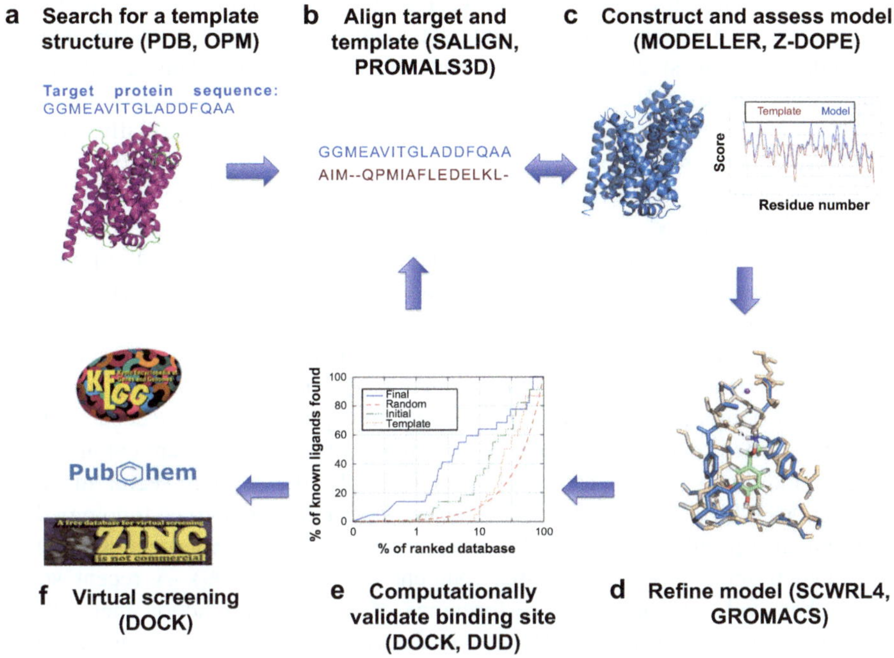

Fig. 2.1 Steps in structure-based ligand prediction

(Huang et al. 2006; Fan et al. 2009; Schlessinger et al. 2011; Lorber and Shoichet 1998). Thus, by selecting models based on their enrichment scores, comparative models are assessed for the utility in predicting ligands. The modeling steps and enrichment calculations can be performed iteratively to optimize the models for virtual screening applications (Evers et al. 2003; Cavasotto et al. 2008; Carlsson et al. 2011; Schlessinger et al. 2012). Virtual screening is then performed where large compound libraries, which can contain millions of purchasable compounds (e.g., ZINC (Irwin and Shoichet 2005)), are docked against the high-ranking models (Fig. 2.1f) (Shoichet 2004; Shoichet et al. 1993). Finally, top-ranked compounds and their orientation relative to the target binding site ("poses") are analyzed manually and selected for experimental validation.

2.2.2 Template Identification and Target–Template Alignment

As for other proteins, to construct a comparative model for an SLC transporter target, most similar known template structure or structures need to be detected and aligned with its sequence (Sali and Blundell 1993). The majority of the SLC families share similar topology (10–14 TMH); thus, a key step toward aligning their sequences optimally involves the accurate identification of these TMH regions (Kernytsky and Rost 2003; Elofsson and von Heijne 2007; Punta et al. 2007; Nugent and Jones 2012). Early TMH prediction methods primarily included "hydropathy plots" derived from "propensity scales," which usually include 20 values corresponding to amino acid properties when found in membrane spanning helices (e.g., the Kyte–Doolite hydrophobicity scale (Kyte and Doolittle 1982)). These methods were shown to be surprisingly accurate because of the compositional bias on the amino acid sequence of membrane proteins imposed by the lipid bilayer environment (Elofsson and von Heijne 2007; Eyre et al. 2004; Punta et al. 2007).

However, state-of-the-art TMH prediction methods, which typically use advanced statistical inference methodologies (e.g., machine-learning algorithms) and usually combine more than one approach, are significantly more accurate than the early TMH prediction methods (i.e., accuracy of 80–90 %) (Punta et al. 2007; Nugent and Jones 2012) (Table 2.1). For example, the support vector machine program MEMSAT-SVM, which was trained on structurally derived topology data and sequence profiles, achieved ~90 % accuracy on a set of experimentally determined membrane protein structures (Nugent and Jones 2009). A recent study comprehensively evaluated the performance of topology prediction methods on various sets including a set of 103 newly determined atomic structures (Tsirigos et al. 2012). This analysis indicated that the best performance is obtained by TOPCONS (Bernsel et al. 2009), a consensus method that combines five other prediction methods (e.g., SCAMPI (Bernsel et al. 2008)) and that the best method for distinguishing membrane from non-membrane proteins is the Hidden Markov Model (HMM)-based method Phobius (Kall et al. 2005).

The identification of the best templates for modeling is relatively straightforward for some human SLC members. In particular, several human SLC families exhibit sequence similarities of 30–40 % to proteins of known structure. For example, the human Na^+-coupled nucleoside transporters (SLC28), the oligopeptide:H^+ co-transporters (SLC15), and the high-affinity glutamate and neutral amino acid transporter family (SLC1) are homologs of the proteins of known structures vcCNT (Johnson et al. 2012), PEPTso (Newstead et al. 2011), and Glt(Ph) (Yernool et al. 2004), respectively. For these families the template identification is feasible even when using relatively simple sequence-based alignment methods such as BLAST (Altschul et al. 1990; Rost et al. 2003).

However, the template identification for other human SLC families and the template–target alignment can be more difficult. Such challenging targets can be transporters that are distantly related to proteins of known structure. For instance, members of the SLC22 family of organic cation/anion/zwitterion transporters share sequence identity of 20 % or lower with the closest template structure—the structure of the fungal phosphate transporter PiPT (Pedersen et al. 2013). Other challenging targets can be those that exhibit significant (yet often distant) sequence similarity to more than one potential template structure. For instance, the SLC7 family of amino acid transporters is related to several transporters with known

Table 2.1 Sequence-based methods

Method[a]	Type[b]	URL[c]	Reference[d]
Transmembrane helix prediction			
MEMSAT-SVM	S, P	http://bioinf.cs.ucl.ac.uk/psipred/	Nugent and Jones (2009)
TOPCONS	S, P	http://topcons.cbr.su.se	Bernsel et al. (2009)
SCAMPI	S	http://scampi.cbr.su.se	Bernsel et al. (2008)
Phobius	S, P	http://phobius.sbc.su.se	Kall et al. (2005)
PHDhtm	S, P	http://www.predictprotein.org	Rost et al. (1995)
TMHMM2	S, P	http://www.cbs.dtu.dk/services/TMHMM	Sonnhammer et al. (1998)
Template detection and alignment			
HHpred	S, P	http://www.toolkit.tuebingen.mpg.de/hhpred	Soding et al. (2005)
MAFFT	S, P	http://www.mafft.cbrc.jp/alignment/server/	Katoh and Toh (2008)
PROMALS3D	S, P	http://prodata.swmed.edu/promals3d/promals3d.php	Pei et al. (2008)
SALIGN	S, P	http://modbase.compbio.ucsf.edu/salign-cgi/index.cgi	Madhusudhan et al. (2009)
AlignMe*	S, P	http://www.bioinfo.mpg.de/AlignMe	Khafizov et al. (2010)
MP-T*	S	http://www.opig.stats.ox.ac.uk/webapps/MPT/php/	Hill and Deane (2013)
PRALINE™*	S	http://www.ibi.vu.nl/programs/pralinewww/	Pirovano et al. (2008)
TM-Coffee*	S	http://www.tcoffee.crg.cat/apps/tcoffee/do:tmcoffee	Chang et al. (2012)

[a]*Method* marks computational sequence-based tools for TMH prediction and template detection/alignment. Template detection and alignment methods specifically designed for membrane proteins are marked with "*"

[b]*Type* gives type of the prediction method; "S" and "P" correspond to a web server and a computer program, respectively

[c]*URL* provides the Internet address for the service

[d]*Reference* gives the primary literature describing the method

structures of the APC family such as the amino acid transporter ApcT from *Methanococcus jannaschii* (Shaffer et al. 2009) and the arginine:agmatine antiporter AdiC from *Escherichia coli* (Gao et al. 2010).

For such challenging targets more state-of-the art fold assignment and threading methods need to be applied and analyzed carefully. They include profile–profile alignment and fold recognition methods that rely on observed or predicted structural information (e.g., HHpred (Soding et al. 2005)), multiple sequence alignment methods (e.g., MAFFT (Katoh and Toh 2008)), as well as methods that combine these approaches (e.g., Promals3D (Pei et al. 2008)). Several alignment methods have been specifically optimized to align membrane proteins (Khafizov et al. 2010; Hill and Deane 2013; Pirovano et al. 2008; Chang et al. 2012) (Table 2.1). Two recent developments include (1) alignment of membrane proteins (AlignMe), which is a user-friendly program that uses information such as residue hydrophobicity and predicted secondary structure to align sequences or hydropathy profiles of membrane proteins (Khafizov et al. 2010; Stamm et al. 2013) and (2) membrane protein threader (MP-T), which is a sequence-structure alignment method based on multiple sequence alignment (Hill and Deane 2013). MP-T uses predicted and observed information such as accessible surface area, membrane positioning, as well as secondary structure derived from homologs of the target protein. MP-T achieved higher accuracy than other alignment methods (e.g., Promals3D) using a variety of measures. For instance, models based on MP-T alignment were more accurate than those based on the best alternative alignment tool, where 25 % of the test set proteins exhibited an increase of ≥4 % in the global distance test total score (GDT_TS), which measures structural similarity between two protein structures with identical amino acid sequences (Zemla et al. 1999; Hill and Deane 2013).

Sequence similarity networks that visualize relationships among protein sequences and structures have shown to be useful in annotating the structures and functions of human SLC transporters (Schlessinger et al. 2010, 2013b). In brief, Schlessinger et al. first performed an "all-against-all" profile–profile comparison with SALIGN (Madhusudhan et al. 2009) among all the human SLC members. In combination with sequence-based clustering, they constructed and visualized sequence-based similarity networks for which varying cutoffs to define related sequences were examined. The final cutoffs, some representing distant sequence relationships (sequence identity of ~15 %), were selected to be permissive enough to detect the majority of known distant relationships among SLC members with minimal false positives. The analysis identified previously unknown relationships among human SLC families. For example, it was predicted that the glucose transporter family (GLUT, SLC2) are related to the organic ion transporter family (SLC22), which was subsequently confirmed experimentally for representative members of these families (Schlessinger et al. 2010). This similarity network also captured the structural similarity among the SLC5, SLC6, and SLC7 families (NSS fold) and predicted that other human families, including the SLC32 (the vesicular inhibitory amino acid transporter family), SLC36 (the proton-coupled amino acid transporter family), and SLC38 (System A and System N sodium-coupled neutral amino acid transporter family) are members of the NSS fold.

Recent discoveries of previously unknown human SLC families (i.e., SLC49-52) and the substantial increase of experimentally determined atomic structures of homologs from various organisms initiated an updated comprehensive analysis of the SLC members (Schlessinger et al. 2013b). For example, it was shown that the SLC52 family of riboflavin transporters is highly connected to the SLC29 of nucleoside transporters, suggesting that they are related in structure, function, or even both.

2.2.3 Model Construction, Refinement, and Assessment

Once a template structure or structures have been identified and aligned to the target sequence, an atomic comparative model can be constructed. The under-representation of experimentally determined membrane protein structures makes the human SLC targets difficult targets for modeling for two main reasons. First, there is only a small number of closely related protein structures; thus, researchers often rely on templates that are only distantly related to their target proteins (e.g., share the same fold). Second, most modeling methods were not optimized for membrane proteins, therefore, unique structural properties (e.g., specific inter-helical interactions (Harrington and Ben-Tal 2009)) are often missed by the current prediction methods. Interestingly, a previous benchmark study revealed that the model accuracy of membrane proteins strongly depends on the sequence similarity between the target and template, similarly to globular proteins (Forrest et al. 2006). For example, comparative models corresponding to the transmembrane regions exhibited <2 Å Cα RMSD to the native structure for targets that share sequence identity of 30 % or more with their templates. Although these results are encouraging, the same study identified the loop regions connecting the transmembrane regions to be considerably less accurate than the transmembrane domains (Forrest et al. 2006).

However, a recent increase in the number of atomic structures of membrane proteins (Pieper et al. 2013), particularly of SLC homologs (Schlessinger et al. 2013b), has expanded our ability to model the structures of many previously unmodelable pharmacologically important SLC transporters, including drug target (e.g., SLC2/GLUTs) and drug ADME (e.g., CNTs) transporter families. Some of these transporters can now be modeled with accuracy that might be sufficient for virtual screening (Schlessinger et al. 2010, 2013b; Pieper et al. 2013). For example, the concentrative nucleoside transporter from *Vibrio cholerae* (vcCNT) (Johnson et al. 2012) structure (resolution of 2.4 Å) was determined in a ligand-bound conformation. vcCNT shares sequence identity of 35–40 % and a highly similar binding site with the human SLC28/CNTs family, whose members transport key anticancer and antiviral drugs, thereby providing an excellent template for modeling of these proteins (Johnson et al. 2012).

Comparative modeling programs typically rely on several common approaches [reviewed in Sanchez and Sali (1998) and Marti-Renom et al. (2000)]. For example, the program MODELLER calculates comparative models based on satisfaction of spatial restraints that are derived from the target–template alignment, atomic statistical potentials, and molecular mechanics (Sali and Blundell 1993) (Table 2.2). MODELLER exists as standalone program and can also be run through the structure analysis and visualization program UCSF CHIMERA (Pettersen et al. 2004) or via the web server ModWeb (Pieper et al. 2011). ModWeb and another highly popular comparative modeling server, Swiss-Model (Kiefer et al. 2009), can take as input protein sequence and perform key steps of comparative modeling automatically, including template identification, target–template alignment, model building,

Table 2.2 Structural modeling methods

Method[a]	Type[b]	URL[c]	Reference[d]
Comparative modeling			
MODELLER/ModWeb	S, P	http://modbase.compbio.ucsf.edu/ModWeb20-html/modweb.html	Sali and Blundell (1993)
Swiss-Model	S, P	http://swissmodel.expasy.org/	Kiefer et al. (2009)
MEDELLER*	S	http://opig.stats.ox.ac.uk/webapps/medeller/home.pl?app=MEDELLER	Kelm et al. (2010)
Model assessment			
ModEval	S, P	http://modbase.compbio.ucsf.edu/modeval/help.cgi?type=help	Eramian et al. (2008)
Verify3D	S, P	http://nihserver.mbi.ucla.edu/Verify_3D/	Bowie et al. (1991)
PROCHECK	S, P	http://nihserver.mbi.ucla.edu/SAVES/	Laskowski et al. (1993)
Model refinement			
ModLoop	S, P	http://www.modbase.compbio.ucsf.edu/modloop/	Fiser and Sali (2003)
ArchPRED	S	http://manaslu.aecom.yu.edu/loopred/	Fernandez-Fuentes et al. (2006)
LoopyMod	P	http://wiki.c2b2.columbia.edu/honiglab_public/index.php/Software:Loopy	Soto et al. (2008)
RosettaMembane*	S, P	https://www.rosettacommons.org/software/	Yarov-Yarovoy et al. (2006)
GROMACS*	P	http://www.gromacs.org/	Hess et al. (2008)
Positioning in the membranous environment			
iMEMBRANE	S	http://opig.stats.ox.ac.uk/webapps/medeller/home.pl?app=iMembrane	Kelm et al. (2009)
PPM	S	http://opm.phar.umich.edu/server.php	Lomize et al. (2012)
TMDET	S	http://tmdet.enzim.hu/?go=form	Tusnady et al. (2005)

[a]*Method* marks freely available methods for model construction, assessment, analysis, and refinement. Comparative modeling and model refinement methods specifically optimized for membrane proteins are marked with "*"
[b]*Type* gives type of the prediction method; "S" and "P" correspond to a web server and computer program, respectively
[c]*URL* provides the Internet address for the service
[d]*Reference* gives the primary literature describing the method

and model assessment. Moreover, several recent programs have been designed to model the structures of membrane proteins (Kelm et al. 2010; Yarov-Yarovoy et al. 2006). For example, MEDELLER works by first identifying a reliable core structure and subsequently building a structural model by extending the core to other transmembrane regions and loops (Kelm et al. 2010). MEDELLER performance is overall better than that of MODELLER for membrane protein targets; however, for targets that share 20–40 % with their templates (i.e., "medium" set), which is the case for most of the modelable human SLC targets (Schlessinger et al. 2010, 2013b), MODELLER is slightly more accurate (RMSD of 5.33 vs 5.84 Å for MEDELLER) (Kelm et al. 2010).

After an initial model has been built, the model can be further refined. In particular, loop modeling, which usually does not require a template structure, can be employed for the non-membranous loop regions that connect the transmembrane helices; several top-performing loop modeling methods are publically available as downloadable programs (e.g., LoopyMod (Soto et al. 2008), PLOP (Jacobson et al. 2004)), via a web server (e.g., ArchPRED (Fernandez-Fuentes et al. 2006)), or both (ModLoop (Fiser and Sali 2003)). In addition, comparative models can be further refined by optimized the sidechain configuration on a fixed backbone conformation (Krivov et al. 2009; Xiang et al. 2007). For example, the program SCWRL4 is based on a graph theory algorithm that solves the combinatorial problem in side chain prediction faster than other approaches (Krivov et al. 2009). Finally, MD simulations (e.g., with GROMACS (Hess et al. 2008)) can be employed to minimize the energy of models thus refining the initial model, where an implicit solvent model can be used to account for the hydrophobic membrane environment (Lindorff-Larsen et al. 2010; Shaw et al. 2010) (Table 2.2).

Evaluation of the model accuracy is important for its appropriate interpretation. For example, the sequence identity between the target and template is related to the model accuracy, where models that share sequence identity of 50 % or more with their templates are expected to have 1 Å root mean square (RMS) error for the main chain atoms, making them potentially suitable for ligand docking (Baker and Sali 2001). Because the sequence identity is a crude measure that is not correlated strongly with model, various advanced methods have been developed for evaluation comparative models (Marti-Renom et al. 2000; Eramian et al. 2008). Perhaps the most popular model evaluation approaches are knowledge-based statistical potentials that are derived from distributions of structural features in experimentally determined protein structures (Melo et al. 2002). For instance, Z-DOPE is an atomic distance-dependent statistical potential that is implemented in MODELLER (Shen and Sali 2006); ModEval is a server that evaluates the accuracy of models using a variety of statistical potentials and a support vector machine predicting the accuracy of the model based on nine features, such as sequence similarity measures and statistical potentials that are extracted from a tailored training set of models unique to the assessed model (Eramian et al. 2008). Furthermore, other approaches evaluate the stereochemical properties of the structural model, such as its bond lengths, angles, torsion angles, planarity, and nonbond distances (e.g., the program PROCHECK) (Laskowski et al. 1993) (Table 2.2).

Finally, the position within the lipid bilayer of the protein structure obtained experimentally or computationally can also be estimated using a variety of methods, including PPM, TMDET, and iMEMBRANE (Tusnady et al. 2005; Lomize et al. 2012; Kelm et al. 2009) (Table 2.1). For example, iMEMBRANE, relies on coarse-grained MD simulations in membranous environment precalculated for proteins related to the target protein (Kelm et al. 2009).

2.2.4 Modeling Different SLC Conformations

An important step toward understanding transport by SLC transporter is the description of the conformations that the transporter undergoes during the transport cycle. Several approaches can be applied to model SLC transporters in different conformational states. First, initially shown by Forrest et al. for LeuT, the internal pseudo-symmetry of the transporter structure can be used, by swapping the symmetry-repeats (repeated halves of the transporter); essentially, one symmetry-repeat is used to model the other resulting in alternative conformation of the transporter (Forrest et al. 2008; Radestock and Forrest 2011; Faraldo-Gomez and Forrest 2011). Such approach has been applied to characterize a variety of transporters, including the glutamate transporter GltPh (Crisman et al. 2009), Na(+)/H (+) NhaA antiporter (Schushan et al. 2012), and LacY (Radestock and Forrest 2011).

Second, some transporters can be modeled based on structures of transporters that share the same fold but are only distantly related to the target (i.e., sequence identity of <10 %) (Madej et al. 2012). For example, vSGLT is currently the best template to model the human SLC5 family of Na^+ glucose co-transporter (e.g., sequence identity of 30 % with SLC5A2), but its structure was determined in only two distinct conformations (i.e., substrate bound (Faham et al. 2008) and substrate free (Watanabe et al. 2010)). However, there are total of 86 related structures of the NSS fold, representing at least six distinct conformations (Lomize et al. 2006) that can be used indirectly to model additional conformations of SLC5 members. In these cases, hybrid approaches for which low resolution data is integrated with modeling can be applied. For example, in a recent study, Madej et al. modeled an *apo*-intermediate conformation of LacY based on the atomic structure of PepT and information derived from other MFS structures including EmrD, FucP, and GlpT, in combination with low-resolution experimental data from double electron–electron resonance (DEER) measurements (Madej et al. 2012).

Third, specialized parallel computer architectures and efficient simulation algorithms have enabled researchers to perform simulations of longer times, thereby capturing important features of the transport process (Enkavi et al. 2013; Faraldo-Gomez and Forrest 2011; Khalili-Araghi et al. 2009; Lindahl and Sansom 2008). For example, employing steered molecular dynamics simulations, followed by experimental validation, Shi et al. characterized the substrate translocation pathway of LeuT, including the identification of a putative, second substrate-binding site located in the extracellular vestibule (Shi et al. 2008).

2.2.5 Molecular Docking to Comparative Models

In molecular docking, small molecules are sampled in large number of configurations and their binding modes ("poses") are evaluated or scored based on their

Table 2.3 Molecular docking software

Method[a]	Type[b]	URL[c]	Reference[d]
DOCK	P, S	http://blaster.docking.org/	Shoichet et al. (1993), Mysinger and Shoichet (2010)
AutoDock Vina	P, S	http://vina.scripps.edu/	Trott and Olson (2010)
GOLD	P, S	http://gold.ccdc.cam.ac.uk/setup_and_run_docking.php	Verdonk et al. (2003)
RosettaLigand	P, S	https://www.rosettacommons.org/software/	Meiler and Baker (2006)

[a]*Method* marks molecular docking/virtual screening methods
[b]*Type* gives type of the prediction method; "S" and "P" correspond to web server and computer program, respectively
[c]*URL* provides the Internet address for the service
[d]*Reference* gives the primary literature describing the method

complementarity to the target protein-binding site (Shoichet 2004; Coupez and Lewis 2006; Brooijmans and Kuntz 2003; Sperandio et al. 2006). Docking algorithms are most different in their conformational space search (sampling), representation of protein–ligand interactions, and binding affinity estimation (scoring). Popular docking programs that are publically available include AutoDock Vina (Trott and Olson 2010), GOLD (Verdonk et al. 2003), RosettaLigand (Meiler and Baker 2006), and DOCK (Shoichet et al. 1993; Mysinger and Shoichet 2010) (Table 2.3). For example, DOCK generates a set of overlapping spheres in contact with the surface of the target-binding site, representing a negative image of the binding site. These spheres are then matched with ligand atoms via a graph-matching algorithm. Finally, the score of each pose is derived from a sum of van der Waals, Poisson–Boltzmann electrostatic, and ligand desolvation penalty terms (Shoichet et al. 1993; Mysinger and Shoichet 2010).

When proteins do not have experimentally determined structures, docking can also be performed against structural models calculated using comparative modeling methods (Jacobson and Sali 2004; Fan et al. 2012; Kaufmann and Meiler 2012). Although most algorithms were not optimized for docking against comparative models, Fan et al. have recently developed two distance-dependent atomic statistical potentials for scoring protein–ligand complexes that were evaluated for docking against comparative models (Fan et al. 2011). In brief, PoseScore is optimized for predicting the binding pose of a known ligand and RankScore for identifying ligands through virtual screening.

The accuracy of virtual screening against structures or models of proteins with known ligands can be estimated by the enrichment for known ligands relative to decoy compounds, calculated by docking against the tested model ("enrichment calculations") (Huang et al. 2006; Fan et al. 2009; Schlessinger et al. 2011). In particular, the enrichment curve can be obtained by plotting the percentage of actual ligands identified (y-axis) within the top ranked subset of all database compounds (x-axis on logarithmic scale) (Huang et al. 2006; Irwin and Shoichet

2005; Fan et al. 2009). The area under the curve (logAUC) of the enrichment plot for $\Delta x = 0.1$ can be calculated using Eq. (2.1):

$$\text{logAUC} = \frac{1}{\log_{10} 100/0.1} \sum_{0.1}^{100} \frac{\text{ligand}_{\text{selected}}(x)}{\text{ligand}_{\text{total}}} \Delta x \quad \text{with} \quad x = \log_{10} \frac{N_n}{N_{\text{total}}} \tag{2.1}$$

where ligand total is the number of known ligands in a database containing N_{total} compounds, and ligand selected is the number of ligands found in a given subset of N_n compounds. For example, a random selection of known ligands from a database consisting of known ligands and decoys yields a logAUC of 14.5; a selection that picks twice as many ligands at any N_{subset} as a random selection yields a logAUC of 24.5 and is considered mediocre; a selection that picks ten times as many ligands at any N_{subset} as a random selection yields a logAUC of 47.7 is considered highly accurate (Fan et al. 2012). Moreover, by selecting models based on their enrichment scores, comparative models can be optimized for protein–ligand complementarity and thus for the utility in predicting additional ligands via virtual screening (Evers et al. 2003; Cavasotto et al. 2008; Carlsson et al. 2011; Schlessinger et al. 2012). In recent years, virtual screening against comparative models has been used to identify small molecule ligands for a variety of targets, including enzymes, GPCRs, and human SLC transporters [reviewed in Jacobson and Sali (2004), Shoichet and Kobilka (2012), Schlessinger et al. (2013a) and Fan et al. (2012)].

It was proposed that the following three criteria are needed to be fulfilled for a successful virtual screening against comparative models of human SLC members (Schlessinger et al. 2013b): (1) high-quality template structure (e.g., resolution of 3 Å or lower), (2) high-sequence similarity between the target and template (e.g., sequence identity of 25 % or higher), particularly in the binding site region, and (3) the conformation of the template structure. In particular, ligand-bound conformations are likely to yield more accurate predictions because they typically represent conformational states with high affinity for ligands and because the binding site location in the model can be predicted based on its location on the template structure (de novo identification of the residues involved in ligand binding is highly nontrivial (Ghersi and Sanchez 2012)) (Fan et al. 2009; Schlessinger et al. 2013b).

2.3 Examples

2.3.1 The Norepinephrine Transporter (NET, SLC6A2)

The solute carrier 6 family (SLC6) includes 20 Na^+- and Cl^--dependent membrane transporters of neurotransmitters, amino acids, and other metabolites that are involved in a variety of biological processes (Chen et al. 2004). The SLC6 is classified into four groups based on their amino acid sequences and functions—monoamine transporters, GABA transporters, amino acid transporters, and "orphan" transporters that include three amino acids transporters and one transporter with unknown function (Chen et al. 2004; Hahn and Blakely 2007). The norepinephrine transporter (NET, SLC6A2) is responsible for the reuptake of extracellular monoamine neurotransmitters such as norepinephrine, serotonin, and dopamine, thereby regulating adrenergic signaling pathways that are related to behavior and cardiovascular effects (Hahn and Blakely 2007; Pacholczyk et al. 1991). NET is a target for dozens of prescription (e.g., Ritalin) and recreational (cocaine) drugs that inhibit reuptake by NET to increase neurotransmitter availability for triggering adrenergic signaling.

Schlessinger et al. hypothesized that NET is involved in interacting with additional known drugs that act via polypharmacology and took modeling/docking approach to identify previously unknown NET ligand drugs (Schlessinger et al. 2011). The best template for modeling NET and other SLC6 members is LeuT which shares sequence identity of 20–25 % with the human SLC6 members as well as a conserved binding site dubbed the "S1 site." The LeuT structure has been determined in different conformations with or without ligands (Yamashita et al. 2005; Singh et al. 2007, 2008; Krishnamurthy et al. 2009). It was also shown that a putative substrate-binding site ("S2") is positioned on LeuT's surface (Shi et al. 2008; Zhao et al. 2010, 2011), and X-ray crystallography structures revealed that inhibitors can stabilize different conformations of LeuT by binding to sites overlapping with the S2 site (Claxton et al. 2010; Nyola et al. 2010). However, the S2 substrate-binding site has not been confirmed experimentally for the human SLC6 members.

Importantly, ligands transported by LeuT (amino acids) are chemically distinct from those transported by NET (positively charged monoamines), suggesting that the virtual screening against the LeuT S1-binding site is not likely to be productive for NET ligand identification and that a NET model is needed. The NET–LeuT alignment was a refined version of previously published comprehensive comparison of the SLC6 family, including eukaryotic and prokaryotic members (Beuming et al. 2006). Particularly, Beuming et al. used a variety of computational tools, including multiple sequence alignment, topology prediction, prediction of interior and lipid-exposed faces of transmembrane domains, secondary structure prediction, and even comparative modeling and previously published experimental data to construct an alignment of the SLC6 members.

NET comparative model was constructed using MODELLER and was subsequently refined by modeling key side chain using SCWRL4 (Krivov et al. 2009) and PyMOL (Fig. 2.2). For example, in the initial NET model, the side chain of Asp75, which corresponds to Gly24 in LeuT, did not face the binding site. Because an aspartate is highly conserved in the corresponding position of other SLC6 monoamine transporters, the dopamine transporter (DAT, SLC6A1) (Beuming et al. 2008) and the serotonin transporter (SERT, SLC6A4) (Kaufmann et al. 2009; Celik et al. 2008), where it makes key polar interactions with ligands (Severinsen et al. 2012; Beuming et al. 2006), Asp75 was remodeled on a fixed backbone. The utility of the model for virtual screening was estimated by the enrichment for the known ligands among the top scoring decoy compounds.

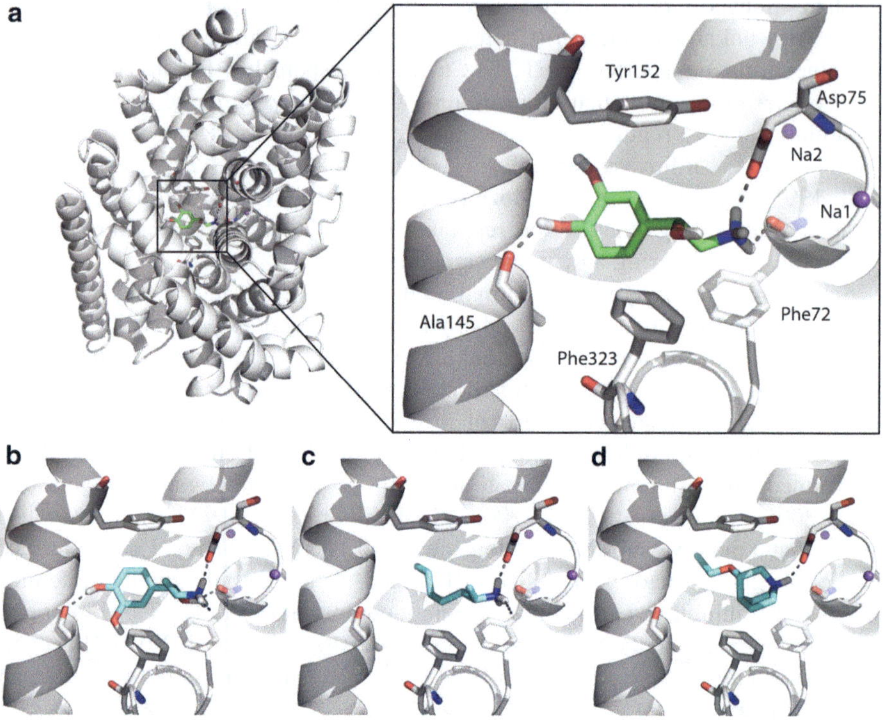

Fig. 2.2 *Predicted NET-binding site and mode of interaction with validated ligands*. (**a**) Predicted structure of the NET–norepinephrine complex. *Inset* shows the NET-binding site in complex with norepinephrine where important residues for binding are displayed as *sticks*. The known substrate norepinephrine is colored in *green*, with oxygen, nitrogen, and hydrogen atoms in *red*, *blue*, and *white*, respectively. Sodium ions are depicted as *purple* spheres. NET's transmembrane helices are shown as *white* ribbons; hydrogen bonds between norepinephrine and NET residues (e.g., Asp75) are shown as *dotted gray lines*. (**b–d**) Predicted binding modes of representative previously unknown prescription drugs that interact with NET (in *cyan sticks*). The drugs are (**b**) levonordefrin, and the chemically novel ligands tuaminoheptane (**c**) and talsaclidine (**d**)

The final model indicates four key binding site features that are likely to be important for NET-ligand recognition. They include (1) the polar interactions involving Asp75, (2) propensity to form interactions involving π-electrons of four aromatic residues (Phe72, Tyr152, Phe317, and Phe323), (3) hydrophobic affect involving small hydrophobic residues (e.g., Ala145 and Val148), and (4) a small binding site that limits the size of the ligands (Fig. 2.2a).

Six thousand four hundred and thirty six prescription drugs and drug-like molecules from the Kyoto Encyclopedia of Genes (KEGG DRUG) were docked against the S1-binding site of the final NET model. The top 200 predicted complexes were visually evaluated for experimental testing. In particular, molecules were selected based on the following five considerations: (1) the predicted pose was similar to predicted poses of known ligands, (2) scaffolds were predicted repeatedly, (3) the novelty of the chemical structure of the ligand, (4) the pharmacological function of the compound, and (5) the likelihood of a false positive prediction related to the limitation of the docking program, including docking poses with high internal energies or unbound polar groups. Ten of the 18 experimentally tested molecules were identified as NET ligands, where five of these were chemically novel (Fig. 2.2b–d). Interestingly, these results reveal polypharmacology of known drugs that were not known previously to bind NET, thereby rationalizing their positive and/or negative effects. For instance, talsaclidine is a muscarinic M1 receptor ligand that failed in clinical trials for treating Alzheimer's disease, due to side effects related to adrenergic effects (e.g., high blood pressure) (Fig. 2.2d) (Adamus et al. 1995). The pharmacodynamic (Giacomini et al. 2010; Schlessinger et al. 2011) and kinetic (Schlessinger et al. 2011) properties of talsaclidine suggest that inhibition of the NET uptake might occur in clinically relevant concentrations and contribute to the positive and negative pharmacological effects of the drug.

Because of NET's role in regulating adrenergic signaling, naturally occurring mutations in NET have been associated with behavioral disorders (e.g., panic disorder) and orthostatic hypotension (Hahn and Blakely 2007). The effect of one such mutation can be rationalized based on its location on the NET structural model. Ala457Pro is strongly associated with orthostatic hypotension and tachycardia (Runkel et al. 2000), likely resulted from high levels of norepinephrine in the plasma (Paczkowski et al. 2002). Ala457 is found in TMH 9, which is located in close proximity to what corresponds to the S2-binding site. Therefore, a mutation of alanine to proline in this position might impact NET structure through (1) distortion of the α-helix geometry, (2) disruption of the helix-helix packing, and (3) decreased flexibility.

2.3.2 *The GABA Transporter 2 (GAT-2, SLC6A13)*

GABA is a neurotransmitter that activates the GABAergic receptors in inhibitory neurons of the mammalian central nervous system (CNS). In humans, there are four SLC6 GABA transporters (GATs) that include three CNS transporters (GAT-1, GAT-3, and BGT-1) and one peripheral transporter (GAT-2). The CNS GATs are targeted by anticonvulsants and relaxant drugs (e.g., tiagabine) that inhibit GABA reuptake, which can increase concentration of GABA available for signaling. GAT-2 (SLC6A13), however, is relatively uncharacterized transporter that is highly expressed in the liver and kidney and was suggested to be important for disposition and distribution of GABAergic drugs (Madsen et al. 2010; Nakashita et al. 1997).

GAT-2 ligand-binding mechanism and pharmacological profile were explored by structure-based discovery approach (Schlessinger et al. 2012). In particular, GAT-2 structure was modeled based on the LeuT X-ray structures in an occluded substrate-bound conformation (Yamashita et al. 2005) and in an outward-facing inhibitor-bound conformation (Singh et al. 2008), using a refined version of the alignment developed by Beuming et al. (Beuming et al. 2006) (Fig. 2.3). Because GAT-2-binding site residues and mode of interaction with ligands have not been extensively characterized, thousands of GAT-2 configurations were sampled using sidechain modeling and energy minimization with MD simulations and were subsequently evaluated for their enrichment of known ligands among a database of known ligands and decoys. Thus, for GAT-2 modeling, enrichment scores were used for optimizing the model for protein–ligand complementarity, differently from NET modeling, where enrichment scores were used for assessing the model. Interestingly, during refinement, models in which the side chain of Glu48 faced the binding site and interacted with known ligands obtained the highest enrichment scores (and vice versa). This suggested that Glu48 plays a key role in ligand binding, which was confirmed experimentally via mutagenesis and kinetic measurements. In addition, the occluded and outward-facing GAT-2 models include minor structural rearrangements (e.g., the extracellular gate of Tyr129 and Phe288) that make additional volume accessible to solvent in the S1-binding site of the outward facing binding model, similarly to the corresponding LeuT template structures (Singh et al. 2008).

Virtual screening of 594,166 small molecule drugs, metabolites, and fragment-like molecules from the ZINC database was performed against models of the two conformations. The virtual screening results indicated that chemically distinct ligands are preferred by the different GAT-2 conformations, despite the small structural differences between their binding site conformations (Fig. 2.3). For example, larger and more hydrophobic molecules such as the muscle relaxant baclofen interact only with the outward facing model and were too large to be accommodated in the occluded conformation-binding site. Some ligands were predicted by using both models (e.g., homotaurine), thereby increasing the confidence in these predictions because the models were constructed independently.

Finally, 12 of the 31 experimentally tested molecules were identified as GAT-2 ligands, including known drugs (e.g., GABOB) and endogenous metabolites (e.g., homotaurine), where six of these ligands were chemically novel ligands of GAT-2 (e.g., baclofen). Surprisingly, the inhibitors identified from the outward-facing model, which is thought to represent an inhibited conformation, were not more potent than those predicted with the occluded model based on a substrate-bound conformation. These results indicate that the disparity between substrates and competitive inhibitors of GAT-2 is more complex than just the differences of simple physicochemical features such as size and hydrophobicity. Analogously, Albers et al. modeled the alanine–serine–cysteine transporter 2 (SLC1A5, ASCT2) based on structures of the archaeal aspartate transporter GltPh in an occluded and an outward-facing conformations (Albers et al. 2012). By using docking of serine derivatives against the two ASCT2 conformations, followed by experimental testing, they revealed that ligands identified via the screen against the occluded model were substrates, whereas ligands identified via the screen against the outward-facing inhibited conformation were inhibitors with larger and more hydrophobic side chains (Albers et al. 2012).

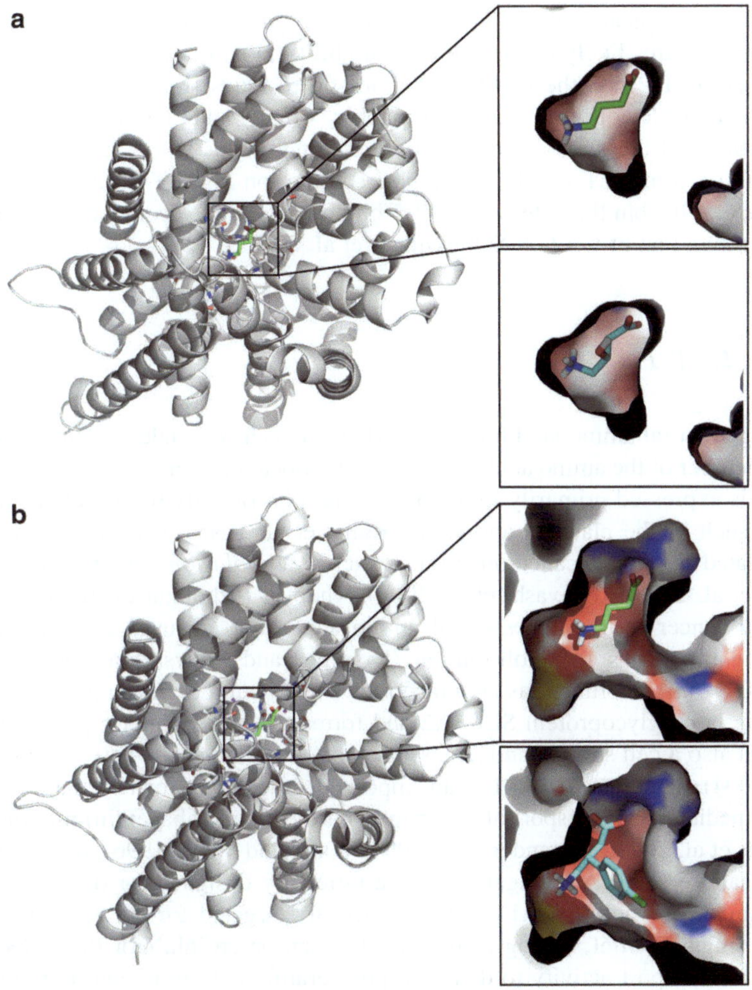

Fig. 2.3 Predicted GAT-2 structures and modes of interaction with ligands. Models of GAT-2 in occluded (**a**) and outward-facing (**b**) conformations. TMH regions are illustrated in *white ribbons*. *Inset* shows surface representation of the binding site in complex with GABA (*green*) and representative experimentally confirmed ligands that were identified from the virtual screening (*cyan*), including the GABAergic drugs vigabatrin (**a**) and baclofen (**b**). Oxygen, nitrogen, and hydrogen atoms are colored *red*, *blue*, and *white*, respectively

Comparison between models of SLC6 members and their mode of interactions with their corresponding ligands can help identifying rules for substrate specificity among members within this important family. Particularly, NET and GAT-2 share sequence identity of 45 % and their overall structures are predicted to be highly similar; however, these transporters have different substrate specificities, where NET substrates include aromatic monoamine neurotransmitters and GAT-2 substrates are typically small (7–8 heavy atoms), linear zwitterions. The following three key features likely determine the variability in substrate specificities among NET and GAT-2: (1) the number of aromatic residues (four in NET vs two in GAT-2) capable of forming hydrophobic effect and interaction involving π-electrons, (2) the number and location of amino acids with charged/polar side chains (Asp75 in NET vs Glu48 and Na1 sodium ion in GAT-2), and (3) the size and shape of the binding site where NET binding site in the occluded conformation is larger than that of GAT-2 (Schlessinger et al. 2013a).

2.3.3 LAT-1

The large neutral amino acid transporter (LAT-1) is a Na^+-independent exchanger and a member of the amino acid/polyamine/organocation transporter (APC) family. LAT-1 is expressed primarily in the blood–brain barrier (BBB) as well as in other tissues such as the placenta and testis (Kanai et al. 1998). LAT-1 is also highly upregulated in various cancerous tumors such as *glioblastoma multiforme* (GBM) (Kaira et al. 2008; Kobayashi et al. 2008), where it is thought to supply the fast-growing cancer cells with essential amino acids that are used as nutrients and signaling molecules for proliferation (Kroemer and Pouyssegur 2008; Nicklin et al. 2009). LAT-1 functions as transporter when bound to the single membrane-spanning helix glycoprotein SLC3A2 and forms the heterodimeric protein CD98, which is also a cell surface antigen associated with lymphocyte activation (Verrey et al. 2004). Therefore, LAT-1 is an important drug target for two main reasons: first, it mediates the transport of prescription drugs (e.g., gabapentin) into the CNS (Roberts et al. 2008; Alexander et al. 1994; Wang and Welty 1996), thus, potential drugs that are also LAT-1 ligands can be rationally designed for optimized BBB permeability. Second, LAT-1 can be specifically targeted for the development of cancer drugs. In brief, a drug against LAT-1 can be an inhibitor that selectively blocks its transport activity to deny the proliferating cell important nutrients or a cytotoxic substrate "hijacking" the transporter to deliver a drug against a different target in the cell (e.g., a metabolic enzyme) (Geier et al. 2013).

Sequence-based prediction methods indicate that LAT-1 includes twelve TMH (Verrey et al. 2004). Currently, there are 16 atomic structures representing four APC family members in the PDB. These structures were determined in five distinct conformations, representing different snapshots of the transport cycle (Lomize et al. 2006; Gao et al. 2009, 2010; Shaffer et al. 2009; Kowalczyk et al. 2011a; Schlessinger et al. 2010; Petrascheck et al. 2007) (Fig. 2.4a). For example, the arginine:agmatine antiporter AdiC from *E. coli* was determined in three conformations, including an "outward-facing," a "substrate-bound," and "intermediate" conformations (Gao et al. 2009, 2010; Kowalczyk et al. 2011a; Lomize et al. 2006). LAT-1 structure was recently modeled in complex with its natural substrates (Fig. 2.4) (Geier et al. 2013). Because several possible templates can be used to model LAT-1, the most optimal templates for modeling the transporter for the

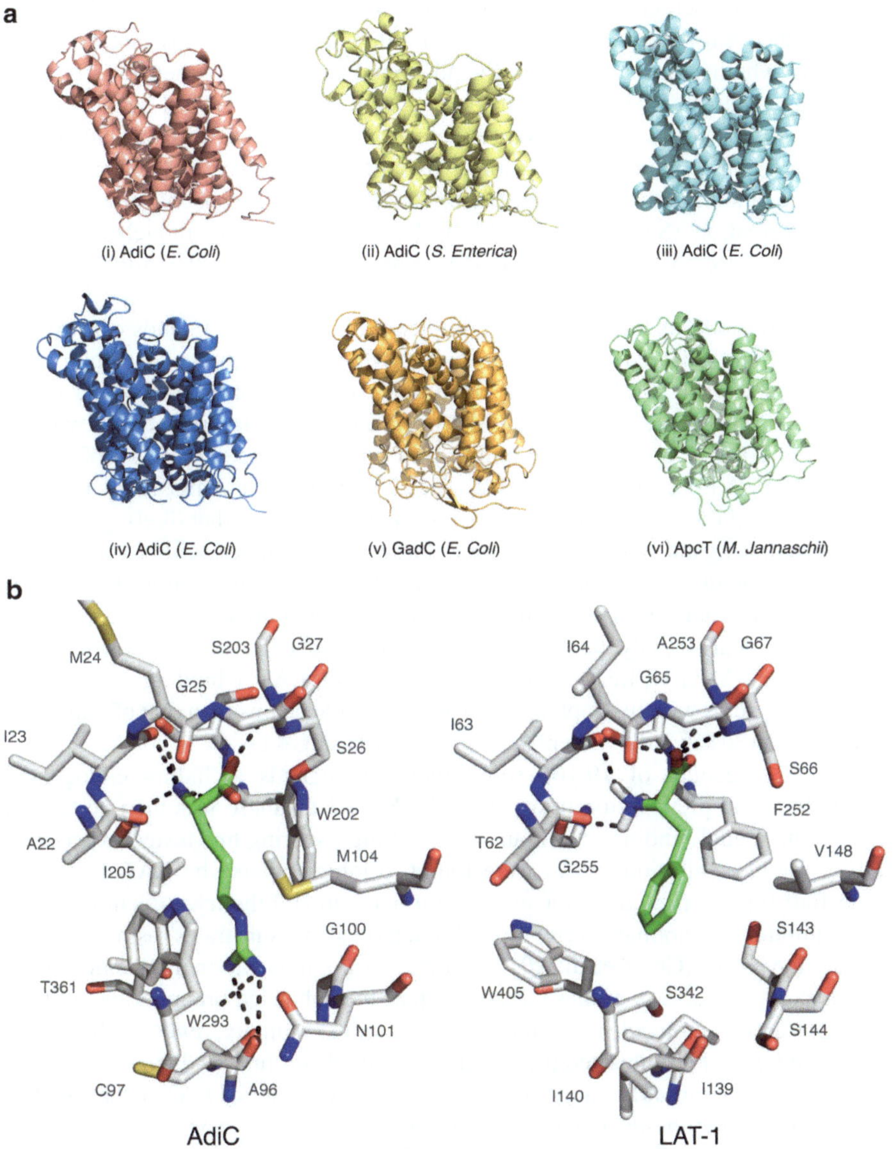

Fig. 2.4 LAT-1 modeling. (**a**) Potential template structures for modeling LAT-1. Structures belonging to the amino acid-polyamine-organocation (APC) family in OPM, including the arginine/agmatine antiporter (AdiC) in outward-facing conformation from (1) *E. coli* (PDB identifier 3lrb (Gao et al. 2009)) and (2) *Salmonella enterica* (3hqk (Fang et al. 2009)); AdiC from *E. coli* in an (3) intermediate outward-facing (3ob6 (Kowalczyk et al. 2011b)) and in a (4) substrate-bound outward-facing (3l1l (Gao et al. 2010)) conformations; (5) the glutamate/GABA antiporter (GadC) from *E. coli* in an inward conformation (4dji (Ma et al. 2012)); and (6) amino acid transporter ApcT from *M. jannaschii* in inward-facing occluded conformation (3gia (Shaffer et al. 2009)). (**b**) Binding sites of LAT-1 and AdiC. The X-ray structure of AdiC (*left panel*) and comparative

utility of virtual screening were selected by taking the following considerations: (1) sequence similarity to LAT-1, (2) structure quality, and (3) conformation, including whether the structure is ligand bound or not, and whether it is in an occluded, inward-facing, or outward-facing conformation. LAT-1 was therefore modeled based on the AdiC structure from *E. coli* in the outward-occluded arginine-bound conformation (Gao et al. 2010) and the structure of the amino acid, polyamine, and organo-cation transporter ApcT from *M. jannaschii* in an inward-*apo* conformation (Shaffer et al. 2009). The most enriching models that were ultimately subjected to virtual screening were those based on the AdiC structure from *E. coli*.

The final model of the LAT-1 in complex with its natural ligands can explain similarities and differences in ligand-binding specificity among the amino acid transporters LAT-1 and AdiC (Fig. 2.4b). In particular, backbone atoms in the binding site residues of AdiC and LAT-1 form conserved polar interactions with the carboxy and amine groups of the amino acid ligands. For example, Thr62, Ile63, and Ile64 in LAT-1 correspond to Ala22, Ile23, and Met24 in AdiC. In contrast, the binding sites of the two transporters have two key differences that likely give rise to their different ligand-binding preference (large neutral amino acids for LAT-1 vs polar amino acids for AdiC). First, LAT-1-binding site is much larger than that in AdiC, which results from substitution of amino acids with smaller side chains in LAT-1 (e.g., Val148, Gly255, and Ser342) with amino acids with larger side chains in AdiC (e.g., Met104, Ile205, and Trp293). Second, LAT-1-binding site includes more amino acids with hydrophobic and aromatic side chains than that of AdiC. For example, Thr361 in AdiC is replaced with Trp405 in LAT-1.

Virtual screening of 19,166 small molecule ligands including endogenous metabolites and prescription drugs from the KEGG database was then performed against the model binding-site. Four of the 12 high-scoring hits tested experimentally using *cis*-inhibition assays were found to inhibit uptake by LAT-1. Two of these four ligands (i.e., the anticancer agent acivicin and the tryptophan hydroxylase inhibitor fenclonine) were also characterized as substrates using a *trans*-stimulation assay (Geier et al. 2013). These results may rationalize some of the pharmacological effects of these drugs, such as the enhanced brain permeability of the CNS-active molecules acivicin and fenclonine. Finally, two of the newly discovered ligand hit inhibited proliferation of GBM cell line by distinct molecular mechanisms. In particular, acivicin is a cytotoxic substrate of LAT-1 that inhibits the activity glutamine-dependent amidotransferases in the biosynthesis of purines and pyrimidines (O'Dwyer et al. 1984); 3-iodo-L-tyrosine is a potent LAT-1 inhibitor that binds LAT-1 with stronger affinity, thereby possibly starving cancer

Fig. 2.4 (continued) model of LAT-1 (*right panel*) are shown in *light gray*. Protein atoms are shown as *sticks*; oxygen, nitrogen, and sulfur atoms are colored in *red*, *purple*, and *yellow*, respectively. Phenylalanine and arginine, substrates of LAT-1 and AdiC, respectively, are depicted as *green sticks*, and their predicted/observed hydrogen bonds with LAT-1 and AdiC, are shown as *dotted gray lines*

cells of nutrients supplied by LAT-1. These results provide useful chemical tools to characterize the role of LAT-1 in cancer metabolism as well as a starting point for optimizing the ligands for desired BBB and GBM cells permeability.

2.4 Conclusions and Future Outlook

Human SLC transporters are important for mediating transport of essential solutes, such as ions, metabolites, and prescription drugs. Recent determination of atomic structures of SLC transporters and improvement in computational structural biology methods have extended the applicability of computational structural biology to characterize many previously unmodelable human SLC transporters (see Chap. 5). We began this chapter by providing an introduction into the structures and functions of the human SLC transporters followed by the description of in silico methods that are often applied to predict the structures and substrate specificities of human SLC transporters. Finally, we illustrated how these computational methods can be applied in combination with experimental testing to characterize three important human SLC transporters.

Computational techniques are expected to continue making significant contribution to structural pharmacology investigation of membrane transporters. Particular active areas of research involve the development and application of hybrid approaches that integrate data from diverse biochemical and biophysical experiments such as NMR spectroscopy, electron microscopy, and FRET spectroscopy (e.g., Integrative Modeling Platform (Russel et al. 2012)) to model structures of unknown conformations with growing accuracy and confidence. Furthermore, the availability of increasing number of membrane protein structures will continue to enable researchers to develop more methods predicting various functional and structural aspects of membrane proteins (e.g., TMH contact prediction (Nugent and Jones 2010)). Third, new computer technologies will continue to emerge and enable researchers to perform MD simulations for longer times, which in turn will improve the state-of-the art force fields and might lead to new discoveries (Borhani and Shaw 2012).

Finally, in recent years there is a growing appreciation for the key role of the SLC transporters in human diseases such as cancer and diabetes (DeBerardinis and Thompson 2012; Wise and Thompson 2010; Vander Heiden et al. 2009; Adekola et al. 2012) and in drug ADME (Zamek-Gliszczynski et al. 2012; Giacomini et al. 2010). For example, most recently, the Food and Drug Administration (FDA) has approved canagliflozin—the first of a new class of diabetes drugs known as Na^{+}-glucose co-transporter 2 (SGLT2, SLC5A2) inhibitors (Traynor 2013). This understanding about the biomedical importance of the SLC transporters, coupled with the improvement in computational structural biology technology and increase in the number of SLC structures, will likely lead to additional future applications of rational structure-based drug design to membrane transporters (similarly to GPCRs). In fact, it is now feasible to target small biological systems such as cancer metabolism pathways and whole organs (e.g., the BBB) by modeling multiple transporter targets in multiple conformations, as well as predict ligands for these targets simultaneously. In the next few years, the number of drugs rationally designed against human SLC members is, thus, expected to grow.

Acknowledgments I am grateful to Greg Madej (UCLA) and Hao Fan (UCSF) for helpful comments on the manuscript. I am also thankful to Andrej Sali, Kathleen Giacomini, Ethan Geier, Sook Wah Yee, Matthias Wittwer, Robert Stroud, Andrew Waight, Bjørn Pedersen, Arik Zur, Ligong Chen, Ursula Pieper, Ben Webb, Massimiliano Bonomi, John Irwin, and Brian Shoichet (all UCSF), as well as to Christof Grewer (SUNY Binghamton), David Smith (University of Michigan Ann Arbor), Nir Ben-Tal (TAU), Lucy Forrest (NIH), Ron Kaback (UCLA), Christine Ziegler (MPIBP Frankfurt), Reinhard Krämer (University of Cologne), and Seok-Yong Lee (Duke) for discussions about SLC transporters, structural modeling, ligand docking, and related research.

References

Abramson J, Wright EM (2009) Structure and function of Na(+)-symporters with inverted repeats. Curr Opin Struct Biol 19(4):425–432. doi:10.1016/j.sbi.2009.06.002

Adamus WS, Leonard JP, Troger W (1995) Phase I clinical trials with WAL 2014, a new muscarinic agonist for the treatment of Alzheimer's disease. Life Sci 56(11–12):883–890

Adekola K, Rosen ST, Shanmugam M (2012) Glucose transporters in cancer metabolism. Curr Opin Oncol 24(6):650–654. doi:10.1097/CCO.0b013e328356da72

Albers T, Marsiglia W, Thomas T, Gameiro A, Grewer C (2012) Defining substrate and blocker activity of alanine–serine–cysteine transporter 2 (ASCT2) ligands with novel serine analogs. Mol Pharmacol 81(3):356–365. doi:10.1124/mol.111.075648

Alexander GM, Schwartzman RJ, Grothusen JR, Gordon SW (1994) Effect of plasma levels of large neutral amino acids and degree of parkinsonism on the blood-to-brain transport of levodopa in naive and MPTP parkinsonian monkeys. Neurology 44(8):1491–1499

Altschul SF, Gish W, Miller W, Myers EW, Lipman DJ (1990) Basic local alignment search tool. J Mol Biol 215(3):403–410. doi:10.1006/jmbi.1990.9999, S0022283680799990 [pii]

Baker D, Sali A (2001) Protein structure prediction and structural genomics. Science 294 (5540):93–96

Berardi MJ, Shih WM, Harrison SC, Chou JJ (2011) Mitochondrial uncoupling protein 2 structure determined by NMR molecular fragment searching. Nature 476(7358):109–113. doi:10.1038/nature10257

Bernsel A, Viklund H, Falk J, Lindahl E, von Heijne G, Elofsson A (2008) Prediction of membrane-protein topology from first principles. Proc Natl Acad Sci U S A 105(20):7177–7181. doi:10.1073/pnas.0711151105, 0711151105 [pii]

Bernsel A, Viklund H, Hennerdal A, Elofsson A (2009) TOPCONS: consensus prediction of membrane protein topology. Nucleic Acids Res 37(Web Server issue):W465–W468. doi:10.1093/nar/gkp363

Beuming T, Kniazeff J, Bergmann ML, Shi L, Gracia L, Raniszewska K, Newman AH, Javitch JA, Weinstein H, Gether U, Loland CJ (2008) The binding sites for cocaine and dopamine in the dopamine transporter overlap. Nat Neurosci 11(7):780–789. doi:10.1038/nn.2146, nn.2146 [pii]

Beuming T, Shi L, Javitch JA, Weinstein H (2006) A comprehensive structure-based alignment of prokaryotic and eukaryotic neurotransmitter/Na^+ symporters (NSS) aids in the use of the LeuT structure to probe NSS structure and function. Mol Pharmacol 70(5):1630–1642. doi:10.1124/mol.106.026120, mol.106.026120 [pii]

Borhani DW, Shaw DE (2012) The future of molecular dynamics simulations in drug discovery. J Comput Aided Mol Des 26(1):15–26. doi:10.1007/s10822-011-9517-y

Bowie JU, Luthy R, Eisenberg D (1991) A method to identify protein sequences that fold into a known three-dimensional structure. Science 253(5016):164–170

Brooijmans N, Kuntz ID (2003) Molecular recognition and docking algorithms. Annu Rev Biophys Biomol Struct 32:335–373. doi:10.1146/annurev.biophys.32.110601.142532

Carlsson J, Coleman RG, Setola V, Irwin JJ, Fan H, Schlessinger A, Sali A, Roth BL, Shoichet BK (2011) Ligand discovery from a dopamine D3 receptor homology model and crystal structure. Nat Chem Biol 7(11):769–778. doi:10.1038/nchembio.662

Cavasotto CN, Orry AJ, Murgolo NJ, Czarniecki MF, Kocsi SA, Hawes BE, O'Neill KA, Hine H, Burton MS, Voigt JH, Abagyan RA, Bayne ML, Monsma FJ Jr (2008) Discovery of novel chemotypes to a G-protein-coupled receptor through ligand-steered homology modeling and structure-based virtual screening. J Med Chem 51(3):581–588. doi:10.1021/jm070759m

Celik L, Sinning S, Severinsen K, Hansen CG, Moller MS, Bols M, Wiborg O, Schiott B (2008) Binding of serotonin to the human serotonin transporter. Molecular modeling and experimental validation. J Am Chem Soc 130(12):3853–3865. doi:10.1021/ja076403h

Chang JM, Di Tommaso P, Taly JF, Notredame C (2012) Accurate multiple sequence alignment of transmembrane proteins with PSI-Coffee. BMC Bioinformatics 13(Suppl 4):S1. doi:10.1186/1471-2105-13-S4-S1

Chen NH, Reith ME, Quick MW (2004) Synaptic uptake and beyond: the sodium- and chloride-dependent neurotransmitter transporter family SLC6. Pflugers Arch 447(5):519–531. doi:10.1007/s00424-003-1064-5

Choi JH, Yee SW, Ramirez AH, Morrissey KM, Jang GH, Joski PJ, Mefford JA, Hesselson SE, Schlessinger A, Jenkins G, Castro RA, Johns SJ, Stryke D, Sali A, Ferrin TE, Witte JS, Kwok PY, Roden DM, Wilke RA, McCarty CA, Davis RL, Giacomini KM (2011) A common 5'-UTR variant in MATE2-K is associated with poor response to metformin. Clin Pharmacol Ther 90(5):674–684. doi:10.1038/clpt.2011.165

Claxton DP, Quick M, Shi L, de Carvalho FD, Weinstein H, Javitch JA, McHaourab HS (2010) Ion/substrate-dependent conformational dynamics of a bacterial homolog of neurotransmitter: sodium symporters. Nat Struct Mol Biol 17(7):822–829. doi:10.1038/nsmb.1854

Coupez B, Lewis RA (2006) Docking and scoring—theoretically easy, practically impossible? Curr Med Chem 13(25):2995–3003

Crisman TJ, Qu S, Kanner BI, Forrest LR (2009) Inward-facing conformation of glutamate transporters as revealed by their inverted-topology structural repeats. Proc Natl Acad Sci U S A 106(49):20752–20757. doi:10.1073/pnas.0908570106

DeBerardinis RJ, Thompson CB (2012) Cellular metabolism and disease: what do metabolic outliers teach us? Cell 148(6):1132–1144. doi:10.1016/j.cell.2012.02.032

Dresser MJ, Gray AT, Giacomini KM (2000) Kinetic and selectivity differences between rodent, rabbit, and human organic cation transporters (OCT1). J Pharmacol Exp Ther 292(3):1146–1152

Elofsson A, von Heijne G (2007) Membrane protein structure: prediction versus reality. Annu Rev Biochem 76:125–140. doi:10.1146/annurev.biochem.76.052705.163539

Enkavi G, Li J, Mahinthichaichan P, Wen PC, Huang Z, Shaikh SA, Tajkhorshid E (2013) Simulation studies of the mechanism of membrane transporters. Methods Mol Biol 924:361–405. doi:10.1007/978-1-62703-017-5_14

Eramian D, Eswar N, Shen MY, Sali A (2008) How well can the accuracy of comparative protein structure models be predicted? Protein Sci 17(11):1881–1893. doi:10.1110/ps.036061.108, ps.036061.108 [pii]

Evers A, Gohlke H, Klebe G (2003) Ligand-supported homology modelling of protein binding-sites using knowledge-based potentials. J Mol Biol 334(2):327–345

Eyre TA, Partridge L, Thornton JM (2004) Computational analysis of alpha-helical membrane protein structure: implications for the prediction of 3D structural models. Protein Eng Des Sel 17(8):613–624. doi:10.1093/protein/gzh072

Faham S, Watanabe A, Besserer GM, Cascio D, Specht A, Hirayama BA, Wright EM, Abramson J (2008) The crystal structure of a sodium galactose transporter reveals mechanistic insights into Na^+/sugar symport. Science 321(5890):810–814

Fan H, Irwin JJ, Sali A (2012) Virtual ligand screening against comparative protein structure models. Methods Mol Biol 819:105–126. doi:10.1007/978-1-61779-465-0_8

Fan H, Irwin JJ, Webb BM, Klebe G, Shoichet BK, Sali A (2009) Molecular docking screens using comparative models of proteins. J Chem Inf Model 49(11):2512–2527. doi:10.1021/ci9003706

Fan H, Schneidman-Duhovny D, Irwin JJ, Dong G, Shoichet BK, Sali A (2011) Statistical potential for modeling and ranking of protein–ligand interactions. J Chem Inf Model 51 (12):3078–3092. doi:10.1021/ci200377u

Fang Y, Jayaram H, Shane T, Kolmakova-Partensky L, Wu F, Williams C, Xiong Y, Miller C (2009) Structure of a prokaryotic virtual proton pump at 3.2 A resolution. Nature 460 (7258):1040–1043. doi:10.1038/nature08201, nature08201 [pii]

Faraldo-Gomez JD, Forrest LR (2011) Modeling and simulation of ion-coupled and ATP-driven membrane proteins. Curr Opin Struct Biol 21(2):173–179. doi:10.1016/j.sbi.2011.01.013

Feng B, Dresser MJ, Shu Y, Johns SJ, Giacomini KM (2001) Arginine 454 and lysine 370 are essential for the anion specificity of the organic anion transporter, rOAT3. Biochemistry 40 (18):5511–5520, bi002841o [pii]

Fernandez-Fuentes N, Zhai J, Fiser A (2006) ArchPRED: a template based loop structure prediction server. Nucleic Acids Res 34(Web Server issue):W173–W176. doi:10.1093/nar/gkl113

Fiser A, Sali A (2003) ModLoop: automated modeling of loops in protein structures. Bioinformatics 19(18):2500–2501

Forrest LR (2013) Structural biology. (Pseudo-)symmetrical transport. Science 339(6118):399–401. doi:10.1126/science.1228465

Forrest LR, Kramer R, Ziegler C (2011) The structural basis of secondary active transport mechanisms. Biochim Biophys Acta 1807(2):167–188. doi:10.1016/j.bbabio.2010.10.014

Forrest LR, Rudnick G (2009) The rocking bundle: a mechanism for ion-coupled solute flux by symmetrical transporters. Physiology (Bethesda) 24:377–386. doi:10.1152/physiol.00030. 2009, 24/6/377 [pii]

Forrest LR, Tang CL, Honig B (2006) On the accuracy of homology modeling and sequence alignment methods applied to membrane proteins. Biophys J 91(2):508–517

Forrest LR, Zhang YW, Jacobs MT, Gesmonde J, Xie L, Honig BH, Rudnick G (2008) Mechanism for alternating access in neurotransmitter transporters. Proc Natl Acad Sci U S A 105 (30):10338–10343

Gao X, Lu F, Zhou L, Dang S, Sun L, Li X, Wang J, Shi Y (2009) Structure and mechanism of an amino acid antiporter. Science 324(5934):1565–1568. doi:10.1126/science.1173654, 1173654 [pii]

Gao X, Zhou L, Jiao X, Lu F, Yan C, Zeng X, Wang J, Shi Y (2010) Mechanism of substrate recognition and transport by an amino acid antiporter. Nature 463(7282):828–832. doi:10. 1038/nature08741

Geier EG, Schlessinger A, Fan H, Gable JE, Irwin JJ, Sali A, Giacomini KM (2013) Structure-based ligand discovery for the large-neutral amino acid transporter 1, LAT-1. Proc Natl Acad Sci U S A 110(14):5480–5485. doi:10.1073/pnas.1218165110

Gether U, Andersen PH, Larsson OM, Schousboe A (2006) Neurotransmitter transporters: molecular function of important drug targets. Trends Pharmacol Sci 27(7):375–383

Ghersi D, Sanchez R (2012) Automated identification of binding sites for phosphorylated ligands in protein structures. Proteins 80(10):2347–2358. doi:10.1002/prot.24117

Giacomini KM, Huang SM, Tweedie DJ, Benet LZ, Brouwer KL, Chu X, Dahlin A, Evers R, Fischer V, Hillgren KM, Hoffmaster KA, Ishikawa T, Keppler D, Kim RB, Lee CA, Niemi M, Polli JW, Sugiyama Y, Swaan PW, Ware JA, Wright SH, Yee SW, Zamek-Gliszczynski MJ,

Zhang L (2010) Membrane transporters in drug development. Nat Rev Drug Discov 9(3):215–236. doi:10.1038/nrd3028

Gruswitz F, Chaudhary S, Ho JD, Schlessinger A, Pezeshki B, Ho CM, Sali A, Westhoff CM, Stroud RM (2010) Function of human Rh based on structure of RhCG at 2.1 A. Proc Natl Acad Sci U S A 107(21):9638–9643. doi:10.1073/pnas.1003587107

Guan L, Kaback HR (2006) Lessons from lactose permease. Annu Rev Biophys Biomol Struct 35:67–91. doi:10.1146/annurev.biophys.35.040405.102005

Guastella J, Nelson N, Nelson H, Czyzyk L, Keynan S, Miedel MC, Davidson N, Lester HA, Kanner BI (1990) Cloning and expression of a rat brain GABA transporter. Science 249 (4974):1303–1306

Hahn MK, Blakely RD (2007) The functional impact of SLC6 transporter genetic variation. Annu Rev Pharmacol Toxicol 47:401–441. doi:10.1146/annurev.pharmtox.47.120505.105242

Harrington SE, Ben-Tal N (2009) Structural determinants of transmembrane helical proteins. Structure 17(8):1092–1103. doi:10.1016/j.str.2009.06.009

Hediger MA, Romero MF, Peng JB, Rolfs A, Takanaga H, Bruford EA (2004) The ABCs of solute carriers: physiological, pathological and therapeutic implications of human membrane transport proteinsIntroduction. Pflugers Arch 447(5):465–468

Hess B, Kutzner C, van der Spoel D, Lindahl E (2008) GROMACS 4: algorithms for highly efficient, load-balanced, and scalable molecular simulation. J Chem Theory Comput 4(3):435–447. doi:10.1021/ct700301q

Hill JR, Deane CM (2013) MP-T: improving membrane protein alignment for structure prediction. Bioinformatics 29(1):54–61. doi:10.1093/bioinformatics/bts640

Hopf TA, Colwell LJ, Sheridan R, Rost B, Sander C, Marks DS (2012) Three-dimensional structures of membrane proteins from genomic sequencing. Cell 149(7):1607–1621. doi:10.1016/j.cell.2012.04.012

Huang N, Shoichet BK, Irwin JJ (2006) Benchmarking sets for molecular docking. J Med Chem 49 (23):6789–6801. doi:10.1021/jm0608356

Irwin JJ, Shoichet BK (2005) ZINC—a free database of commercially available compounds for virtual screening. J Chem Inf Model 45(1):177–182

Jacobson M, Sali A (2004) Comparative protein structure modeling and its applications to drug discovery. In: Overington J (ed) Annual reports in medicinal chemistry, vol 39. Inpharmatica Ltd, London, pp 259–276

Jacobson MP, Pincus DL, Rapp CS, Day TJ, Honig B, Shaw DE, Friesner RA (2004) A hierarchical approach to all-atom protein loop prediction. Proteins 55(2):351–367. doi:10.1002/prot.10613

Jardetzky O (1966) Simple allosteric model for membrane pumps. Nature 211(5052):969–970

Johnson ZL, Cheong CG, Lee SY (2012) Crystal structure of a concentrative nucleoside transporter from *Vibrio cholerae* at 2.4 A. Nature 483(7390):489–493. doi:10.1038/nature10882

Kaira K, Oriuchi N, Imai H, Shimizu K, Yanagitani N, Sunaga N, Hisada T, Tanaka S, Ishizuka T, Kanai Y, Endou H, Nakajima T, Mori M (2008) Prognostic significance of L-type amino acid transporter 1 expression in resectable stage I–III nonsmall cell lung cancer. Br J Cancer 98 (4):742–748. doi:10.1038/sj.bjc.6604235

Kall L, Krogh A, Sonnhammer EL (2005) An HMM posterior decoder for sequence feature prediction that includes homology information. Bioinformatics 21(Suppl 1):i251–i257. doi:10.1093/bioinformatics/bti1014

Kanai Y, Segawa H, Miyamoto K, Uchino H, Takeda E, Endou H (1998) Expression cloning and characterization of a transporter for large neutral amino acids activated by the heavy chain of 4 F2 antigen (CD98). J Biol Chem 273(37):23629–23632

Kanner BI, Zomot E (2008) Sodium-coupled neurotransmitter transporters. Chem Rev 108 (5):1654–1668. doi:10.1021/cr078246a

Katoh K, Toh H (2008) Recent developments in the MAFFT multiple sequence alignment program. Brief Bioinform 9(4):286–298. doi:10.1093/bib/bbn013

Kaufmann KW, Dawson ES, Henry LK, Field JR, Blakely RD, Meiler J (2009) Structural determinants of species-selective substrate recognition in human and Drosophila serotonin transporters revealed through computational docking studies. Proteins 74(3):630–642. doi:10.1002/prot.22178

Kaufmann KW, Meiler J (2012) Using RosettaLigand for small molecule docking into comparative models. PLoS One 7(12):e50769. doi:10.1371/journal.pone.0050769

Kelm S, Shi J, Deane CM (2009) iMembrane: homology-based membrane-insertion of proteins. Bioinformatics 25(8):1086–1088. doi:10.1093/bioinformatics/btp102

Kelm S, Shi J, Deane CM (2010) MEDELLER: homology-based coordinate generation for membrane proteins. Bioinformatics 26(22):2833–2840. doi:10.1093/bioinformatics/btq554

Kernytsky A, Rost B (2003) Static benchmarking of membrane helix predictions. Nucleic Acids Res 31(13):3642–3644

Khafizov K, Staritzbichler R, Stamm M, Forrest LR (2010) A study of the evolution of inverted-topology repeats from LeuT-fold transporters using AlignMe. Biochemistry 49(50):10702–10713. doi:10.1021/bi101256x

Khalili-Araghi F, Gumbart J, Wen PC, Sotomayor M, Tajkhorshid E, Schulten K (2009) Molecular dynamics simulations of membrane channels and transporters. Curr Opin Struct Biol 19 (2):128–137. doi:10.1016/j.sbi.2009.02.011

Kiefer F, Arnold K, Kunzli M, Bordoli L, Schwede T (2009) The SWISS-MODEL repository and associated resources. Nucleic Acids Res 37(Database issue):D387–D392. doi:10.1093/nar/gkn750

Kobayashi K, Ohnishi A, Promsuk J, Shimizu S, Kanai Y, Shiokawa Y, Nagane M (2008) Enhanced tumor growth elicited by L-type amino acid transporter 1 in human malignant glioma cells. Neurosurgery 62(2):493–503. doi:10.1227/01.neu.0000316018.51292.19, discussion 503–494

Kowalczyk L, Ratera M, Paladino A, Bartoccioni P, Errasti-Murugarren E, Valencia E, Portella G, Bial S, Zorzano A, Fita I, Orozco M, Carpena X, Vazquez-Ibar JL, Palacin M (2011a) Molecular basis of substrate-induced permeation by an amino acid antiporter. Proc Natl Acad Sci U S A 108(10):3935–3940. doi:10.1073/pnas.1018081108

Kowalczyk L, Ratera M, Paladino A, Bartoccioni P, Errasti-Murugarren E, Valencia E, Portella G, Bial S, Zorzano A, Fita I, Orozco M, Carpena X, Vazquez-Ibar JL, Palacin M (2011b) Molecular basis of substrate-induced permeation by an amino acid antiporter. Proc Natl Acad Sci U S A 108(10):3935–3940. doi:10.1073/pnas.1018081108

Krishnamurthy H, Gouaux E (2012) X-ray structures of LeuT in substrate-free outward-open and apo inward-open states. Nature 481(7382):469–474. doi:10.1038/nature10737

Krishnamurthy H, Piscitelli CL, Gouaux E (2009) Unlocking the molecular secrets of sodium-coupled transporters. Nature 459(7245):347–355. doi:10.1038/nature08143, nature08143 [pii]

Krivov GG, Shapovalov MV, Dunbrack RL Jr (2009) Improved prediction of protein side-chain conformations with SCWRL4. Proteins 77(4):778–795. doi:10.1002/prot.22488

Kroemer G, Pouyssegur J (2008) Tumor cell metabolism: cancer's Achilles' heel. Cancer Cell 13 (6):472–482. doi:10.1016/j.ccr.2008.05.005

Kyte J, Doolittle RF (1982) A simple method for displaying the hydropathic character of a protein. J Mol Biol 157(1):105–132

Laskowski RA, Moss DS, Thornton JM (1993) Main-chain bond lengths and bond angles in protein structures. J Mol Biol 231(4):1049–1067. doi:10.1006/jmbi.1993.1351

Lindahl E, Sansom MS (2008) Membrane proteins: molecular dynamics simulations. Curr Opin Struct Biol 18(4):425–431. doi:10.1016/j.sbi.2008.02.003

Lindorff-Larsen K, Piana S, Palmo K, Maragakis P, Klepeis JL, Dror RO, Shaw DE (2010) Improved side-chain torsion potentials for the Amber ff99SB protein force field. Proteins 78 (8):1950–1958. doi:10.1002/prot.22711

Lomize MA, Lomize AL, Pogozheva ID, Mosberg HI (2006) OPM: orientations of proteins in membranes database. Bioinformatics 22(5):623–625. doi:10.1093/bioinformatics/btk023, btk023 [pii]

Lomize MA, Pogozheva ID, Joo H, Mosberg HI, Lomize AL (2012) OPM database and PPM web server: resources for positioning of proteins in membranes. Nucleic Acids Res 40(Database issue):D370–D376. doi:10.1093/nar/gkr703

Lorber DM, Shoichet BK (1998) Flexible ligand docking using conformational ensembles. Protein Sci 7(4):938–950. doi:10.1002/pro.5560070411

Lu F, Li S, Jiang Y, Jiang J, Fan H, Lu G, Deng D, Dang S, Zhang X, Wang J, Yan N (2011) Structure and mechanism of the uracil transporter UraA. Nature 472(7342):243–246. doi:10.1038/nature09885

Ma D, Lu P, Yan C, Fan C, Yin P, Wang J, Shi Y (2012) Structure and mechanism of a glutamate–GABA antiporter. Nature 483(7391):632–636. doi:10.1038/nature10917

Madej MG, Dang S, Yan N, Kaback HR (2013) Evolutionary mix-and-match with MFS transporters. Proc Natl Acad Sci U S A 110(15):5870–5874. doi:10.1073/pnas.1303538110

Madej MG, Soro SN, Kaback HR (2012) Apo-intermediate in the transport cycle of lactose permease (LacY). Proc Natl Acad Sci U S A 109(44):E2970–E2978. doi:10.1073/pnas.1211183109

Madhusudhan MS, Webb BM, Marti-Renom MA, Eswar N, Sali A (2009) Alignment of multiple protein structures based on sequence and structure features. Protein Eng Des Sel 22(9):569–574. doi:10.1093/protein/gzp040, gzp040 [pii]

Madsen KK, White HS, Schousboe A (2010) Neuronal and non-neuronal GABA transporters as targets for antiepileptic drugs. Pharmacol Ther 125(3):394–401. doi:10.1016/j.pharmthera.2009.11.007

Marks DS, Colwell LJ, Sheridan R, Hopf TA, Pagnani A, Zecchina R, Sander C (2011) Protein 3D structure computed from evolutionary sequence variation. PLoS One 6(12):e28766. doi:10.1371/journal.pone.0028766

Marti-Renom MA, Stuart AC, Fiser A, Sanchez R, Melo F, Sali A (2000) Comparative protein structure modeling of genes and genomes. Annu Rev Biophys Biomol Struct 29:291–325. doi:10.1146/annurev.biophys.29.1.291

Meiler J, Baker D (2006) ROSETTALIGAND: protein-small molecule docking with full side-chain flexibility. Proteins 65(3):538–548. doi:10.1002/prot.21086

Melo F, Sanchez R, Sali A (2002) Statistical potentials for fold assessment. Protein Sci 11(2):430–448

Murzin AG, Brenner SE, Hubbard T, Chothia C (1995) SCOP: a structural classification of proteins database for the investigation of sequences and structures. J Mol Biol 247(4):536–540. doi:10.1006/jmbi.1995.0159, S0022283685701593 [pii]

Mysinger MM, Shoichet BK (2010) Rapid context-dependent ligand desolvation in molecular docking. J Chem Inf Model 50(9):1561–1573. doi:10.1021/ci100214a

Nakashita M, Sasaki K, Sakai N, Saito N (1997) Effects of tricyclic and tetracyclic antidepressants on the three subtypes of GABA transporter. Neurosci Res 29(1):87–91

Newstead S, Drew D, Cameron AD, Postis VL, Xia X, Fowler PW, Ingram JC, Carpenter EP, Sansom MS, McPherson MJ, Baldwin SA, Iwata S (2011) Crystal structure of a prokaryotic homologue of the mammalian oligopeptide–proton symporters, PepT1 and PepT2. EMBO J 30(2):417–426. doi:10.1038/emboj.2010.309

Nicklin P, Bergman P, Zhang B, Triantafellow E, Wang H, Nyfeler B, Yang H, Hild M, Kung C, Wilson C, Myer VE, MacKeigan JP, Porter JA, Wang YK, Cantley LC, Finan PM, Murphy LO (2009) Bidirectional transport of amino acids regulates mTOR and autophagy. Cell 136(3):521–534. doi:10.1016/j.cell.2008.11.044

Nugent T, Jones DT (2009) Transmembrane protein topology prediction using support vector machines. BMC Bioinforma 10:159. doi:10.1186/1471-2105-10-159

Nugent T, Jones DT (2010) Predicting transmembrane helix packing arrangements using residue contacts and a force-directed algorithm. PLoS Comput Biol 6(3):e1000714. doi:10.1371/journal.pcbi.1000714

Nugent T, Jones DT (2012) Membrane protein structural bioinformatics. J Struct Biol 179(3):327–337. doi:10.1016/j.jsb.2011.10.008

Nyola A, Karpowich NK, Zhen J, Marden J, Reith ME, Wang DN (2010) Substrate and drug binding sites in LeuT. Curr Opin Struct Biol 20(4):415–422. doi:10.1016/j.sbi.2010.05.007
O'Dwyer PJ, Alonso MT, Leyland-Jones B (1984) Acivicin: a new glutamine antagonist in clinical trials. J Clin Oncol 2(9):1064–1071
Pacholczyk T, Blakely RD, Amara SG (1991) Expression cloning of a cocaine- and antidepressant-sensitive human noradrenaline transporter. Nature 350(6316):350–354. doi:10.1038/350350a0
Paczkowski FA, Bonisch H, Bryan-Lluka LJ (2002) Pharmacological properties of the naturally occurring Ala(457)Pro variant of the human norepinephrine transporter. Pharmacogenetics 12 (2):165–173
Pedersen BP, Kumar H, Waight AB, Risenmay AJ, Roe-Zurz Z, Chau BH, Schlessinger A, Bonomi M, Harries W, Sali A, Johri AK, Stroud RM (2013) Crystal structure of a eukaryotic phosphate transporter. Nature. doi:10.1038/nature12042
Pei J, Kim BH, Grishin NV (2008) PROMALS3D: a tool for multiple protein sequence and structure alignments. Nucleic Acids Res 36(7):2295–2300. doi:10.1093/nar/gkn072
Perez C, Ziegler C (2013) Mechanistic aspects of sodium-binding sites in LeuT-like fold symporters. Biol Chem. doi:10.1515/hsz-2012-0336
Petrascheck M, Ye X, Buck LB (2007) An antidepressant that extends lifespan in adult Caenorhabditis elegans. Nature 450(7169):553–556. doi:10.1038/nature05991, nature05991 [pii]
Pettersen EF, Goddard TD, Huang CC, Couch GS, Greenblatt DM, Meng EC, Ferrin TE (2004) UCSF Chimera—a visualization system for exploratory research and analysis. J Comput Chem 25(13):1605–1612. doi:10.1002/jcc.20084
Pieper U, Schlessinger A, Kloppmann E, Chang GA, Chou JJ, Dumont ME, Fox BG, Fromme P, Hendrickson WA, Malkowski MG, Rees DC, Stokes DL, Stowell MH, Wiener MC, Rost B, Stroud RM, Stevens RC, Sali A (2013) Coordinating the impact of structural genomics on the human alpha-helical transmembrane proteome. Nat Struct Mol Biol 20(2):135–138. doi:10.1038/nsmb.2508
Pieper U, Webb BM, Barkan DT, Schneidman-Duhovny D, Schlessinger A, Braberg H, Yang Z, Meng EC, Pettersen EF, Huang CC, Datta RS, Sampathkumar P, Madhusudhan MS, Sjolander K, Ferrin TE, Burley SK, Sali A (2011) ModBase, a database of annotated comparative protein structure models, and associated resources. Nucleic Acids Res 39(Database issue):D465–D474. doi:10.1093/nar/gkq1091
Pirovano W, Feenstra KA, Heringa J (2008) PRALINETM: a strategy for improved multiple alignment of transmembrane proteins. Bioinformatics 24(4):492–497. doi:10.1093/bioinformatics/btm636
Povey S, Lovering R, Bruford E, Wright M, Lush M, Wain H (2001) The HUGO gene nomenclature committee (HGNC). Hum Genet 109(6):678–680. doi:10.1007/s00439-001-0615-0
Punta M, Forrest LR, Bigelow H, Kernytsky A, Liu J, Rost B (2007) Membrane protein prediction methods. Methods 41(4):460–474. doi:10.1016/j.ymeth.2006.07.026
Radestock S, Forrest LR (2011) The alternating-access mechanism of MFS transporters arises from inverted-topology repeats. J Mol Biol 407(5):698–715. doi:10.1016/j.jmb.2011.02.008
Ressl S, Terwisscha van Scheltinga AC, Vonrhein C, Ott V, Ziegler C (2009) Molecular basis of transport and regulation in the Na(+)/betaine symporter BetP. Nature 458(7234):47–52. doi:10.1038/nature07819, nature07819 [pii]
Roberts LM, Black DS, Raman C, Woodford K, Zhou M, Haggerty JE, Yan AT, Cwirla SE, Grindstaff KK (2008) Subcellular localization of transporters along the rat blood–brain barrier and blood–cerebral-spinal fluid barrier by in vivo biotinylation. Neuroscience 155(2):423–438. doi:10.1016/j.neuroscience.2008.06.015
Rohl CA, Strauss CE, Misura KM, Baker D (2004) Protein structure prediction using Rosetta. Methods Enzymol 383:66–93. doi:10.1016/S0076-6879(04)83004-0
Rost B, Casadio R, Fariselli P, Sander C (1995) Transmembrane helices predicted at 95 % accuracy. Protein Sci 4(3):521–533

Rost B, Liu J, Nair R, Wrzeszczynski KO, Ofran Y (2003) Automatic prediction of protein function. Cell Mol Life Sci 60(12):2637–2650. doi:10.1007/s00018-003-3114-8

Runkel F, Bruss M, Nothen MM, Stober G, Propping P, Bonisch H (2000) Pharmacological properties of naturally occurring variants of the human norepinephrine transporter. Pharmacogenetics 10(5):397–405

Russel D, Lasker K, Webb B, Velazquez-Muriel J, Tjioe E, Schneidman-Duhovny D, Peterson B, Sali A (2012) Putting the pieces together: integrative modeling platform software for structure determination of macromolecular assemblies. PLoS Biol 10(1):e1001244. doi:10.1371/journal.pbio.1001244

Saier MH Jr (2000) A functional-phylogenetic classification system for transmembrane solute transporters. Microbiol Mol Biol Rev 64(2):354–411

Saier MH Jr, Yen MR, Noto K, Tamang DG, Elkan C (2009) The transporter classification database: recent advances. Nucleic Acids Res 37(Database issue):D274–D278. doi:10.1093/nar/gkn862, doi:gkn862 [pii]

Sali A, Blundell TL (1993) Comparative protein modelling by satisfaction of spatial restraints. J Mol Biol 234(3):779–815

Sanchez R, Sali A (1998) Large-scale protein structure modeling of the *Saccharomyces cerevisiae* genome. Proc Natl Acad Sci U S A 95(23):13597–13602

Schlessinger A, Geier E, Fan H, Irwin JJ, Shoichet BK, Giacomini KM, Sali A (2011) Structure-based discovery of prescription drugs that interact with the norepinephrine transporter, NET. Proc Natl Acad Sci U S A 108(38):15810–15815. doi:10.1073/pnas.1106030108

Schlessinger A, Khuri N, Giacomini KM, Sali A (2013a) Molecular modeling and ligand docking for solute carrier (SLC) transporters. Curr Top Med Chem 13(7):843–856

Schlessinger A, Matsson P, Shima JE, Pieper U, Yee SW, Kelly L, Apeltsin L, Stroud RM, Ferrin TE, Giacomini KM, Sali A (2010) Comparison of human solute carriers. Protein Sci 19 (3):412–428. doi:10.1002/pro.320

Schlessinger A, Wittwer MB, Dahlin A, Khuri N, Bonomi M, Fan H, Giacomini KM, Sali A (2012) High selectivity of the gamma-aminobutyric acid transporter 2 (GAT-2, SLC6A13) revealed by structure-based approach. J Biol Chem 287(45):37745–37756. doi:10.1074/jbc.M112.388157

Schlessinger A, Yee SW, Sali A, Giacomini KM (2013) SLC classification: an update. Clin Pharmacol Ther 94(1):19–23. doi:10.1038/clpt.2013.73

Schulze S, Koster S, Geldmacher U, Terwisscha van Scheltinga AC, Kuhlbrandt W (2010) Structural basis of Na(+)-independent and cooperative substrate/product antiport in CaiT. Nature 467(7312):233–236. doi:10.1038/nature09310

Schushan M, Rimon A, Haliloglu T, Forrest LR, Padan E, Ben-Tal N (2012) A model-structure of a periplasm-facing state of the NhaA antiporter suggests the molecular underpinnings of pH-induced conformational changes. J Biol Chem 287(22):18249–18261. doi:10.1074/jbc.M111.336446

Severinsen K, Kraft JF, Koldso H, Vinberg KA, Rothman RB, Partilla JS, Wiborg O, Blough B, Schiott B, Sinning S (2012) Binding of the amphetamine-like 1-phenyl-piperazine to monoamine transporters. ACS Chem Neurosci 3(9):693–705. doi:10.1021/cn300040f

Shaffer PL, Goehring A, Shankaranarayanan A, Gouaux E (2009) Structure and mechanism of a Na^+-independent amino acid transporter. Science 325(5943):1010–1014. doi:10.1126/science.1176088, 1176088 [pii]

Shaw DE, Maragakis P, Lindorff-Larsen K, Piana S, Dror RO, Eastwood MP, Bank JA, Jumper JM, Salmon JK, Shan Y, Wriggers W (2010) Atomic-level characterization of the structural dynamics of proteins. Science 330(6002):341–346. doi:10.1126/science.1187409

Shen MY, Sali A (2006) Statistical potential for assessment and prediction of protein structures. Protein Sci 15(11):2507–2524

Shi L, Quick M, Zhao Y, Weinstein H, Javitch JA (2008) The mechanism of a neurotransmitter: sodium symporter—inward release of Na^+ and substrate is triggered by substrate in a second binding site. Mol Cell 30(6):667–677. doi:10.1016/j.molcel.2008.05.008, S1097-2765(08) 00359-6 [pii]

Shi L, Weinstein H (2010) Conformational rearrangements to the intracellular open states of the LeuT and ApcT transporters are modulated by common mechanisms. Biophys J 99(12):L103–L105. doi:10.1016/j.bpj.2010.10.003

Shoichet BK (2004) Virtual screening of chemical libraries. Nature 432(7019):862–865

Shoichet BK, Kobilka BK (2012) Structure-based drug screening for G-protein-coupled receptors. Trends Pharmacol Sci 33(5):268–272. doi:10.1016/j.tips.2012.03.007

Shoichet BK, Stroud RM, Santi DV, Kuntz ID, Perry KM (1993) Structure-based discovery of inhibitors of thymidylate synthase. Science 259(5100):1445–1450

Shu Y, Brown C, Castro RA, Shi RJ, Lin ET, Owen RP, Sheardown SA, Yue L, Burchard EG, Brett CM, Giacomini KM (2008) Effect of genetic variation in the organic cation transporter 1, OCT1, on metformin pharmacokinetics. Clin Pharmacol Ther 83(2):273–280. doi:10.1038/sj.clpt.6100275, 6100275 [pii]

Shu Y, Leabman MK, Feng B, Mangravite LM, Huang CC, Stryke D, Kawamoto M, Johns SJ, DeYoung J, Carlson E, Ferrin TE, Herskowitz I, Giacomini KM (2003) Evolutionary conservation predicts function of variants of the human organic cation transporter, OCT1. Proc Natl Acad Sci U S A 100(10):5902–5907. doi:10.1073/pnas.0730858100, 0730858100 [pii]

Shu Y, Sheardown SA, Brown C, Owen RP, Zhang S, Castro RA, Ianculescu AG, Yue L, Lo JC, Burchard EG, Brett CM, Giacomini KM (2007) Effect of genetic variation in the organic cation transporter 1 (OCT1) on metformin action. J Clin Invest 117(5):1422–1431

Singh SK, Piscitelli CL, Yamashita A, Gouaux E (2008) A competitive inhibitor traps LeuT in an open-to-out conformation. Science 322(5908):1655–1661. doi:10.1126/science.1166777, 322/5908/1655 [pii]

Singh SK, Yamashita A, Gouaux E (2007) Antidepressant binding site in a bacterial homologue of neurotransmitter transporters. Nature 448(7156):952–956

Smith DE, Clemencon B, Hediger MA (2013) Proton-coupled oligopeptide transporter family SLC15: physiological, pharmacological and pathological implications. Mol Aspects Med 34 (2–3):323–336. doi:10.1016/j.mam.2012.11.003

Soding J, Biegert A, Lupas AN (2005) The HHpred interactive server for protein homology detection and structure prediction. Nucleic Acids Res 33(Web Server issue):W244–W248. doi:10.1093/nar/gki408

Sonnhammer EL, von Heijne G, Krogh A (1998) A hidden Markov model for predicting transmembrane helices in protein sequences. Proc Int Conf Intell Syst Mol Biol 6:175–182

Soto CS, Fasnacht M, Zhu J, Forrest L, Honig B (2008) Loop modeling: sampling, filtering, and scoring. Proteins 70(3):834–843. doi:10.1002/prot.21612

Sperandio O, Miteva MA, Delfaud F, Villoutreix BO (2006) Receptor-based computational screening of compound databases: the main docking-scoring engines. Curr Protein Pept Sci 7 (5):369–393

Stamm M, Staritzbichler R, Khafizov K, Forrest LR (2013) Alignment of helical membrane protein sequences using alignme. PLoS One 8(3):e57731. doi:10.1371/journal.pone.0057731

Traynor K (2013) Canagliflozin approved for type 2 diabetes. Am J Health Syst Pharm 70(10):834. doi:10.2146/news130035

Trott O, Olson AJ (2010) AutoDock Vina: improving the speed and accuracy of docking with a new scoring function, efficient optimization, and multithreading. J Comput Chem 31(2):455–461. doi:10.1002/jcc.21334

Tsirigos KD, Hennerdal A, Kall L, Elofsson A (2012) A guideline to proteome-wide alpha-helical membrane protein topology predictions. Proteomics 12(14):2282–2294. doi:10.1002/pmic.201100495

Tusnady GE, Dosztanyi Z, Simon I (2005) TMDET: web server for detecting transmembrane regions of proteins by using their 3D coordinates. Bioinformatics 21(7):1276–1277. doi:10.1093/bioinformatics/bti121

Vander Heiden MG, Cantley LC, Thompson CB (2009) Understanding the Warburg effect: the metabolic requirements of cell proliferation. Science 324(5930):1029–1033. doi:10.1126/science.1160809

Verdonk ML, Cole JC, Hartshorn MJ, Murray CW, Taylor RD (2003) Improved protein–ligand docking using GOLD. Proteins 52(4):609–623. doi:10.1002/prot.10465

Verrey F, Closs EI, Wagner CA, Palacin M, Endou H, Kanai Y (2004) CATs and HATs: the SLC7 family of amino acid transporters. Pflugers Arch 447(5):532–542. doi:10.1007/s00424-003-1086-z

Wang Y, Welty DF (1996) The simultaneous estimation of the influx and efflux blood–brain barrier permeabilities of gabapentin using a microdialysis-pharmacokinetic approach. Pharm Res 13(3):398–403

Watanabe A, Choe S, Chaptal V, Rosenberg JM, Wright EM, Grabe M, Abramson J (2010) The mechanism of sodium and substrate release from the binding pocket of vSGLT. Nature 468 (7326):988–991. doi:10.1038/nature09580

Weyand S, Shimamura T, Yajima S, Suzuki S, Mirza O, Krusong K, Carpenter EP, Rutherford NG, Hadden JM, O'Reilly J, Ma P, Saidijam M, Patching SG, Hope RJ, Norbertczak HT, Roach PC, Iwata S, Henderson PJ, Cameron AD (2008) Structure and molecular mechanism of a nucleobase-cation-symport-1 family transporter. Science 322(5902):709–713. doi:10.1126/science.1164440, 1164440 [pii]

Wise DR, Thompson CB (2010) Glutamine addiction: a new therapeutic target in cancer. Trends Biochem Sci 35(8):427–433. doi:10.1016/j.tibs.2010.05.003

Xiang Z, Steinbach PJ, Jacobson MP, Friesner RA, Honig B (2007) Prediction of side-chain conformations on protein surfaces. Proteins 66(4):814–823. doi:10.1002/prot.21099

Yamashita A, Singh SK, Kawate T, Jin Y, Gouaux E (2005) Crystal structure of a bacterial homologue of Na^+/Cl^- dependent neurotransmitter transporters. Nature 437(7056):215–223

Yarov-Yarovoy V, Schonbrun J, Baker D (2006) Multipass membrane protein structure prediction using Rosetta. Proteins 62(4):1010–1025. doi:10.1002/prot.20817

Yernool D, Boudker O, Jin Y, Gouaux E (2004) Structure of a glutamate transporter homologue from *Pyrococcus horikoshii*. Nature 431(7010):811–818

Zamek-Gliszczynski MJ, Hoffmaster KA, Tweedie DJ, Giacomini KM, Hillgren KM (2012) Highlights from the international transporter consortium second workshop. Clin Pharmacol Ther 92(5):553–556. doi:10.1038/clpt.2012.126

Zemla A, Venclovas C, Moult J, Fidelis K (1999) Processing and analysis of CASP3 protein structure predictions. Proteins Suppl 3:22–29

Zhao Y, Terry D, Shi L, Weinstein H, Blanchard SC, Javitch JA (2010) Single-molecule dynamics of gating in a neurotransmitter transporter homologue. Nature 465(7295):188–193. doi:10.1038/nature09057

Zhao Y, Terry DS, Shi L, Quick M, Weinstein H, Blanchard SC, Javitch JA (2011) Substrate-modulated gating dynamics in a Na(+)-coupled neurotransmitter transporter homologue. Nature. doi:10.1038/nature09971

Chapter 3
Structures of the Prokaryotic Galactose Transporter vSGLT and Their Implications on Alternating Access Mechanism in Human SGLT1

Jeff Abramson, Aviv Paz, and Armand S. Vartanian

Abstract Secondary active transporters couple electrochemical energy to the transport of solutes across biological membrane. The sodium solute symporters (SSS) comprise a large family of secondary active transporters fundamental to the homeostasis of a variety of solutes including sugars, inositols, anions, short chain fatty acids, choline, and vitamins. This chapter will explicate the structural framework of the sodium galactose transporter from *Vibrio parahaemolyticus* (vSGLT) and focus on mechanistic hypothesis drawn from these structures, some of which were directly transferable to the human homologues of vSGLT, which are members of the medically important SLC5 family. Compiling functional, structural, and bioinformatics data generated from these studies and incorporating unique conformations from structurally similar proteins, we propose a modified mechanism for sodium-dependent transport.

Keywords Glucose transport • Sugar transport • Homology modeling • Symporters • Membrane protein expression systems • Lipidic crystallization

---

J. Abramson (✉) • A.S. Vartanian
Department of Physiology, David Geffen School of Medicine, University of California, Los Angeles, CA, USA

Institute for Stem Cell Biology and Regenerative Medicine, NCBS-TIFR, Bangalore, Karnataka - 560065, India
e-mail: JAbramson@mednet.ucla.edu

A. Paz
Department of Physiology, David Geffen School of Medicine, University of California, Los Angeles, CA, USA

R. Krämer and C. Ziegler (eds.), *Membrane Transport Mechanism*, Springer Series in Biophysics 17, DOI 10.1007/978-3-642-53839-1_3,

3.1 Solute Sodium Symporter Family

The solute sodium symporter (SSS) family (TC# 2.A.21) constitutes a large family of co-transporters with over 250 members spanning all kingdoms of life. These proteins facilitate the co-transport of Na^+ with a diverse group of substrates ranging from sugars, inositols, anions, short chain fatty acids, choline, and vitamins (Wright et al. 2004). Crystal structures from a diverse group of transporter families (BCCT—TC# 2.A.15; SSS—TC# 2.A.21; NSS—TC# 2.A.22; and NCS1—TC# 2.A.39) show a similar structural fold indicative of a common evolutionary origin. Recently, on the basis of this similar structural fold and a minor sequence similarity originating from ancient intragenic duplications, these transporter families have been reclassified into the amino acid-polyamine-organocation (APC) superfamily, making it the second largest family of transporters (Wong et al. 2012).

The SSS family has 12 human members comprising the SLC5 family, which are expressed in a wide variety of tissues. All are sodium-coupled transporters for sugars (SGLT1, SGLT2, SGLT4, and SGLT5), inositols (SMIT1 and SMIT2), anions (NIS), short chain fatty acids (SMCT1 and SMCT2), choline (CHT), and vitamins (SMVT), with the exception of hSGLT3 that functions as a glucose sensor (Table 3.1). These 12 genes share a sequence identity of 20–70 % and many members of this family play a decisive role in human health and physiology (Wright 2013).

3.2 Disease and Therapeutic Applications

Several trafficking mutations in the hSGLT1 gene cause glucose galactose malabsorption (GGM), a metabolic disorder characterized by a defect in intestinal glucose and galactose absorption (Wright et al. 2011). Newborn children presenting GGM suffer from diarrhea leading to severe and fatal dehydration unless dietary sugars are completely removed. Additionally, SGLT1 is the transporter targeted in oral rehydration therapy (ORT) for the treatment of acute diarrhea in patients infected with cholera. The World Health Organization estimates that every year ORT saves the lives of over one million children who are suffering from infectious diarrhea. The physiological importance of SGLT1 likely extends beyond the small intestine as it is functionally expressed in a wide variety of cell types throughout the body including the kidney, heart, liver brain, prostate, and tumor cells (Wright 2013).

SGLT2 is the primary renal Na^+ glucose symporter, which is responsible for reabsorption of glucose from glomerular filtrate. Mutations in the hSGLT2 gene cause familial renal glucosuria (FRG), an autosomal recessive condition characterized by elevated urinary glucose excretion that surprisingly bears no other systemic consequences. The absence of any other clinical phenotype in FRG patients has encouraged the pharmaceutical industry to develop hSGLT2 inhibitors to lower blood glucose levels in diabetes patients (Wright et al. 2011). Most recently, an

Table 3.1 SCL5A family members

Human gene name	Protein name	Substrate	Transport type/coupling ion	Stoichiometry	Tissue	Disease
SLC5A1	SGLT1	Glucose, galactose, urea, and H_2O	$C/Na^+(H^+)$ F/Na^+ (H^+)	2:1	Intestine, trachea, kidney, heart, brain, testis, and prostate	Glucose galactose malabsorption
SLC5A2	SGLT2	Glucose	C/Na^+	1:1	Kidney, brain, liver, thyroid, heart and salivary glands	Familial renal glucosuria
SCL5A3	SMIT1	Myo-inositol (Glucose)	C/Na^+	2:1	Brain, heart, kidney, and lungs	Down's syndrome
SLC5A4	SGLT3	$Na^+(H^+)$	Glucose activated Na^+ (H^+) channel	2:1	Intestine, muscle, kidney, uterus, and testis	
SLC5A5	NIS	I^- (ClO_4^-, SCN^-, NO_3^-, Br^-	C/Na^+ uniporter, Na^+ channel, urea, H_2O	2:1	Thyroid, lactating breast, colon, stomach, and ovaries	Iodide transport defect, thyroid hormonogenesis
SLC5A6	SMVT	Pantothenate, biotin, 1- and lipoate	C/Na^+	>1	Brain, heart, kidney, placenta, and lungs	
SLC5A7	CHT1	Choline	$C/NA^+/Cl^-$	–	Medulla and spinal cord	
SLC5A8	SMCT1	Short-chain fatty acids	C/Na^+	2:1	Intestine, kidney, brain, retina, muscle	
SLC5A9	SGLT4	Glucose, mannose, fructose	C/Na^+	?	Intestine, kidney, liver, brain, heart, lung, and uterus	
SLC5A10	SGLT5	Glucose, mannose, fructose	C/Na^+	?	Kidney cortex	
SLC5A11	SMIT2	Myo-inositol, chiro-inositol	C/Na^+	2:1	throid, brain, heart, muscle, spleen, liver, and lung	
SLC5A12	SMCT2	Short-chain fatty acids	C/Na^+	2:1	Intestine, brain, retina, and muscle	

Substrate, transport type, stoichiometry, tissue locations and linked disease are shown for all 12 members of the SLC5A family. Substrate specificity, ion, and stoichiometry were determined using heterologous expression systems. The "?" denotes a unknown stoichiometry

hSGLT2 inhibitor (INVOKANA) has received FDA approval for the treatment of type II diabetes. This potent inhibitor reduces renal glucose reabsorption, thereby increasing urinary glucose excretion and effectively lowers blood glucose levels. It should also be noted, and has not been thoroughly investigated, that significant increases in SGLT2 expression have been observed in lung cancer metastasis (Ishikawa et al. 2001), suggesting that SGLT2 inhibitors may be useful for the treatment of some types of cancers.

Another SLC5 member, hNIS (sodium-iodide symporter), can carry autosomal recessive mutations resulting in a rare congenital iodide transport defect (ITD) leading to hypothyroidism, low thyroid iodide uptake, and goiter. In addition, hNIS plays a role in the uptake of radioiodide used for the diagnosis and treatment of both thyroid and breast cancer (Bizhanova and Kopp 2009).

Additional members of the SLC5 family related to disease are hSMIT1 associated with the pathogenesis of Down syndrome, and hSMCT1, a tumor suppressor gene, the expression of which is silenced in tumors of colon, thyroid, stomach, kidney, and brain (Ganapathy et al. 2008). In total, SLC5 proteins are essential components for cell communication, function, and survival. Alterations in their inherent functions result in a number disorders; thus, an intricate understanding of their structure and dynamics remains a critical objective for basic and medical research.

3.3 The Structures of SGLTs

While no structure of a SLC5 family member has been solved, a well-expressing bacterial homolog of the sodium glucose symporter from *Vibrio parahaemolyticus* (vSGLT) has been determined in two distinct conformations (Faham et al. 2008; Watanabe et al. 2010) (PDB ID 3DH4 & 2XQ2). vSGLT has a high sequence identity (32 % amino acid identity) with hSGLT1 and like its human counterpart, it contains 14 transmembrane helical segments (TM) with both N- and C-termini exposed to the extracellular side of the membrane (Fig. 3.1a). The substrate-binding site, which harbors a single galactose molecule in the 3DH4 structure, is in the center of the protein, approximately halfway across the lipid bilayer. The core domain of ten TMS has an inverted repeat topology, where TM(1)–TM(5) and TM (6)–TM(10) are related by an approximate twofold rotational axis parallel to the membrane plane (Fig. 3.1b, c) with additional supporting helices located on the N- and C-termini. The architecture of this core domain resembles that of the leucine transporter (LeuT) (Yamashita et al. 2005), although there is no apparent sequence homology between the two proteins.

Since the time this structure was solved in 2008, there has been a surge of crystal structures (Faham et al. 2008; Weyand et al. 2008; Ressl et al. 2009; Schulze et al. 2010; Shaffer et al. 2009; Gao et al. 2009; Zhou et al. 2007) displaying this

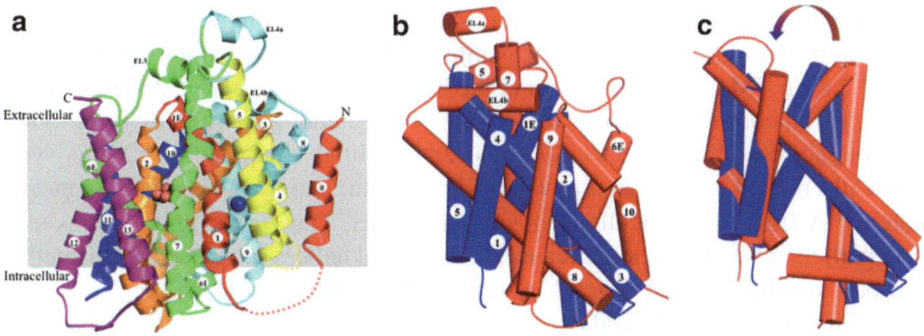

Fig. 3.1 Overall structure of vSGLT. (**a**) vSGLT structure presented in the membrane plane is colored as *rainbow* from N terminus to C terminus. Bound galactose is shown as *black* and *red spheres* for the carbon and oxygen atoms, with the proposed Na^+ ion presented as a *blue sphere* in the Na2 site. (**b**) The ten helical structural core of vSGLT is formed from inverted repeats of five TM helices: TM1–TM5 in *blue* and TM6–TM10 in *red*. (**c**) Superpositioning of the two inverted repeats (RMSD 3.9 Å), coloring as in (**b**)

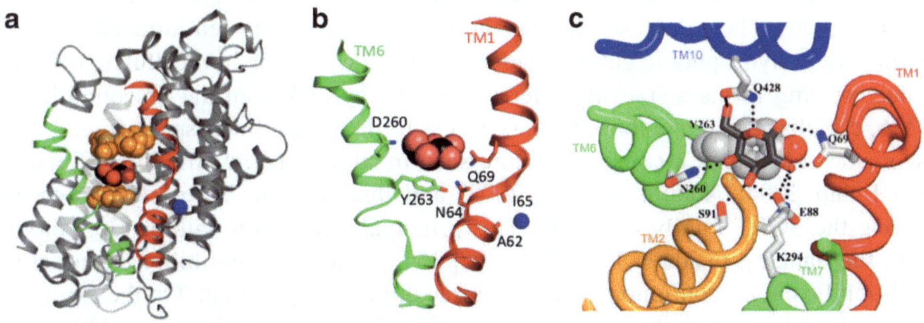

Fig. 3.2 Galactose binding and gating. (**a**) An overview of the galactose- and Na^+-binding sites with the side chains comprising the outer (F424, Y87 M73) and inner gates (Y263) shown as *orange spheres*. The unwound helices TM1 and TM6 are colored *red* and *green*, respectively. (**b**) The unwound helices TM 1 (in *red*) and symmetrically related TM 6 (in *green*) coordinate the galactose (*black* and *red spheres*) and the sodium ion (*blue sphere*). This pattern of discontinuous helices has been implicated in transport mechanisms of a number of co-transporters. (**c**) The galactose-binding site viewed from the extracellular side with the extracellular hydrophobic gate residues removed. Hydrogen bonding side chains are displayed as *sticks* colored by atom type. Hydrogen bonds are depicted in *black dashed lines*. The coloring scheme follows that of Fig. 3.1a

5-helix inverted repeat motif (5HIR) in genetically diverse proteins that differ widely in their mode, type, and number of driving ligands (Lolkema et al. 2008; Abramson and Wright 2009) (see Chap. 4).

In the inward-occluded structure of vSGLT (PDB ID: 3DH4), a galactose is located approximately halfway across the membrane bilayer in the center of the core domain (Fig. 3.2). There is a large cavity that extends from just below the galactose-binding site to the intracellular space representing an inward-facing conformation. A striking feature of the vSGLT and all proteins with 5HIR is the presence of two discontinuous transmembrane helices in the center of the protein (Fig. 3.2a, b). TM1 has a disruption in the helical structure approximately halfway across the membrane bilayer, adjacent to the bound galactose dividing TM1 into roughly equivalent intracellular and extracellular components. TM6 is symmetrically related to TM1 and unwinds closer to the intracellular region, approximately 7 Å from the intracellular membrane leaflet. These structural features may have particular functional significance, as unwound helices have been implicated in transport mechanisms of several co-transporters (Hunte et al. 2005; Yamashita et al. 2005; Yernool et al. 2004).

3.4 Galactose-Binding Site

Galactose is bound approximately halfway across the membrane bilayer where it is coordinated by extensive sidechain interactions from TMs 1,2,6,7, and 10 (Fig. 3.2c). The galactose-binding site is sandwiched between intracellular and extracellular hydrophobic gating residues (Fig. 3.2a). On the intracellular side, the galactose ring forms a stacking interaction with Tyr263 from the unwound helix TM6, establishing a gate that prevents access to the intracellular space (Fig. 3.2c). Aromatic stacking is a common feature seen in sugar-binding structures (Sujatha and Balaji 2004; Abramson et al. 2003) and this tyrosine is strictly conserved among the SGLT family members. This principal interaction with Tyr263, along with the flanking residues Tyr262 and Trp264, prevents accessibility to the large hydrophilic cavity contiguous with the intracellular compartment. On the extracellular side, a triad of hydrophobic residues (Met73 on TM1E, Tyr87 on TM2, and Phe424 on TM10) stabilizes the occluded galactose site, constituting the second hydrophobic gate (Fig. 3.2a). On this face of the cavity, protein mass from the helices as well as from the extracellular loops prevents accessibility to the extracellular surface. The residues forming the hydrophobic gates are highly conserved in the members of the SSS family that bind sugars, indicating a conserved mechanism for positioning and occluding the ligand once it is bound.

All hydroxyl groups of the galactose ring are stabilized through hydrogen bonds from residues (Fig. 3.2c): Asn69 to C1–OH and C2–OH, Glu88 to C2–OH, the gamma oxygen of Ser91 to C3–OH, and Asn260 and Gln428 to C4–OH and C6–OH. There is also a hydrogen bond between the carbonyl oxygen of Tyr87 and C3–OH of galactose. Because Tyr87 is part of the extracellular hydrophobic plug, this additional hydrogen bond may have mechanistic implications for galactose binding and stabilization in the central domain. Lys294 forms a hydrogen bond with the C2–OH in a coordination similar to that observed between substrate and protein in the structures of the glucose/galactose-binding protein (PDB: 1GCA) and lactose permease (PDB: 1PV7). Indeed, hydrogen bonding with a positively charged side chain appears to be a common feature of sugar-binding proteins (Sujatha and Balaji

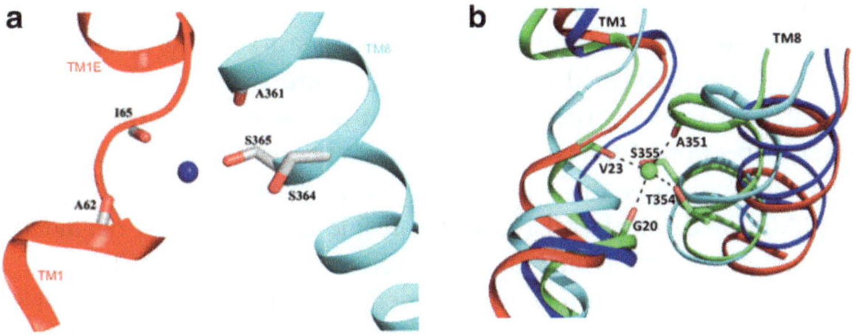

Fig. 3.3 The Na2 site. (**a**) Residues in the putative Na^+-binding site (the Na2 site) of vSGLT are displayed as *sticks*. (**b**) An overlay of the Na2 sites in various LeuT-fold transporters. vSGLT (in *red*) and BetP (in *blue*) reside in inward-facing conformations (presumably in the absence of sodium) and Mhp1 (in *cyan*) and LeuT (in *green*) reside in outward-facing conformations with sodium coordinating residues from LeuT shown as *sticks*

2004; Zou et al. 1993; Abramson et al. 2003). With the exception of Ser91, which is not conserved among the SGLT family of proteins, all of the galactose-binding residues discussed above were shown to be critical for sodium-dependent transport (Faham et al. 2008; Watanabe et al. 2010). In summary, the substrate is supported by extensive sidechain interactions from the central domain and held in an occluded conformation through intracellular and extracellular hydrophobic plugs.

3.5 Sodium-Binding Site

Sodium is the ion-driving substrate translocation in a number of APC family members; however, the sodium to substrate stoichiometry may differ among these sodium-dependent transporters. For instance, LeuT and BetP have a 2:1 stoichiometry (Shi et al. 2008; Quick et al. 2009; Ressl et al. 2009) whereas vSGLT and Mhp1 have a 1:1 stoichiometry (Veenstra et al. 2004; Weyand et al. 2008). In the founding structure of LeuT (at 1.65 Å), two likely sodium sites (Na1 and Na2) were identified (Yamashita et al. 2005). The positions of both bound ions are in the vicinity of the unwound segments of TM1 and TM6 with a high degree of conservation (Noskov and Roux 2008). In both LeuT and BetP, the sodium ion at the Na1 site is directly coupled to the carboxy oxygen of their corresponding substrates. For both vSGLT and Mhp1, there is no Na1 site; however, members of the APC superfamily with a 2:1 stoichiometry likely harbor a sodium ion in vicinity of the Na1 site.

The Na2 sodium site is conserved among all sodium-dependent APC family members. For vSGLT, a sodium-coordination site resides at the intersection of TM1 and TM8, approximately 8 Å away from the substrate-binding site (Fig. 3.3a). In this region, the carbonyl oxygens from Ala62 and Ile65 in the unwound segment of TM1 are in position to coordinate the Na^+ ion. An additional interaction could be established with the hydroxyl oxygen of Ser365 of TM8, a residue strictly conserved throughout the SSS family. This site bears a striking resemblance to the second sodium site identified in the LeuT structure (Yamashita et al. 2005) and present in all four known structures of sodium-dependent APC members.

A closer inspection of the Na2 site reveals some distinct alterations in TM8 that appear to be dependent on the conformation of the transporter. Both TM1 and TM8 of LeuT (2A65) and Mhp1 (2JLN), which reside in the outward-occluded conformation, align nicely with one another. Conversely, BetP (2WIT) and vSGLT (3DH4), which reside in an inward-occluded conformations, align well with one another but are shifted ~3 Å away from LeuT (2A65) and Mhp1 (2JLN), making a more open environment around the Na2 site (Fig. 3.3b). This can further be observed by a comparison of the sodium coordinating distances, where both LeuT (2.2–2.5 Å) and Mhp1 (2.1–2.7 Å) are tighter than those for vSGLT (3.1–3.8 Å). This raises the possibility that sodium is not bound in the vSGLT crystal structure (3DH4) which is reinforced by molecular dynamic studies (Watanabe et al. 2010). Simulations on the Na2 site in LeuT indicate a tightly bound sodium ion (Celik et al. 2008), whereas in vSGLT sodium rapidly escapes from the Na2 site into the inner aqueous vestibule and, after a transient interaction with D186, flows into the cytoplasm (Li and Tajkhorshid 2009; Watanabe et al. 2010). This conserved aspartate has been shown to be involved in cation selectivity in hSGLT1 (Quick et al. 2001).

3.6 Galactose Release Mechanism

Additional MD simulations of >50 ns (Watanabe et al. 2010; Li et al. 2013) reveal that after Na^+ release, galactose undergoes significant fluctuations within the binding pocket. Approximately ~50 ns later, the sugar leaves the binding site and enters the intracellular space. Importantly, the simulation shows that the substrate moves past the inner hydrophobic gate, which is coordinated, by N64 and the gating residue Y263. N64 is located in the unwound segment of TM1 and likely forms an important link between the Na2- and the galactose-binding sites, thus propagating the signal of sodium binding/release to the sugar-binding site. These simulations are in remarkable agreement with the inward-open structure of vSGLT (2XQ2).

A comparison between the inward-occluded and inward-open structures of vSGLT reveals many similarities, but there are distinct structural differences that result from ligand release (Fig. 3.4). Substrate release is a result of a subtle movement of subdomains. This transition from inward occluded to inward open is likely triggered by the release of sodium from the Na2 site and the resulting reorganization of the hydrogen bonding network surrounding the unwound segment of TM1. As the intracellular half of TM1 begins to flex by approximately 13°, Asn64—located on the unwound segment of TM1—alters its coordination pattern (Fig. 3.5). In the absence of Na^+ and galactose, Asn64 coordinates both Tyr263 and

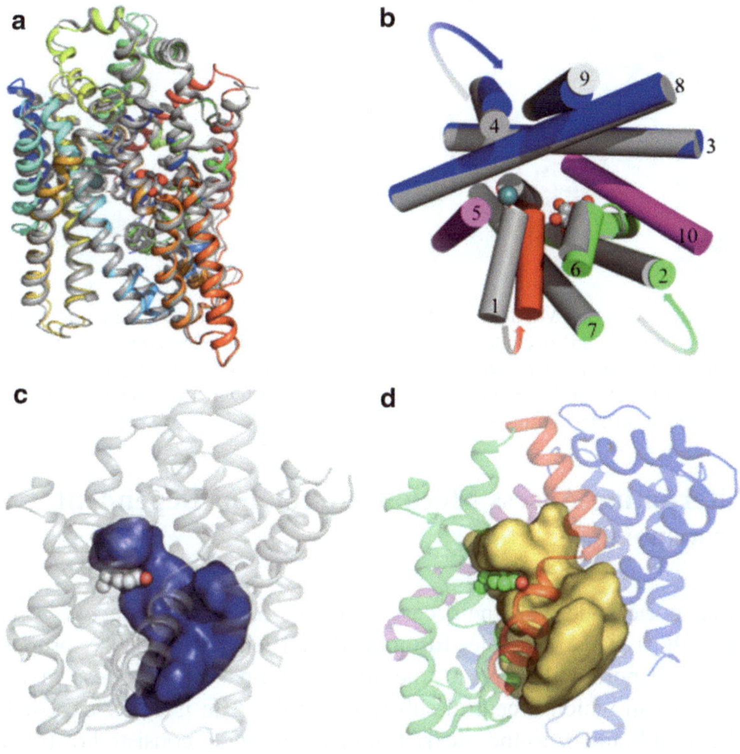

Fig. 3.4 The transition from inward occluded to inward open. (**a**) An overlay of the inward-open (*colored*) and inward-occluded (*gray*) conformations. (**b**) Conformational changes in the inward-open structure reveal a ~13° kink in the unwound segment of TM1 and rigid-body rotations of the hash motif (TMs 3,4,8,9) and sugar bundle (TMs 1,2,6,7) by 3° in opposite directions. This exposes the substrate-binding site to the intracellular milieu. (**c**) Accessibility cavity of the inward-occluded conformation (*colored blue*). (**d**) Accessibility cavity of the inward-open conformation (*colored gold*). The conformational changes of TM1, hash motif, and sugar bundle increase the intracellular cavity by ~1,400 Å^3 in the inward-open conformation, facilitating galactose release

Glu88 (Fig. 3.5b). However, in the inward-occluded structure, Asn64 and Glu88 are coordinated to the galactose hydroxyls at position O2 and O3, respectively. This transition increases the volume of the accessible cavity by over 1,400 Å^3 as each domain moves approximately 3° in opposite directions (Fig. 3.4c, d). This overall 6° of domain movement presumably alters the interaction between the protein and its substrate by allowing H_2O to flow into the expanded cavity. As water competes for available hydrogen bonds in the cavity, galactose binding is weakened and finally releases into the cell. Yet, these details are derived from static intermediates and computer simulations. To verify these conclusions, additional structural conformations and biophysical techniques to monitor transitions as well as the binding, translocation, and release of substrates are required.

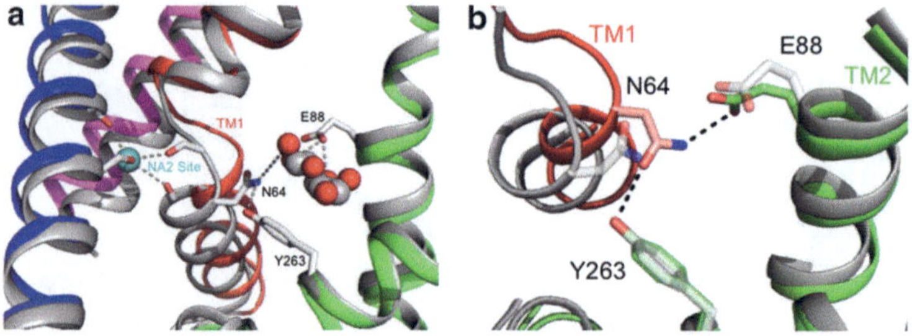

Fig. 3.5 The transition of TM1 links the Na2 site to the galactose-binding site. (**a**) An overlay of the inward-open and inward-occluded (*gray*) conformations. The release of sodium causes a kink in TM1 which in turn disrupts the galactose-binding site. (**b**) In the absence of galactose, Asn64 hydrogen bonds to Glu88 and Tyr263, maintaining an open pathway from the intracellular space to the now exposed substrate-binding site

3.7 Alternating Access Mechanism for Ion-Dependent Transport

Active transporters deliver their cargo via an *alternating access mechanism* (Jardetzky 1966). This principle, formulated in 1966 by Jardetzky, proposes that a solute is transported through a "membrane pump" by a process of allosteric rearrangements in which a central substrate-binding site is either exposed to one side of the membrane or to the other. The original model consisted of two distinct conformations and has since been expanded to six-states largely based on functional data (Sala-Rabanal et al. 2012; Guan and Kaback 2006). More recently, Forrest (Forrest et al. 2011) and Jeschke (Jeschke 2013) have extended this scheme into an eight state model which incorporate crystal structure data.

The symmetric architecture of the APC family members makes them ideal alternating access machines. The inverted repeat motif is a highly economical and efficient way for a transporter to alternate the exposure of a centrally located binding site to both sides of the membrane with small movements and minimal energy invested in the conformational changes. All of these considerations, combined with accessibility measurements, have led Forrest and Rudnick to propose the rocking bundle mechanism, which proclaims a rocking motion of concerted helices, is sufficient to open and close the ligand pathway. In particular, the hash motif (TMs 3,4,8,9) moves relative to the bundle domain (TMs 1,2,6,7) in excess of a 25° rotation along a screw axis. This movement is not a strict rigid body movement, as a number of TMs independently flex to occlude or open pathways to the substrate-binding sites. Jeschke analyzed 14 structures from 7 different proteins containing the 5HIR core, supporting the rocking bundle model (Forrest et al. 2008) as the primary mechanism transitioning from the outward-open to inward-open conformations. Although the rocking bundle mechanism is a valid guideline for the primary conformational change, the current models for achieving sodium-dependent transport may require additional modifications in order to separate the

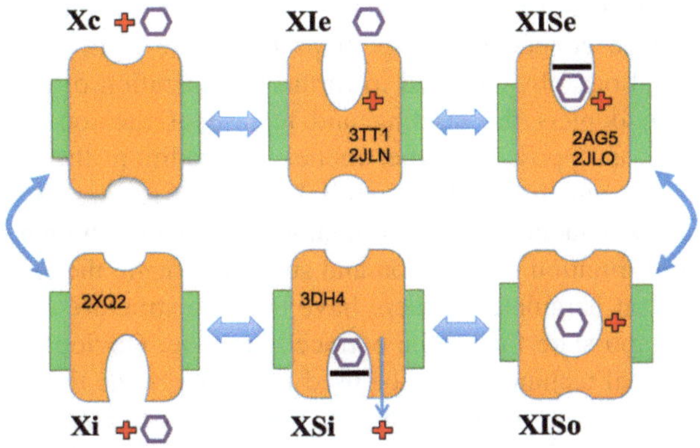

Fig. 3.6 Conformational states for sodium-dependent transporters. In the initial conformation, the carrier free of substrate and ion resides in a closed state (Xc). External ion binding to the transporter exposes the substrate-binding site to the extracellular environment (XIe). Extracellular substrate binds to its now-exposed binding site and this in turn closes the external gate, without changes to the external vestibule (XISe). The next step involves the closure of the external vestibule to form a fully occluded state (XISo). Substrate release begins as the carrier transitions to expose the substrate-vestibule to the intracellular environment (XSi). Once sodium is released, the internal gate opens to permit substrate release (Xe). To complete the transport cycle, the internal vestibule is closed and the transporter returns to its original unbound state (Xc). The pdb codes for different sodium-dependent transporters are placed inside the cartoon representing the appropriate conformational state

binding and release of driving ion (I) and substrate (S) to accurately capture the structural framework of the transport cycle.

Here, we propose a slightly modified scheme (Fig. 3.6) that incorporates the structural models of sodium-dependent transporters with the 5HIR fold which follows the "first-on" "first-off" principles for I and S consecutive binding and release proposed decades ago (Krupka 1993). In our model, the protein resides in various conformations (X) either with a cavity open to the extracellular environment (e), closed (c), or with a cavity open to the intracellular environment (i). In the initial conformation, the carrier is free of substrate and ion residing in a fully closed state (Xc). As seen from functional studies on SGLT1 (Sala-Rabanal et al. 2012), external sodium is required to expose the substrate-binding site to the extracellular environment (XIe). Extracellular substrate binds to its now exposed binding site and this in turn closes the external gate, without changes to the external vestibule (XISe). The next step involves the closure of the external vestibule to form a fully occluded state (XISo). Substrate release begins as the carrier transitions to expose the substrate-vestibule to the intracellular environment. This displacement—following the rocking bundle mechanism—of TM1 and TM8 disrupts the Na2 site and weakens sodium binding, resulting in its release while maintaining an occluded substrate via the internal gate (XSi). Once sodium is released, the internal gate opens to permit substrate release (Xe). To complete the transport cycle, the internal vestibule is closed and the transporter returns to its original unbound state (Xc).

Clearly the mechanism proposed here is oversimplified. Many additional conformations could be conceived and perhaps under the right conditions will be captured by crystallography and lead to further modification of this model. However, this scheme depicts the binding and release of the ion and substrate as sequential events, as opposed to other models. In addition, the scheme includes the conformation in which the 3DH4 vSGLT structure resides (XSi: an occluded substrate with a cavity facing the intracellular side, with no sodium bound) and does not include a conformation without ion and substrate facing the extracellular side. This is an important parameter as there have been no structure or functional data depicting a fully apo state facing the extracellular side. Obviously, further structures and biophysical techniques can be used to further refine this model.

However, capturing different snapshots of the many conformations for any given transporter to provide a high-resolution description of each conformation proves to be a challenging task (see Chap. 4). For example, we could crystallize and solve the structure of vSGLT only in the inward-occluded (Faham et al. 2008) and inward-open conformations (Watanabe et al. 2010). As another example, LeuT was crystallized in 2005 in the outward facing conformation (Yamashita et al. 2005) and the substrate-free inward-facing conformation was reported only in 2012 (Krishnamurthy and Gouaux 2012). To obtain this structure, Gouaux and colleagues identified a mutant that stabilized the inward-facing conformation and further developed a conformation-specific antibody to aid crystallization. It should be noted that LeuT may have a unique mechanism of transporting substrate as TM1 is dramatically rotated by ~45°, away from the transport path, along the unwound segment of TM1 and does not entirely conform to the rocking bundle mechanism.

It is paramount to extend the structural knowledge obtained by crystallography with complementary biophysical techniques to gather a more complete understanding of the different conformation and transitions between them. So far EPR, DEER, fluorescence, and various computational techniques have been used to extend our understanding of the alternating access mechanism beyond the static snapshots obtained by crystallography (Smirnova et al. 2007; Majumdar et al. 2007; Claxton et al. 2010) (see Chaps. 1, 2, 6, and 11).

3.8 Connecting Bacterial vSGLT Structures to hSGLT1

While currently no human SGLT structures have been resolved, the similarities between hSGLT1 and vSGLT provided a satisfactory template for homology modeling. A detailed study based on this model was recently reported and demonstrated two overwhelming properties (Sala-Rabanal et al. 2012). The first, established a mechanism for hSGLT1 looking at conserved ligand-binding and gating residues identified from the structure of vSGLT. In doing so, Sala-Rabanal et al. made 12 point mutations and comprehensively investigated $K_{0.5}$, I_{max} (for both sodium and sugar), and apparent turnover. They were able to show that (a) mutations of sugar-binding residues dramatically reduce apparent sugar affinity;

(b) mutations of one of the two sodium-binding residues dramatically reduce apparent sodium affinity; and (c) the gating residues—forming the barrier between sugar and the external environment—have a role in coupling sugar to Na^+ transport.

The second overall accomplishment of this study establishes conformational changes of hSGLT1 as Na^+ and sugar are transported. Using site-directed cysteine labeling to monitor accessibility, the authors confirm a number of conformations in the previously proposed six-state model for alternating access. First, external Na^+ opens the sugar-binding vestibule, effectively raising the apparent affinity of the transporter for sugar due to a higher accessibility of glucose to the binding site. After sugar binds, the outer gates closes, and the sugar-binding sites become inaccessible or occluded to the external membrane surface. These accessibility results are consistent with homology modeling and provide insight into the dynamics of conformational changes underlying Na^+ and sugar transport for the external environment. What remains to be concluded are the steps leading to Na^+ and sugar release.

3.9 Future Perspectives

Currently, no mammalian APC structures have been resolved and this is certainly not due to lack of effort. Membrane proteins are notoriously difficult to work with and the rate of successful crystallization is remarkably low compared to soluble proteins. TargetTrack (http://sbkb.org/tt/index.html), a target registration database, provides information on the experimental progress of protein targets selected for structural determination by a number of high-throughput structural biology projects for which the SLC5 family is included. Tracking the clones produced for membrane proteins, only 0.36 % yielded high-quality diffracting crystals compared to 5.11 % for soluble proteins. However, the membrane protein crystallography field is maturing and many new advances in experimental methods are being developed increasing the likelihood of success.

The first hurdle in the process of structure determination is acquiring a suitable protein for crystallization. When the protein cannot be obtained from a natural source, the only option is to clone the protein using heterologous expression systems. Before any cloning is preformed, *in silico* analysis of the gene should be made to identify disordered regions, transmembrane helices, domain boundaries, and glycosylation sites. Each of these parameters may be critical for obtaining high quantities of stable monodispersed protein conducive for crystallization (Malawski et al. 2006). These modified constructs are made, in addition to the full-length sequences, to increase a projects' chance of success.

In a similar manner, the homology approach, pioneered by the Rees lab (Chang et al. 1998), is one of the most useful and robust techniques for identifying a "crystallizable" protein. Using this technique, expression trials of a given protein target are performed on every available gene homolog in the hope of finding the most appropriate one for crystallography. The rationale for this approach is that minor sequence differentiation inherent in homologous genes from multiple organisms may completely change the expression pattern and protein stability. The only compromise, when using a wide range of sources, is that the successful target may not be well characterized. Fortunately, the high sequence similarity and conservation of essential residues is often retained in homologues proteins, which usually wields meaningful experimental results that are transferable to the human protein target. As more cDNA sequences from various organisms are made available, the power of this approach will likely increase, but it remains a successful approach for eukaryotic structure determination (Jaroszewski et al. 2008).

The two approaches described above result in having tens to hundreds of different constructs for each protein target that require rapid screening. New expression techniques and detection assays are greatly increasing the ability to screen many constructs in a rapid manner. For example, utilizing *Saccharomyces cerevisiae* as an expression host in conjunction with customized GFP-fusion plasmids (Drew and Kim 2012; Drew et al. 2002, 2008), a target protein expression level can be rapidly monitored by whole cell GFP fluorescence. All viable constructs can further be processed to determine the proteins stability and monodispersity in various detergents by using Fluorescence Size Exclusion Chromatography (FSEC) (Hattori et al. 2012; Kawate and Gouaux 2006; Hsieh et al. 2010). Using this methodology multiple constructs can be tested for expression and stability in different detergents to validate their potential for crystallization trials and has been successfully used for structures determination (Jasti et al. 2007; Kawate et al. 2009; Yernool et al. 2004; Sobolevsky et al. 2009; Krishnamurthy and Gouaux 2012).

Yeast systems are typically implemented due to their ease of handling and the ability to rapidly screen constructs; however, more advanced insect and mammalian expression systems are also being used as crucial expression tools. While each system is inherently more difficult and costly to work with, insect and mammalian expression systems could potentially produce protein when other systems fail. Insect expression typically utilizes SF9 or Hi5 cells and the Invitrogen Bac-toBac expression system. Once a construct has been cloned, a baculovirus is made to infect the insect cells for expression of the construct. Expression in insect cells can be screened using various methods. The high sensitivity of western blots or fluorescent antibodies allows for expression screening to be performed on small growth samples either by FACS or gel analysis. These small-scale expression tests allow for the effective and rapid screening of many clones in parallel, without the need and costs of large-scale expression. Since milligram quantities are needed for crystallization trials, the best viruses could be scaled up relatively easily. This methodology has been very successful for producing human membrane protein structures, in particular GPCRs (Cherezov et al. 2007; Rosenbaum et al. 2007, 2011; Rasmussen et al. 2011a, b; Shimamura et al. 2011; Granier et al. 2012; Haga et al. 2012; Hanson et al. 2012; Hino et al. 2012; Manglik et al. 2012; Zhang et al. 2012).

Figure 3.7a shows the distribution of successful expression sources for unique X-ray structures of eukaryotic membrane proteins. Overall, the percentage of

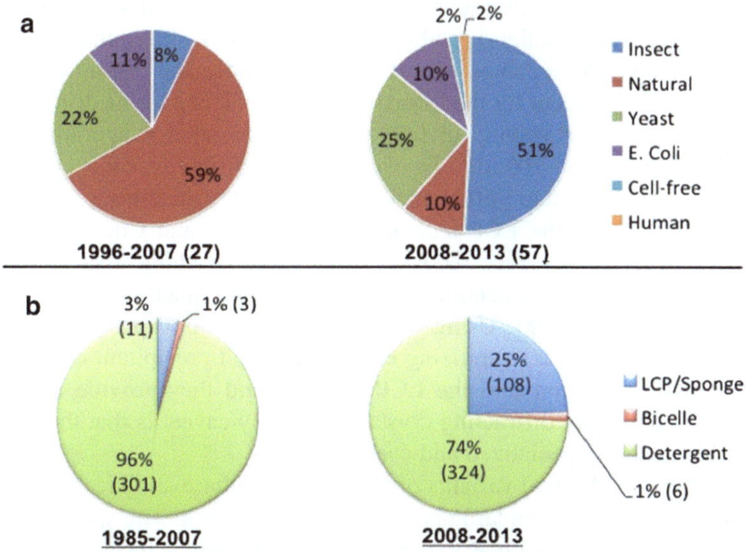

Fig. 3.7 The distribution of expression sources and crystallization techniques for published transmembrane proteins. (**a**) The distribution of different expression sources for unique eukaryotic proteins is shown for the year range of 1996–2007 to the *left* and for 2008–2013 to the *right*. (**b**) The distribution of crystallization techniques used for all membrane proteins is shown for the year range of 1985–2007 to the *left* and for the range of 2008–2013 to the *right*. The numbers in *parenthesis* represent the total number of structures published for each specific timeframe

structures solved using yeast, bacteria, and human cell expression systems remain virtually identical. However, the ratio of structures solved from proteins isolated from natural sources has considerably decreased (from 59 % for the range of 1996–2007 to 10 % over the past 5 years) and substantially more structures are derived from proteins isolated from insect cells (from 8 % for the range of 1996–2007 to 51 % over the past 5 years).

Once the protein has been successfully isolated, the challenge of crystallizing the protein ensues. Traditionally, and still widely more successful (Fig. 3.7b), membrane proteins are crystallized in detergents, but lipidic techniques are becoming increasingly popular. Lipidic crystallization replaces the solubilizing detergent with lipids immediately prior to crystallization. These lipids often more closely mimic the native membrane environment and thus should enhance the proteins stability. Another advantage of lipidic techniques is that the removal of detergent increases the surface area available to generate crystal contacts and thus, theoretically, increases the chance of obtaining crystals. These crystals generate a type I crystal packing in which contacts are formed between both polar and nonpolar regions of the protein. Frequently, these crystals are better ordered than type II crystals resulting in better crystal diffraction.

The most popular lipidic technique (Fig. 3.7b) is the lipidic cubic phase (LCP) also known as *in meso* crystallization (Caffrey 2011; Cherezov 2011). LCP crystallization proceeds by thoroughly mixing the protein solution with a host cubic phase lipid (usually monoolein) producing a viscous sticky mesophase. Because of this issue, the mixture becomes very difficult to handle and crystal trials require special care and equipment for the formation of the LCP and for its dispensing. One of the critical issues in setting up LCP plates is the rapid evaporation of small sample size (typically < 100 nL). To overcome this difficulty, robots are designed to move very rapidly and are often enclosed in humidity chambers.

To further enhance the LCP technique, the Stevens and Cherezov labs have developed a LCP fluorescence recovery after photobleaching assay that enables prescreening crystallization conditions with microgram quantities of purified protein (Cherezov et al. 2008). Using this technique, one is capable of identifying suitable protein constructs, stabilizing compounds, and precipitant conditions that promote protein diffusion into the LCP medium and thus provide an increased probability of obtaining diffracting crystals. The only caveat is that the equipment used for this assay is specialized and costly.

There are two additional variants of LCP, sponge phase and bicelles, which are both more conducive to standard crystallization equipment. The sponge phase is a liquid variant of the LCP with an expanded mesophase where the aqueous channels are able to accommodate larger membrane proteins (Wadsten et al. 2006). Bicelles are small bilayer discs made from a mixture of lipids and amphiphiles (usually a DMPC:CHAPSO mixture) used for both NMR and crystallography (Faham and Bowie 2002; Ujwal and Abramson 2012). Bicelles are particularly advantageous to handle as they form a nonviscous liquid phase at 4 °C enabling simple protein integration. Unlike LCP, the sponge phase and bicelles are suitable for conventional hanging- and sitting-drop crystallization experiments conducive to standard crystallization robotics.

Membrane protein crystallography has always presented difficult challenges but the current interest, funding, and enthusiasm bode well for the field. Solving new structures and deciphering mechanistic details of pharmacologically relevant proteins drive researchers to pursue and expand the current methodology to further increase the throughput. With this much forward momentum it is only a matter of time until SLC5 protein structures are resolved.

References

Abramson J, Wright EM (2009) Structure and function of Na(+)-symporters with inverted repeats. Curr Opin Struct Biol 19:425–432

Abramson J, Smirnova I, Kasho V, Verner G, Kaback HR, Iwata S (2003) Structure and mechanism of the lactose permease of Escherichia coli. Science 301:610–615

Bizhanova A, KOPP P (2009) Minireview: the sodium-iodide symporter NIS and pendrin in iodide homeostasis of the thyroid. Endocrinology 150:1084–1090

Caffrey M (2011) Crystallizing membrane proteins for structure-function studies using lipidic mesophases. Biochem Soc Trans 39:725–732

Celik L, Schiøtt B, Tajkhorshid E (2008) Substrate binding and formation of an occluded state in the leucine transporter. Biophys J 94:1600–1612

Chang G, Spencer RH, Lee AT, Barclay MT, Rees DC (1998) Structure of the MscL homolog from Mycobacterium tuberculosis: a gated mechanosensitive ion channel. Science 282:2220–2226

Cherezov V (2011) Lipidic cubic phase technologies for membrane protein structural studies. Curr Opin Struct Biol 21:559–566

Cherezov V, Rosenbaum DM, Hanson MA, Rasmussen SG, Thian FS, Kobilka TS, Choi HJ, Kuhn P, Weis WI, Kobilka BK, Stevens RC (2007) High-resolution crystal structure of an engineered human beta2-adrenergic G protein-coupled receptor. Science 318:1258–1265

Cherezov V, Liu J, Griffith M, Hanson MA, Stevens RC (2008) LCP-FRAP assay for pre-screening membrane proteins for in meso crystallization. Cryst Growth Des 8:4307–4315

Claxton DP, Quick M, Shi L, DE Carvalho FD, Weinstein H, Javitch JA, Mchaourab HS (2010) Ion/substrate-dependent conformational dynamics of a bacterial homolog of neurotransmitter: sodium symporters. Nat Struct Mol Biol 17:822–829

Drew D, Kim H (2012) Large-scale production of membrane proteins in Saccharomyces cerevisiae: using a green fluorescent protein fusion strategy in the production of membrane proteins. Methods Mol Biol 866:209–216

Drew D, Sjostrand D, Nilsson J, Urbig T, Chin CN, DE Gier JW, von Heijne G (2002) Rapid topology mapping of Escherichia coli inner-membrane proteins by prediction and PhoA/GFP fusion analysis. Proc Natl Acad Sci USA 99:2690–2695

Drew D, Newstead S, Sonoda Y, Kim H, von Heijne G, Iwata S (2008) GFP-based optimization scheme for the overexpression and purification of eukaryotic membrane proteins in Saccharomyces cerevisiae. Nat Protoc 3:784–798

Faham S, Bowie JU (2002) Bicelle crystallization: a new method for crystallizing membrane proteins yields a monomeric bacteriorhodopsin structure. J Mol Biol 316:1–6

Faham S, Watanabe A, Besserer GM, Cascio D, Specht A, Hirayama BA, Wright EM, Abramson J (2008) The crystal structure of a sodium galactose transporter reveals mechanistic insights into Na+/sugar symport. Science 321:810–814

Forrest LR, Zhang Y-W, Jacobs MT, Gesmonde J, Xie L, Honig BH, Rudnick G (2008) Mechanism for alternating access in neurotransmitter transporters. Proc Natl Acad Sci USA 105:10338–10343

Forrest LR, Kramer R, Ziegler C (2011) The structural basis of secondary active transport mechanisms. Biochim Biophys Acta 1807:167–188

Ganapathy V, Thangaraju M, Gopal E, Martin PM, Itagaki S, Miyauchi S, Prasad PD (2008) Sodium-coupled monocarboxylate transporters in normal tissues and in cancer. AAPS J 10:193–199

Gao X, Lu F, Zhou L, Dang S, Sun L, Li X, Wang J, Shi Y (2009) Structure and mechanism of an amino acid antiporter. Science 324:1565–1568

Granier S, Manglik A, Kruse AC, Kobilka TS, Thian FS, Weis WI, Kobilka BK (2012) Structure of the delta-opioid receptor bound to naltrindole. Nature 485:400–404

Guan L, Kaback HR (2006) Lessons from lactose permease. Annu Rev Biophys Biomol Struct 35:67–91

Haga K, Kruse AC, Asada H, Yurugi-Kobayashi T, Shiroishi M, Zhang C, Weis WI, Okada T, Kobilka BK, Haga T, Kobayashi T (2012) Structure of the human M2 muscarinic acetylcholine receptor bound to an antagonist. Nature 482:547–551

Hanson MA, Roth CB, Jo E, Griffith MT, Scott FL, Reinhart G, Desale H, Clemons B, Cahalan SM, Schuerer SC, Sanna MG, Han GW, Kuhn P, Rosen H, Stevens RC (2012) Crystal structure of a lipid G protein-coupled receptor. Science 335:851–855

Hattori M, Hibbs RE, GOUAUX E (2012) A fluorescence-detection size-exclusion chromatography-based thermostability assay for membrane protein precrystallization screening. Structure 20:1293–1299

Hino T, Arakawa T, Iwanari H, Yurugi-Kobayashi T, Ikeda-Suno C, Nakada-Nakura Y, Kusano-Arai O, Weyand S, Shimamura T, Nomura N, Cameron AD, Kobayashi T, Hamakubo T, Iwata S, Murata T (2012) G-protein-coupled receptor inactivation by an allosteric inverse-agonist antibody. Nature 482:237–240

Hsieh JM, Besserer GM, Madej MG, Bui HQ, Kwon S, Abramson J (2010) Bridging the gap: a GFP-based strategy for overexpression and purification of membrane proteins with intra and extracellular C-termini. Protein Sci 19:868–880

Hunte C, Screpanti E, Venturi M, Rimon A, Padan E, Michel H (2005) Structure of a Na+/H+ antiporter and insights into mechanism of action and regulation by pH. Nature 435:1197–1202

Ishikawa N, Oguri T, Isobe T, Fujitaka K, Kohno N (2001) SGLT gene expression in primary lung cancers and their metastatic lesions. Jpn J Cancer Res 92:874–879

Jardetzky O (1966) Simple allosteric model for membrane pumps. Nature 211:969–970

Jaroszewski L, Slabinski L, Wooley J, Deacon AM, Lesley SA, Wilson IA, Godzik A (2008) Genome pool strategy for structural coverage of protein families. Structure 16:1659–1667

Jasti J, Furukawa H, Gonzales EB, Gouaux E (2007) Structure of acid-sensing ion channel 1 at 1.9 A resolution and low pH. Nature 449:316–323

Jeschke G (2013) A comparative study of structures and structural transitions of secondary transporters with the LeuT fold. Eur Biophys J 42:181–197

Kawate T, Gouaux E (2006) Fluorescence-detection size-exclusion chromatography for precrystallization screening of integral membrane proteins. Structure 14:673–681

Kawate T, Michel JC, Birdsong WT, Gouaux E (2009) Crystal structure of the ATP-gated P2X (4) ion channel in the closed state. Nature 460:592–598

Krishnamurthy H, Gouaux E (2012) X-ray structures of LeuT in substrate-free outward-open and apo inward-open states. Nature 481:469–474

Krupka RM (1993) Coupling mechanisms in active transport. Biochim Biophys Acta 1183:105–113

Li J, Tajkhorshid E (2009) Ion-releasing state of a secondary membrane transporter. Biophys J 97: L29–L31

Li J, Shaikh SA, Enkavi G, Wen PC, Huang Z, Tajkhorshid E (2013) Transient formation of water-conducting states in membrane transporters. Proc Natl Acad Sci USA 110:7696–7701

Lolkema JS, Dobrowolski A, Slotboom DJ (2008) Evolution of antiparallel two-domain membrane proteins: tracing multiple gene duplication events in the DUF606 family. J Mol Biol 378:596–606

Majumdar DS, Smirnova I, Kasho V, Nir E, Kong X, Weiss S, Kaback HR (2007) Single-molecule FRET reveals sugar-induced conformational dynamics in LacY. Proc Natl Acad Sci USA 104:12640–12645

Malawski GA, Hillig RC, Monteclaro F, Eberspaecher U, Schmitz AA, Crusius K, Huber M, Egner U, Donner P, Muller-Tiemann B (2006) Identifying protein construct variants with increased crystallization propensity–a case study. Protein Sci 15:2718–2728

Manglik A, Kruse AC, Kobilka TS, Thian FS, Mathiesen JM, Sunahara RK, Pardo L, Weis WI, Kobilka BK, Granier S (2012) Crystal structure of the micro-opioid receptor bound to a morphinan antagonist. Nature 485:321–326

Noskov SY, Roux B (2008) Control of ion selectivity in LeuT: two Na+ binding sites with two different mechanisms. J Mol Biol 377:804–818

Quick M, Loo DD, Wright EM (2001) Neutralization of a conserved amino acid residue in the human Na+/glucose transporter (hSGLT1) generates a glucose-gated H+ channel. J Biol Chem 276:1728–1734

Quick M, Winther AM, Shi L, Nissen P, Weinstein H, Javitch JA (2009) Binding of an octylglucoside detergent molecule in the second substrate (S2) site of LeuT establishes an inhibitor-bound conformation. Proc Natl Acad Sci USA 106:5563–5568

Rasmussen SG, Choi HJ, Fung JJ, Pardon E, Casarosa P, Chae PS, Devree BT, Rosenbaum DM, Thian FS, Kobilka TS, Schnapp A, Konetzki I, Sunahara RK, Gellman SH, Pautsch A,

Steyaert J, Weis WI, Kobilka BK (2011a) Structure of a nanobody-stabilized active state of the beta(2) adrenoceptor. Nature 469:175–180

Rasmussen SG, Devree BT, Zou Y, Kruse AC, Chung KY, Kobilka TS, Thian FS, Chae PS, Pardon E, Calinski D, Mathiesen JM, Shah ST, Lyons JA, Caffrey M, Gellman SH, Steyaert J, Skiniotis G, Weis WI, Sunahara RK, Kobilka BK (2011b) Crystal structure of the beta2 adrenergic receptor-Gs protein complex. Nature 477:549–555

Ressl S, Terwisscha van Scheltinga AC, Vonrhein C, Ott V, Ziegler C (2009) Molecular basis of transport and regulation in the Na(+)/betaine symporter BetP. Nature 458:47–52

Rosenbaum DM, Cherezov V, Hanson MA, Rasmussen SG, Thian FS, Kobilka TS, Choi HJ, Yao XJ, Weis WI, Stevens RC, Kobilka BK (2007) GPCR engineering yields high-resolution structural insights into beta2-adrenergic receptor function. Science 318:1266–1273

Rosenbaum DM, Zhang C, Lyons JA, Holl R, Aragao D, Arlow DH, Rasmussen SG, Choi HJ, Devree BT, Sunahara RK, Chae PS, Gellman SH, Dror RO, Shaw DE, Weis WI, Caffrey M, Gmeiner P, Kobilka BK (2011) Structure and function of an irreversible agonist-beta(2) adrenoceptor complex. Nature 469:236–240

Sala-Rabanal M, Hirayama BA, Loo DD, Chaptal V, Abramson J, Wright EM (2012) Bridging the gap between structure and kinetics of human SGLT1. Am J Physiol Cell Physiol 302:C1293–C1305

Schulze S, Koster S, Geldmacher U, Terwisscha Van Scheltinga AC, Kuhlbrandt W (2010) Structural basis of Na(+)-independent and cooperative substrate/product antiport in CaiT. Nature 467:233–236

Shaffer PL, Goehring A, Shankaranarayanan A, Gouaux E (2009) Structure and mechanism of a Na+-independent amino acid transporter. Science 325:1010–1014

Shi L, Quick M, Zhao Y, Weinstein H, Javitch JA (2008) The mechanism of a neurotransmitter: sodium symporter–inward release of Na+ and substrate is triggered by substrate in a second binding site. Mol Cell 30:667–677

Shimamura T, Shiroishi M, Weyand S, Tsujimoto H, Winter G, Katritch V, Abagyan R, Cherezov V, Liu W, Han GW, Kobayashi T, Stevens RC, Iwata S (2011) Structure of the human histamine H1 receptor complex with doxepin. Nature 475:65–70

Smirnova I, Kasho V, Choe JY, Altenbach C, Hubbell WL, Kaback HR (2007) Sugar binding induces an outward facing conformation of LacY. Proc Natl Acad Sci USA 104:16504–16509

Sobolevsky AI, Rosconi MP, Gouaux E (2009) X-ray structure, symmetry and mechanism of an AMPA-subtype glutamate receptor. Nature 462:745–756

Sujatha MS, Balaji PV (2004) Identification of common structural features of binding sites in galactose-specific proteins. Proteins 55:44–65

Ujwal R, Abramson J (2012) High-throughput crystallization of membrane proteins using the lipidic bicelle method. J Vis Exp (59):e3383

Veenstra M, Lanza S, Hirayama BA, Turk E, Wright EM (2004) Local conformational changes in the Vibrio Na+/galactose cotransporter. Biochemistry 43:3620–3627

Wadsten P, Wohri AB, Snijder A, Katona G, Gardiner AT, Cogdell RJ, Neutze R, Engstrom S (2006) Lipidic sponge phase crystallization of membrane proteins. J Mol Biol 364:44–53

Watanabe A, Choe S, Chaptal V, Rosenberg JM, Wright EM, Grabe M, Abramson J (2010) The mechanism of sodium and substrate release from the binding pocket of vSGLT. Nature 468:988–991

Weyand S, Shimamura T, Yajima S, Suzuki S, Mirza O, Krusong K, Carpenter EP, Rutherford NG, Hadden JM, O'reilly J, Ma P, Saidijam M, Patching SG, Hope RJ, Norbertczak HT, Roach PC, Iwata S, Henderson PJ, Cameron AD (2008) Structure and molecular mechanism of a nucleobase-cation-symport-1 family transporter. Science 322:709–713

Wong FH, Chen JS, Reddy V, Day JL, Shlykov MA, Wakabayashi ST, Saier MH Jr (2012) The amino acid-polyamine-organocation superfamily. J Mol Microbiol Biotechnol 22:105–113

Wright EM (2013) Glucose transport families SLC5 and SLC50. Mol Aspects Med 34:183–196

Wright EM, Loo DD, Hirayama BA, Turk E (2004) Surprising versatility of Na+-glucose cotransporters: SLC5. Physiology (Bethesda) 19:370–376

Wright EM, Loo DD, Hirayama BA (2011) Biology of human sodium glucose transporters. Physiol Rev 91:733–794

Yamashita A, Singh SK, Kawate T, Jin Y, Gouaux E (2005) Crystal structure of a bacterial homologue of Na+/Cl–dependent neurotransmitter transporters. Nature 437:215–223

Yernool D, Boudker O, Jin Y, Gouaux E (2004) Structure of a glutamate transporter homologue from Pyrococcus horikoshii. Nature 431:811–818

Zhang C, Srinivasan Y, Arlow DH, Fung JJ, Palmer D, Zheng Y, Green HF, Pandey A, Dror RO, Shaw DE, Weis WI, Coughlin SR, Kobilka BK (2012) High-resolution crystal structure of human protease-activated receptor 1. Nature 492:387–392

Zhou Z, Zhen J, Karpowich NK, Goetz RM, Law CJ, Reith ME, Wang DN (2007) LeuT-desipramine structure reveals how antidepressants block neurotransmitter reuptake. Science 317:1390–1393

Zou JY, Flocco MM, Mowbray SL (1993) The 1.7 A refined X-ray structure of the periplasmic glucose/galactose receptor from Salmonella typhimurium. J Mol Biol 233:739–752

Chapter 4
Moving Crystallographic Snapshots: A Mechanism for Transport Regulation in BetP

Christine Ziegler and Reinhard Krämer

Abstract The molecular understanding of secondary transport, in particular how transport activity is regulated, is one of the cutting-edge questions in biological science. A number of secondary transporters show regulation of transport activity, which plays an important role in stress-induced cellular responses. Transport activity is regulated in response to various external stimuli, which often are difficult to identify. Therefore, only a few regulated transporters are described to date; one of the best characterized amongst them is the Na^+-coupled betaine symporter from *Corynebacterium glutamicum*, $BetP_{Cg}$. The biochemical background of stimulus sensing by activity regulation of $BetP_{Cg}$ has been elucidated in detail both in cells and proteoliposomes. First insight into a molecular mechanism of regulated transport in BetP was obtained by the combination of two-dimensional (2D) and three-dimensional (3D) crystallization, functional measurements, spectroscopy, and bioinformatics. In the last 5 years several atomic structures of trimeric BetP were solved in different conformational states and under both activating and inactivating conditions. Thereby, the transport cycle of BetP was described in molecular detail; however, the dynamics of osmoregulation is still far from being understood. One major limiting factor on the way to a molecular mechanism is restrictions imposed by the crystalline environment, which may populate or exclude some of the functional important conformations. The example of BetP demonstrates how difficult it is to trap activation and regulation of a transporter in a crystal structure. In this chapter we will critically discuss the efforts in obtaining meaningful structural and functional data and how they are combined in a dynamic description of transport and regulation of a secondary carrier.

C. Ziegler (✉)
University of Regensburg and MPI of Biophysics, Frankfurt, Germany
e-mail: Christine.Ziegler@biophys.mpg.de

R. Krämer
University of Cologne, Cologne, Germany

R. Krämer and C. Ziegler (eds.), *Membrane Transport Mechanism*, Springer Series in Biophysics 17, DOI 10.1007/978-3-642-53839-1_4,

Keywords Alternating access • Secondary transport • Membrane transport • Conformational states • Osmoregulation • Activation • Sodium coupling • Oligomerization • Stress response • Signal transduction

4.1 Introduction

4.1.1 To the Rescue of Water Activity: Impact of Regulated Osmolyte Transporters

The secondary transporter BetP from the Gram-positive soil bacterium *Corynebacterium glutamicum* is a member of the Betaine:Carnitine:Choline:Transporter (BCCT) family (Ziegler et al. 2010). Its physiologic function is to accumulate the compatible solute glycine betaine to high concentrations in the cytoplasm in response to hyperosmotic conditions in the cell's surroundings (Farwick et al. 1995; Peter et al. 1998b), which is a common type of stress for soil bacteria. Together with ProP, an MFS family transporter from *Escherichia coli* (Culham et al. 1993; Tsatskis et al. 2005; Wood 2011), and OpuA, an ABC type carrier from *Lactococcus lactis* (Biemans-Oldehinkel et al. 2006; van der Heide and Poolman 2000), BetP is in functional terms the best-studied compatible solute carrier. Moreover it is the only of these three well-studied systems for which atomic structures are available which strongly contributed to its mechanistic understanding. The driving force for effective accumulation of its substrate is the electrochemical Na^+ potential since import of betaine is coupled to that of Na^+ with a stoichiometry of 1:2 (Farwick et al. 1995). Transport mechanism and regulation of BetP were analyzed in detail in intact cells of *C. glutamicum*, after heterologous expression in *E. coli*, and upon functional reconstitution in proteoliposomes (Rübenhagen et al. 2000, 2001). The combined results of these studies led to the concept of a transporter with a transport mechanism similar to other Na^+-coupled secondary active solute uptake systems and with a rather complex pattern of regulation different from most other secondary transporters. The fact that all functional aspects of BetP, in particular its complex regulation behavior, are not only observed in its native surrounding, i.e., the plasma membrane of *C. glutamicum*, but are maintained when the transporter is expressed in a heterologous host or reconstituted in proteoliposomes, proves that all these functions are intrinsic attributes of BetP and not due to additional factors or (protein) components (Fig. 4.1) (Rübenhagen et al. 2001).

Expression of *betP* is regulated on the level of transcription by MtrAB, an osmoreactive two-component system in *C. glutamicum* (Möker et al. 2004). This aspect is not covered in this chapter, which focuses on structure–function investigations. Direct regulation of BetP's activity in response to osmotic stress is complex in terms of both physiology and mechanism. In intact cells and in the absence of osmotic stress, BetP is barely active. When the extent of stress exceeds a given threshold, BetP is instantly activated and starts to import betaine into the cell. The

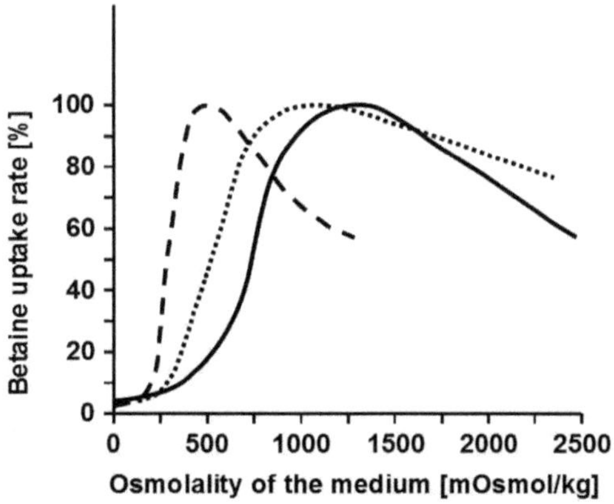

Fig. 4.1 Activation of BetP-mediated betaine uptake by hyperosmotic conditions in *C. glutamicum* (*solid line*), upon heterologous expression in *E. coli* (*broken line*) and after reconstitution in proteoliposomes (*dotted line*). Transport activity was normalized to maximum activity = 100 %

initial betaine uptake activity by BetP directly depends on the extent of hyperosmotic stress resulting in a characteristic pattern with a distinct optimum of response (Fig. 4.1) (Peter et al. 1998a). Accumulation of betaine mediated by BetP ceases again when the cell reaches osmotic compensation according to the increased internal osmolality mainly as a result of betaine uptake. The functional analysis of BetP activation was performed in proteoliposomes and in intact cells. Consequently, two different aspects of BetP regulation on the level of activity can be observed: activation upon external osmotic challenge and inactivation when the cell reaches osmotic compensation. These two regulation events can be experimentally discriminated on the basis of mechanistic aspects (Botzenhardt et al. 2004). Only the former mechanism, activation in response to an osmotic upshift, will be dealt with in this chapter in the context of structure–function analysis.

4.1.2 Sensing of Osmotic Stress from the Perspective of a Secondary Transporter

A large number of possible stimuli from the external and internal water space as well as from the membrane surrounding can account for triggering BetP activation in response to an osmotic upshift (Bremer and Krämer 2000; Wood 1999; Ziegler et al. 2010). The first stimulus which was identified turned out to be the rise in the cytoplasmic K^+ concentration caused initially by instant volume reduction upon osmotic upshift (in proteoliposomes and cells) and subsequently by active K^+ influx (in cells only) (Rübenhagen et al. 2001). The nature of the K^+ stimulus acting on BetP is mechanistically of particular interest. Activation is specific, i.e., K^+, Rb^+, and Cs^+ are stimulating, but not Na^+ (Schiller et al. 2004a). The apparent affinity for K^+ in BetP activation is around 200 mM and thus very low, but it is perfectly within the expected physiologic range of cytoplasmic concentrations (Rübenhagen et al. 2001). Activation was shown to be highly cooperative with a Hill coefficient of around $n = 5$ (Maximov and Krämer, unpublished observation).

Bacteria in general and *C. glutamicum* in particular also host other osmoregulated carriers being members of the BCCT family, e.g., EctP and LcoP (Peter et al. 1998b; Ziegler et al. 2010), or of other transporter families, e.g., ProP (Wood 2011). Interestingly, at least in the case of the carriers from *C. glutamicum* and for members of the BCCT family, also the peculiar mechanistic contribution of the C-terminal extension to both sensing and regulation seems to be conserved (Ott et al. 2008; Peter et al. 1998a; Schiller et al. 2004b).

Recent results have shown that, besides the K^+ related stimulus, a second stimulus mediated by changes in the membrane surrounding is required in *C. glutamicum* cells for effective stimulation of BetP activity (unpublished results). As a consequence, in intact cells the presence of a cytoplasmic K^+ concentration exceeding 200 mM turned out to be essential but not sufficient for full activation of BetP. In addition, a change in the osmotic gradient between the exterior and the interior of the cell has to be imposed to trigger full activity of BetP. This second trigger was shown to be qualitatively related to a change in physical properties of the membrane in which BetP is embedded. Interestingly, this stimulus was not detected in proteoliposomes (unpublished results). This observation again shows that functional reconstitution in proteoliposomes is an extremely helpful tool for a detailed functional analysis of defined aspects of transport systems, but parallel characterization under in vivo conditions is required in addition in order to guarantee a complete mechanistic description.

4.1.3 Teaming Up in Sensing: Role of Terminal Domains and Lipids

BetP is a polypeptide comprised of 595 amino acid residues, 12 transmembrane segments, and two long terminal domains both at the N- and the C-end of the protein (Ott et al. 2008; Peter et al. 1998a; Schiller et al. 2004b). On a mechanistic level, a quantitative description of the contribution of the N- and the C-terminal domains to transport regulation was achieved by analyzing site-directed mutant forms and truncation constructs, as well as by interaction studies using surface plasmon resonance spectroscopy (Ott et al. 2008). Deletion of the terminal domains results in loss of function. Stepwise truncation of the N-terminal domain led to a shift of the regulation pattern towards a higher concentration of K^+ required for activation, whereas alteration of the C-terminal domain resulted in complex changes of BetP regulation. Stepwise truncation of the C-terminal domain affected the response to K^+. Single amino acid replacements, in particular by proline at strategic positions, abolished regulation (Ott et al. 2008; Schiller et al. 2004b). Interestingly, the extent of functional change in the regulatory behavior as a result of alterations in the C-terminal domain critically depends on the nature of the lipid surrounding of BetP. Amino acid replacements always showed a significantly reduced effect in *C. glutamicum* as compared to that in *E. coli* (Ott et al. 2008). This is reflected by a fundamentally different phospholipid composition in the two organisms. Whereas the plasma membrane of *C. glutamicum* contains solely negatively charged phospholipids (phosphatidyl glycerol, phosphatidyl inositol, and cardiolipin) and only two types of fatty acids (palmitic and oleic acid), *E. coli* harbors only around 20 % negatively charged phospholipids, phosphatidyl ethanolamine is dominating, and the fatty acid pattern is more complex (Ozcan et al. 2007). Taken together, the influence of phospholipids on BetP function was found to be twofold. Besides the striking dependence of BetP on the composition of the surrounding membrane lipids, investigations in proteoliposomes further demonstrated that the activity of BetP directly depends on the surface charge of the surrounding phospholipid bilayer (Schiller et al. 2006).

Based solely on these functional data, a basic model for BetP regulation has been suggested (Fig. 4.2) (Ott et al. 2008). This concept takes into account the following contributions of different domains of BetP and the membrane surrounding. (a) The C-terminal domain interacts with cytoplasmic loops of BetP, with the N-terminal domain and with the membrane surface. (b) Interaction of the C-terminal domain and the membrane surface is of electrostatic nature. (c) An increased K^+ concentration at the cytoplasmic side alters the nature of these interactions in a cooperative manner. (d) During its regulatory action, the C-terminal domain alters its conformation, which is described as a kind of molecular switch triggering a number of further consequences in BetP structure and function. (e) The conformational change (switch) of the C-terminal domain ultimately results in the onset of betaine transport. (f) Under default conditions, i.e., in the absence of osmotic stress, an intact C-terminal domain inhibits the transport activity of BetP.

4.1.4 Why Do We Need Structure to Understand Function of BetP?

The knowledge on transport regulation of BetP and the diverse factors that contribute to it, e.g., the binding of K^+ ions, the interaction with negatively charged lipids, the proper helical folding of the C-terminal domains, and their interaction with cytoplasmic loops and lipids, all is based on thorough biochemical data, which have revealed BetP's unique regulatory properties. Apparently *C. glutamicum* has evolved this fine-tuned transporter to take advantage of betaine under hyperosmotic conditions exploiting changes in the membrane state and in internal K^+ concentration as measure for the amount of external stress. The observed functional characteristics during BetP activation are strictly related to measurable stimuli and structural domains allowing for an already rather precise description of the mechanism. This is a very rare situation in biology as in most cases regulation is the result of a complex network of interfering events including interactions of different

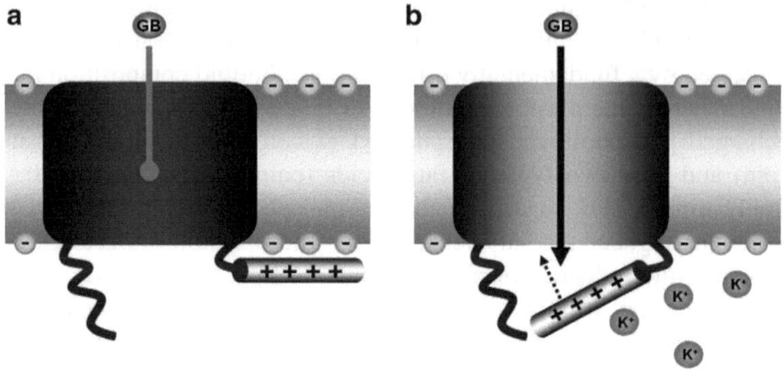

Fig. 4.2 Model of BetP activation mechanism based on biochemical data (Ott et al. 2008; Peter et al. 1998a; Rübenhagen et al. 2001; Schiller et al. 2004b). Putative contributions of the N- and the C-terminal domain of BetP, the membrane surface, and luminal K^+ characteristic for the inactive (**a**) and active (**b**) state are included in the model

proteins. This is not the case for BetP, where at least one of the activation stimuli is not only well defined, but also quantitatively measurable. Common and diverging aspects in the regulation mechanism of BetP, ProP, and Opu A were identified and contributed to the better understanding of osmoregulation in their respective organisms (van der Heide and Poolman 2000; Wood 2011; Ziegler et al. 2010).

What kind of further information can be expected from an atomic structure of BetP that justifies the tedious and risky structure determination procedure? Mainly, it is very difficult to provide a chronological description of what happens during regulation based exclusively on biochemical data as complete and sophisticated they might be. Certainly, biophysical measurements and spectroscopy could help on the track to understand protein dynamics, but they provide a set of physical parameters that have to be translated into a biological process. The most direct information is the visualization of the protein by a structure; however, structural data risk creating the illusion of dynamics, while in fact these are just static snapshots.

The gap between a functional concept based on biochemical data and the mechanistic understanding of BetP is particularly wide because of the complex regulation mechanism of this transport system. Many questions could not be addressed adequately without the availability of a structure regarding, for example, (a) the oligomeric state of BetP, (b) the exclusive substrate specificity, (c) the substrate and co-substrate translocation as well as the coupling between betaine and two sodium ions, (d) the mechanistic explanation of the mode of action of the K^+ stimulus, (e) an understanding of the functionally significant interaction of BetP with lipids, (f) a detailed view how the two terminal domains of BetP exert their regulatory function in close contact with each other and with cytoplasmic loops of the carrier, and finally (g) an idea how the regulatory functions of BetP are intertwined with its catalytic properties. Only the availability of a series of detailed 2D and 3D structures of BetP was an appropriate basis for a better mechanistic understanding of at least some of the fundamental questions listed above concerning the transport regulation.

4.2 Towards a Functional Meaningful Atomistic Structure of BetP$_{Cg}$

4.2.1 Useful Thoughts Before Starting Crystallization

To understand the intriguing transport and regulation properties of BetP$_{Cg}$ extensive structural information was indispensable. First crystallization attempts were initiated in 2001, but it took more than 8 years to obtain the first atomic structure of BetP$_{Cg}$ (Ressl et al. 2009). Interestingly, the wild type of BetP$_{Cg}$ crystallized quite readily yielding fast crystals diffracting to 10–8 Å, but the optimization of those crystals to atomic resolution was long and tedious mainly due to fact that no appropriate homologues for BetP$_{Cg}$ were available. The crystallization of BetP$_{Cg}$ is an instructive example for possible pitfalls and drawbacks in a transporter structure determination demonstrating the risky nature of such a project. In retrospective we can only recommend the younger readership to choose a target protein for which crystallography can be coupled with functional transport and binding studies. This strategy will provide not only a steady scientific outcome, but also valuable information on the way towards the structure.

4.2.2 BetP$_{Cg}$ Crystallization Strategy

BetP$_{Cg}$ turned out to be stable for several days, which for structural studies is one of the most important prerequisites. Topology predictions by the XTALPRED server (http://ffas.burnham.org/XtalPred-cgi/xtal.pl) confirmed the high overall stability in TM helices, but also revealed unordered regions mainly in the N- and C-terminal domains, which we recognized as a major problem for crystallization (Fig. 4.3a). Unordered structural elements always represent an unfortunate starting point for structure determination. It is very likely that these unordered elements are stabilized in the membrane by lipid–protein interactions as it was observed in a structure of BetP$_{Cg}$ in complex with POPG lipids (Fig. 4.3b). While XTALPRED ranks TM2 in BetP$_{Cg}$ as helix with low complexity, it turned out that this helix is very stable and well ordered due to the interaction with a lipid head group at the periplasmic side (Koshy et al. 2013). The most obvious strategy is to truncate those parts and/or stabilize them by the introduction of additional scaffolds, e.g., lysozyme, antibody fragments, or so-called DARPins (Zou et al. 2012; Seeger et al. 2013; Park et al. 2012; Day et al. 2007). Although this method has proven to be successful for GPCRs and also for channels and transporters (Fig. 4.1b, Brohawn et al. 2013; Huber

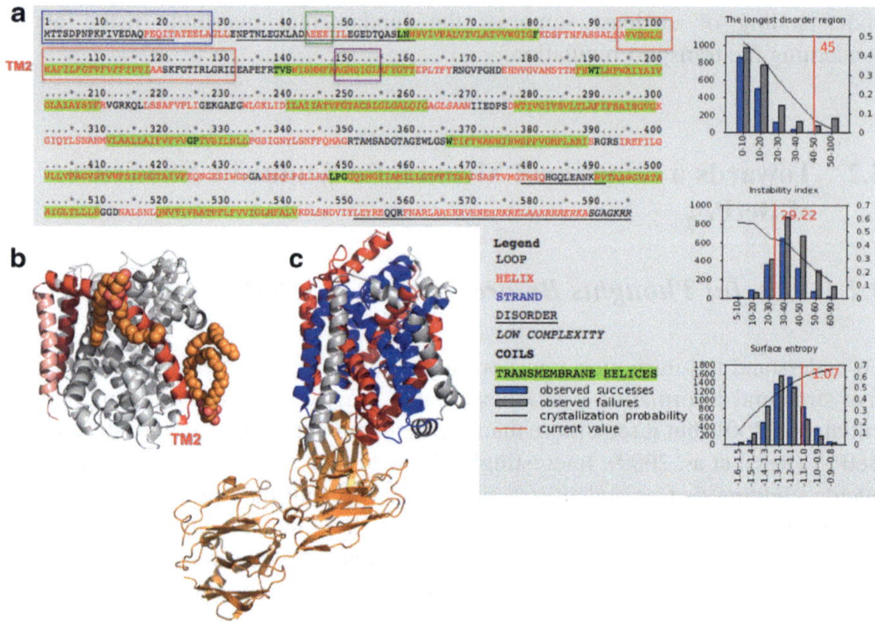

Fig. 4.3 What you should know about your crystallization target protein. (**a**) The web server XTALPRED compares biochemical and biophysical features of the target protein with corresponding probability distributions obtained from a target registration database that provides information on the experimental progress of structure determination projects (PepcDB: http://pepcdb.pdb.org). XTALPRED reveals the empirical distributions of crystallization probability and the positions of the protein in those distributions. (**b**) POPG lipids (*orange* in *sphere representation*) attached to the outer rim of the BetP monomer (pdb entry code 4C7R) (Koshy et al. 2013). (**c**) Outward-facing state of the neurotransmitter:sodium:symporter homologue LeuT from Aquifex aeolicus (pdb entry 3TT3), which was crystallized in complex with an antibody FAB fragment (Krishnamurthy and Gouaux 2012). LeuT and $BetP_{Cg}$ share the same overall folds of two inverted structural repeats of 5TM each. Repeat 1 comprises TM1–TM5 and is colored in *red*, while repeat 2 comprises TM6–TM10 colored in *blue*. The antibody Fab fragment serves as crystallization scaffold; however, there is some probability that the interaction of the Fab fragment modifies the intracellular ionic interaction network in LeuT, which was described previously (Kniazeff et al. 2008)

et al. 2007; Mittal et al. 2012), there are shortcomings due to possible impact on function. In a recent structure of the neurotransmitter transporter homologue LeuT (Krishnamurthy and Gouaux 2012) a functional important ionic network on the cytoplasmic side was modified by co-crystallization with a Fab fragment (Fig. 4.3c) resulting in an artificially wider open inward-facing state. In $BetP_{Cg}$ the unordered regions at least in the C-terminal domain could not be modified by truncation due to their involvement in regulation (Peter et al. 1998a). A stabilizing modification with a comparatively minor functional effect was the truncation of the N-terminal domain. This mutant protein was chosen for crystallization (Fig. 4.1a, blue box).

Truncations were also applied in crystallization of the sodium-coupled galactose transporter vSGLT (Faham et al. 2008) and will be discussed in Chap. 3. Another strategy to overcome such kind of problems is the use of a more stable homologue from another organism. However, with the increasing amount of secondary transporter and channel structures in the past 10 years, it became obvious that a particular protein structure often has limitations in interpreting the specific functions of homologues. This holds true especially for the understanding of eukaryotic transporters based on the structures of their prokaryotic homologues, which is discussed for several transporters in Chaps. 1, Chap. 3, and Chap. 7. Finally, a new development in structural biology of membrane proteins in general is the use of micro- and nanocrystals that require less material and often show improved diffraction potential. These miniscule crystals require advanced microfocus beamlines and high intensities, e.g., provided by free electron laser X-rays (X-FEL) (Hunter and Fromme 2011).

4.2.3 Back to the Membrane: Pitfalls and Paybacks in the 2D Crystallization of BetP

In a 2D crystal, the closely packed proteins form ordered two-dimensional arrays that can be described by crystal lattice constants (Fig. 4.4a, inset) (Tsai et al. 2007, 2011). As they comprise 1–2 lipid bilayers, they are thin enough to be investigated directly in an electron microscope circumventing the phase problem that we encounter in 3D crystals and X-ray diffraction (Hauptman 1997). Based on the fact that the membrane protein is observed in a native environment raised big hopes in structural biology (Kühlbrandt 2013; Ubarretxena-Belandia and Stokes 2012; Pope and Unger 2012). However, it turned out that the resolution obtained from 2D crystals of secondary transporters was constantly inferior to that reported for 3D crystals of the same protein (Tsai and Ziegler 2010), which pitched the enthusiasm for this method. On the other hand, many examples have proven the usefulness of 2D crystallization, which can be illustrated for the sodium/proton antiporter NhaA from *E. coli* (Padan et al. 2009; Appel et al. 2009). Only the combination of 2D and 3D crystallization enabled an undistorted glimpse on the functional atomic architecture of this important transporter.

The structure determination of $BetP_{Cg}$ provides another example for the usefulness of 2D crystals on the one hand and the limitations in resolution on the other. Already from the very first projection structures to 7.5 Å it became obvious that $BetP_{Cg}$ forms a stable trimer in the membrane, characterized by the intriguing fact that each protomer within the trimer adopts a slightly but significantly different architecture (Fig. 4.4b) (Tsai et al. 2011). However, it was not possible to increase the resolution of the BetP 2D crystals beyond 7.5 Å. The reason for this limitation in resolution is the specific architecture of the trimer (Fig. 4.4b), which provides a strong interaction at the trimeric interface, while the perimeter of the trimer does

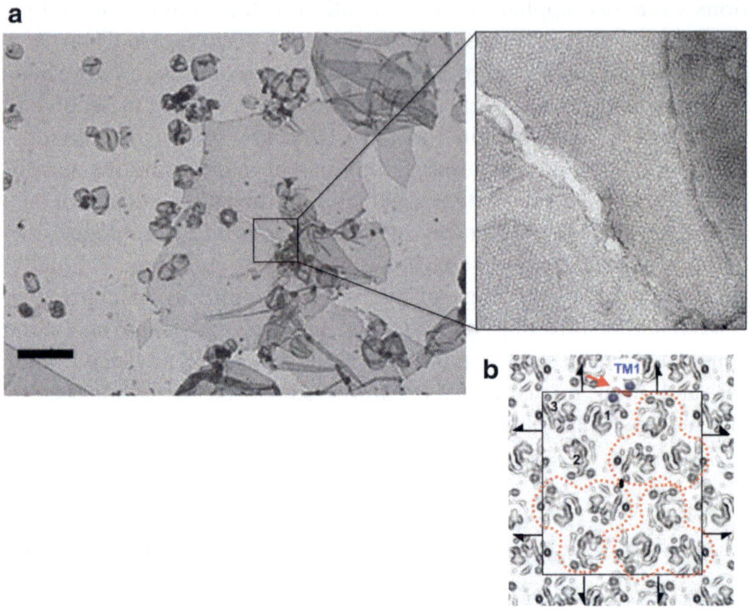

Fig. 4.4 2D crystals in the electron microscope. (**a**) Micrograph of a negatively stained 2D crystalline array of $BetP_{Cg}$ in a low magnification and in a high magnification (*inset*). (**b**) Projection structure of $BetP_{Cg}$ to 8 Å. A *red arrow* indicates the crystal contacts between two protomers of similar orientation in the membrane (Tsai et al. 2011)

not form strong crystal contacts. Despite several attempts under varying conditions, BetP appeared only closely packed in the μm-sized 2D crystals, which due to missing crystal contacts limits the diffracting power significantly. This heterogeneity in packing, which was less a problem in projection, turned out to be devastating for the 3D map determination, which could only be calculated to 8 Å in plane and 16 Å vertical to the plane. Therefore, the asymmetric trimer in 2D crystals could only be interpreted based on the first atomic structure of $BetP_{Cg}$ solved from 3D crystals (Ressl et al. 2009, see next section), on the basis of which the structural differences between individual protomers in 2D crystals could be assigned to different transporter states (Tsai et al. 2011). However, important information was obtained from 2D crystals about potential crystal contacts, e.g., at the N-terminal domain (Fig. 4.4b) and the role of lipids to stabilize the BetP trimer. Combining 2D and 3D crystallization of $BetP_{Cg}$ also helped in distinguishing between protein characteristics and crystallization artifacts. Moreover, 2D crystals can be used in numerous spectroscopic techniques, e.g., for EPR, FRET, and NMR studies. In fact, we have carried out single molecule force spectroscopy by AFM (ref) on $BetP_{Cg}$ as well as FTIR studies (Korkmaz et al. 2013), which have contributed strongly to the understanding of the dynamic behavior of $BetP_{Cg}$, which will not be a topic of this chapter. The simultaneous appearance of conformational asymmetry in 2D and 3D crystals of wild-type $BetP_{Cg}$ and mutant forms opened the discussion towards exciting insight into the folding and regulated transport function of trimeric transporters.

4.2.4 Lessons to Learn from Optimization of BetP$_{Cg}$ 3D Crystals

An iterative process including protein modification by mutagenesis as well as detergent exchange from DDM to Cymal-5 was required to optimize BetP$_{Cg}$ 3D crystals from anistropic low diffraction (Fig. 4.5a) to the best resolution of 2.6 Å (Fig. 4.5b). As a side effect of the detergent exchange it turned out that BetP$_{Cg}$ became significantly de-lipidated as shown by TLC and mass spectrometric analysis (Koshy et al. 2013; Tsai et al. 2007). In addition to partial truncation of the N-terminal domain to remove putative unordered regions (see above), the concept of surface engineering proposed by (Derewenda and Vekilov 2006) was applied and the triple glutamate cluster at position 44–46 was exchanged against alanine (Fig. 4.3a, green box). This mutant is still regulated in *C. glutamicum* cells (Ott et al. 2008; Ressl et al. 2009) although with a shifted profile towards higher osmolality as it was previously observed for other N-terminal truncation mutants (Peter et al. 1998a). This mutant protein was characterized by a higher helical order in the remaining N-terminal domain (Fig. 4.5c), which contributed an additional crystal contact (Koshy et al. 2013). The structure was finally solved by a SAD (single anomalous diffraction) experiment with seleno-methionine-labeled protein crystals diffracting to 3.3 Å. The obtained structure revealed that BetP$_{Cg}$ shares a common fold with several secondary transporters unrelated in sequence (Forrest et al. 2011). This fold was denominated as "LeuT-like fold," acknowledging the fact that it was first observed for LeuT, a member of the neurotransmitter:sodium: symporter (NSS) family (Yamashita et al. 2005). This fold is characterized by the presence of two structurally similar repeats of five transmembrane helices each, which are not sequence related. One repeat is inversely inserted into the membrane with respect to the other repeat and both repeats are closely intertwined. The resulting architecture comprised by this arrangement is a 4 TM bundle formed by the first two helices and a 6 TM scaffold with the remaining 3 helices from each repeat. Both bundle and C-shaped scaffold line the central substrate-binding site (Fig. 4.3c). The medical impact of the NSS family (Bröer and Gether 2012), the superior resolution to 1.65 Å of the first structure (Yamashita et al. 2005), the presence of both an outward-facing and an inward-facing state (Krishnamurthy and Gouaux 2012), and the wealth of biophysical (spectroscopic EPR and FRET) (Zhao et al. 2010, 2011; Claxton et al. 2010) and bioinformatics data (Cheng and Bahar 2013; Zhao et al. 2012, Zdravkovic et al. 2012) turned LeuT to a paradigm for secondary transport (Forrest et al. 2011). Very recently, the structure of the first eukaryotic NSS homologue, the dopamine transporter from *Drosophila*

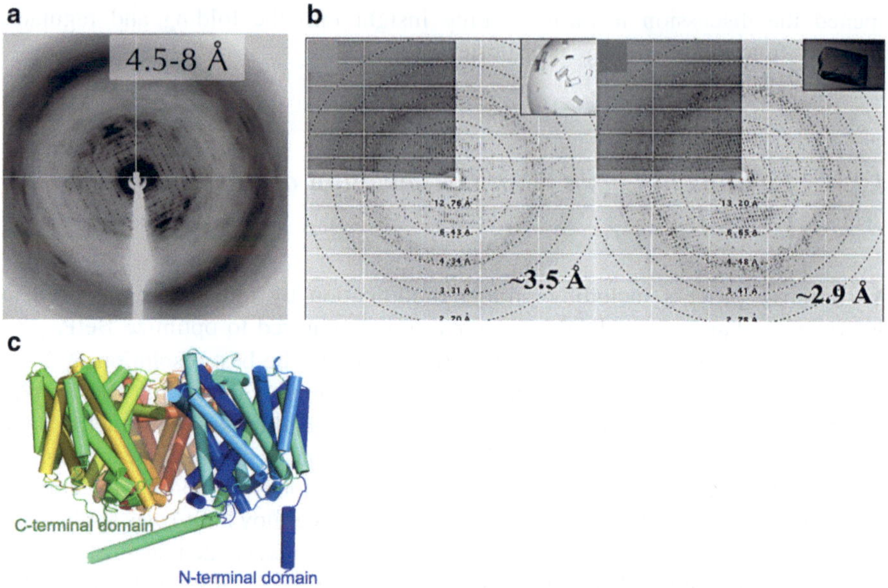

Fig. 4.5 Diffraction pattern from 3D crystals of BetP$_{Cg}$. (**a**) Anistropic diffraction pattern of a WT/DDM crystal. Diffuse scattering is often a side effect of the low order in a third dimension. (**b**) Improved diffraction pattern of the crystallization mutant of BetP$_{Cg}$ in Cymal-5. Variations in resolution are result of differences and un-reproducibility in the purification protocol. (**c**) Helical folding of the crystallization mutant due to the triple mutation of the glutamate cluster at position 44–46 (pdb entry code 4AIN)

melanogaster, was solved by the Gouaux group confirming the major part of conclusions drawn from the prokaryotic structure of LeuT (Penmatsa et al. 2013).

Although sharing the same fold with LeuT and other transporters, the significance of BetP$_{Cg}$ due to its unique features in transport regulation added another interesting aspect to the phenomenon of secondary transport (Forrest et al. 2011). In addition, the conformational asymmetry is an intriguing feature of BetP, which was also observed in EmrE (Lehner 2008), in AcrB (Seeger et al. 2006), and lately in Glt$_{Ph}$ (Verdon and Boudker 2012). It is still not understood what causes different conformations of individual protomers within an oligomeric assembly. There are two possible scenarios, which might lead to an asymmetric conformational distribution, a functional coupling between protomers as it was shown for AcrB (Seeger et al. 2006) or a structural stabilization that can be related to differences in membrane insertion of individual protomers as it might be the case for the asymmetric dimer of EmrE (Steiner-Mordoch et al. 2008). The first has to be addressed by thorough analysis of the cooperative binding and/or transport characteristics. As membrane insertion and assembly of membrane proteins are a complex and still rather unexploited field the second scenario is difficult to address experimentally (Otzen and Andersen 2013; Booth 2012). In the context of conformational asymmetry in BetP, to date, cooperative transport characteristics were not observed for betaine transport, although cooperative betaine binding in the presence of K^+ was detected by Trp fluorescence (Ge et al. 2011). Conformational asymmetry was observed both in the wild-type protein and the deregulated C45 mutant carrying a truncated C-terminal domain, indicating that this asymmetry is not related to regulation. As the C45 mutant could never be crystallized in 3D, but in 2D, again the combination of both methods provided important insights helping to rule out crystallization artifacts as source of the conformational asymmetry in $BetP_{Cg}$. Although being not understood, this asymmetry provided the unique possibility to detect several distinct conformations of BetP in 2D and 3D crystals. As a result, the total of eight unique conformations represents up to now the most complete atomistic description of the alternating access cycle of any secondary transporter (Perez et al. 2012).

4.2.5 From a Crystallographic Snapshot Towards a Molecular Transport Mechanism

To date the following states of $BetP_{Cg}$ are described by atomic structures with resolutions between 3.4 and 2.6 Å: three outward-facing states, three closed states, and four inward-facing states (Fig. 4.6a). The comparison of these states provides a convincing picture describing how betaine binds to and is released from a tryptophan prism shaped by the BCCT family signature motif (Ziegler et al. 2010) forming the central substrate-binding site (S1) (Fig. 4.6b). Due to side-chain and main chain conformational changes the prism opens and closes alternatively to both sides of the membrane. It can be assumed that these changes are orchestrated by binding and release of the two sodium ions (Na1 and Na2). A Na2 sodium ion could be observed to date in only one of the closed states, while Na1 sodium was never detected in a $BetP_{Cg}$ structure. Instead it was predicted based on symmetry considerations within the LeuT-like fold and was validated by MD simulations as well as by binding and transport studies (Khafizov et al. 2012). Concerning the role of sodium ions it was an important advantage that structures of other sodium symporters of similar fold, e.g., of vSGLT (Faham et al. 2008), LeuT (Yamashita et al. 2005), and Mhp1 (Weyand et al. 2008), were solved at about the same time as the $BetP_{Cg}$ structure (Ressl et al. 2009). The careful analysis of the different structures had a tremendous impact on understanding the role of sodium coupling in LeuT-like fold transporters (Perez and Ziegler 2013).

From all states, the closed state (Cc; see Fig. 4.6a), which is significantly distinct from the outward- and inward-facing states, occluding the S1 site from both sides of the membrane by protein bulk, is one of the most interesting and unique features of the BetP structure. In the closed state in complex with betaine and sodium, we observe the tightest coordination of betaine. In the corresponding apo Cc state without substrate the three Trp residues coordinate a water molecule instead of a betaine, which in this state would not fit into the S1 site. A third Cc state showed a

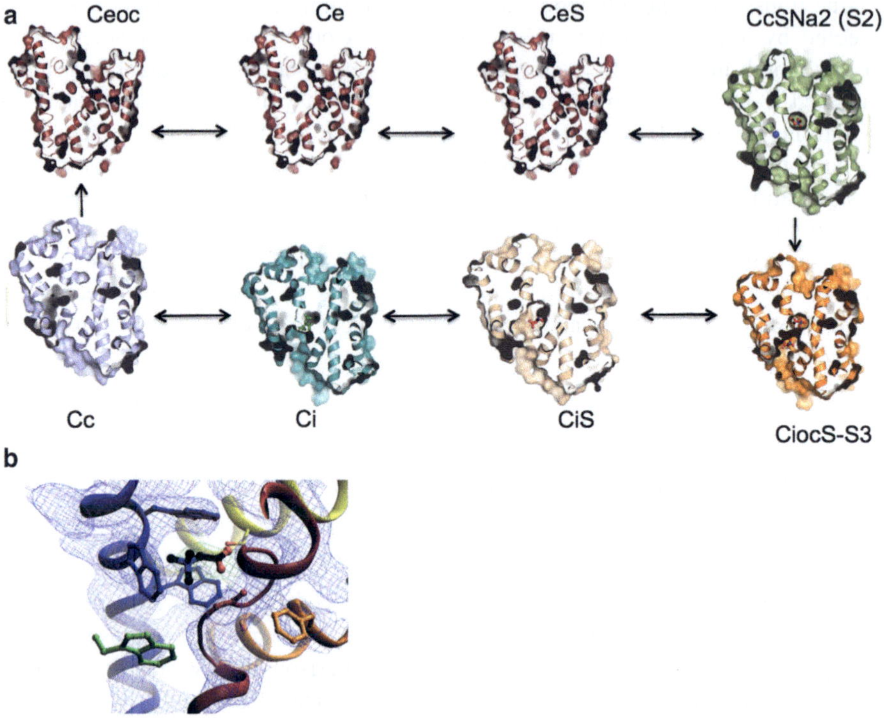

Fig. 4.6 Alternating access cycle and activation cycle of $BetP_{Cg}$. (**a**) Binding and release of betaine can be described by a sequence of eight different states. (**b**) Betaine is bound to the *Trp prism* or *box* in the *center* of the BetP monomer in the closed state. During the transition from outward- to inward-facing state, the Trp residues change their relative positions to open up the *Trp box* to different sides resulting in an *open box* in the outward-facing state and a ladder-like arrangement in the inward-facing state (Perez et al. 2012)

second betaine molecule bound to a distinct periplasmic binding site. This second binding site was first described for the $BetP_{Cg}$ homologue CaiT and interpreted based on binding studies as possible regulatory site (Schulze et al. 2010), although it was shown recently that the site was not stable in MD simulations (Zomot and Bahar 2012). The existence of a second binding site in LeuT-like fold transporters was intensively discussed in the past (Khelashvili et al. 2012). In the case of $BetP_{Cg}$ this site was discovered to be K^+ dependent (Ge et al. 2011), which was first suggested by the aforementioned AFM studies and later confirmed by functional data. To date, the functional meaning of the second periplasmic binding site in $BetP_{Cg}$ is still unknown.

The sequence of states observed for $BetP_{Cg}$ gives important insights into fundamental questions about the transport mechanism in LeuT-fold-like transporters, which can be summarized as follows: (1) Binding and release of substrate and co-substrate are facilitated by subtle side-chain conformational changes, which are rather symmetric; in other words, the binding event corresponds to a reverse release event. (2) Binding and release of the co-substrate (sodium ions in the case of BetP) modify accessibility and final formation or disruption of the substrate-binding site, but each ion has a different effect. (3) Binding of Na2 sodium opens the Trp prism towards the periplasm, while binding of substrate together with Na1 sodium most likely closes the prism and initiates isomerization to the inward-facing state. *Vice versa* release of Na2 again opens the S1 side to the cytoplasm and triggers release of betaine and Na1 sodium closes the site again for isomerization to the outward-facing state (Perez et al. 2012). The fact that we observed the Ci state only when a citrate buffer molecule is wedged into the cytoplasmic funnel supports this assumption. (4) These binding and release events trigger alternating conformational changes in symmetry-related helical segments demonstrating the symmetric process of secondary transport. (5) The structural basis of the described conformational changes results from strategically conserved glycine and proline residues in this fold, which allows for the flexing of helical segments to open and close the transporter binding sites.

Taken together, the sequence of crystallographic snapshots of different conformations in $BetP_{Cg}$ gave a surprisingly accurate picture of what happens during transport. Individual steps were validated by numerous mutagenesis studies as well as binding and transport assays. Stability of binding sites was also validated by MD simulations, which, at least in the case of BetP, proved to be an indispensable tool to judge the meaningfulness of the crystal structure. At present, new crystal structures are rarely published without appropriate MD simulations accounting for the increasing awareness of crystallization artifacts and model inaccuracies, e.g., for the description of substrate coordination sites.

4.2.6 *Transport Regulation in $BetP_{Cg}$: Limits of a Crystal Structure*

One of the main aims of structural investigation of $BetP_{Cg}$ was to elucidate the molecular aspects of the "molecular switch model" (Ott et al. 2008) (Fig. 4.2). This model assumes that in the inactive state the osmosensory C-terminal domain interacts with the negatively charged membrane surface resulting in transport inhibition. During activation, either by K^+ increase or by changes in the membrane state, the C-terminal domain is assumed to detach from the membrane and interact with cytoplasmic parts of BetP, e.g., loop 8 and the N-terminal domain, resulting in release of the alternating access cycle to function.

The first important information from structural studies was the trimeric state of BetP and its putative role in regulation (Krämer and Ziegler 2009; Perez et al. 2011a, b). As a consequence of the trimeric state C-terminal domains within a trimer apparently have the possibility to interact with the cytoplasmic side of an adjacent protomer at least in the detergent-solubilized form in 3D crystals (Fig. 4.5c). In the crystal structure, the C-terminal domain forms a straight, extended α-helix that protrudes into the cytoplasm pointing away from its own protomer and enclosing a 30° angle with the membrane plane (Fig. 4.5c). The C-terminal domain of one protomer interacts in 3D crystals with loop 2 and the C-terminal domain of the adjacent protomer. Moreover, from a later structure of the $BetP_{Cg}$ trimer in complex with eight specifically bound PG lipids (Koshy et al. 2013) we extrapolated that it would be very difficult for a C-terminal domain adopting the conformation described by the molecular switch model to interact with bulk lipids. However, interaction with the membrane surface is possible in a putative different conformation of the C-terminal domain. In such a conformation the C-terminal domain might not interact with the cytoplasmic loops 2 and 8, as predicted by the molecular switch model. Furthermore, interactions of the N-terminal domain cannot be studied in the present crystal structures as the mutant protein used for crystallization is lacking the first 29 amino acid residues and the residual N-terminal domain is forced into a helical conformation by a triple mutation (Ressl et al. 2009). Interactions of the C-terminal domain with the N-terminal domain, which were predicted by peptide array studies (Ott et al. 2008), might be very well possible in the conformation observed in crystal structures.

In the context of regulation, it turned out that the use of this mutant, as beneficial it was to describe the alternating access mechanism, makes it more difficult to detect important regulatory interactions. An important next step will be to compare structures in the presence and absence of K^+ to describe differences between active and inactive states. As the activation threshold of 220 mM is very high due to the low affinity of $BetP_{Cg}$ for potassium, crystallization has to be carried out at high K^+ concentrations. This raises some obstacles for crystallization as well as data collection, as preliminary data showed increased radiation sensitivity under these conditions. A next, also very important, step will be to include lipids into the crystallization procedure by using lipid cubic phase (LCP) (Liu and Cherezov 2011) and or bicelles (Wang et al. 2012). LCP is a methodical approach where the phase transition of special synthetic lipids is exploited to preform a crystallization matrix in which the membrane protein can be dissolved. It is described in more detail in Chap. 3.

In summary, the example of $BetP_{Cg}$ demonstrates beautifully how crystallographic snapshots can be used to observe and explain a complete alternating access cycle of a secondary transporter. But it also signifies the limits in interpretation due to the presence of crystal contacts, as we observe for one of the C-terminal domains in the trimeric protein. To overcome these limits, novel and creative crystallization strategies have to be developed and applied. In perspective, new purification and crystallization methods that allow maintaining the important interactions with lipids might be also the key to a structure determination of mammal transporters.

Acknowledgment The authors would like to thank Lucy Forrest, Markus Becker, and Camilo Perez for inspiring discussions, which have contributed to topics discussed in this chapter. C.Z. want to thank Caroline Koshy and Belinda Faust for important inputs on transporter structure determination. The structural work on BetP (C.Z.) was supported by the DFG Collaborative

Research Center 807: Transport and Communication across Biological Membranes, and the functional work (R.K.) by the DFG grant KR 693/10-2.

References

Appel M, Hizlan D, Vinothkumar KR, Ziegler C, Kühlbrandt W (2009) Conformations of NhaA, the Na+/H + exchanger from Escherichia coli, in the pH-activated and ion-translocating states. J Mol Biol 388:659–672

Biemans-Oldehinkel E, Mahmood NA, Poolman B (2006) A sensor for intracellular ionic strength. Proc Natl Acad Sci U S A 103:10624–10629

Booth PJ (2012) A successful change of circumstance: a transition state for membrane protein folding. Curr Opin Struct Biol 22:469–475

Botzenhardt J, Morbach S, Krämer R (2004) Activity regulation of the betaine transporter BetP of Corynebacterium glutamicum in response to osmotic compensation. Biochim Biophys Acta 1667:229–240

Bremer E, Krämer R (2000) Coping with osmotic challenges: osmoregulation through accumulation and release of compatible solutes in bacteria. In: Storz G, Hengge-Aronis R (eds) Bacterial stress responses. ASM Press, Washington, DC, pp 79–97

Bröer S, Gether U (2012) The solute carrier 6 family of transporters. Br J Pharmacol 167:256–278

Brohawn SG, Campbell EB, MacKinnon R (2013) Domain-swapped chain connectivity and gated membrane access in a Fab-mediated crystal of the human TRAAK K + channel. Proc Natl Acad Sci U S A 110:2129–2134

Cheng MH, Bahar I (2013) Coupled Global and Local Changes Direct Substrate Translocation by Neurotransmitter-Sodium Symporter Ortholog LeuT. Biophys J 105:630–639

Claxton DP, Quick M, Shi L, de Carvalho FD, Weinstein H, Javitch JA, McHaourab HS (2010) Ion/substrate-dependent conformational dynamics of a bacterial homolog of neurotransmitter: sodium symporters. Nat Struct Mol Biol 17:822–829

Culham DE, Lasby B, Marangoni AG, Milner JL, Steer BA, van Nues RW, Wood JM (1993) Isolation and sequencing of Escherichia coli gene proP reveals unusual structural features of the osmoregulatory proline/betaine transporter, ProP. J Mol Biol 229:268–276

Day PW, Rasmussen SG, Parnot C, Fung JJ, Masood A, Kobilka TS, Yao XJ, Choi HJ, Weis WI, Rohrer DK, Kobilka BK (2007) A monoclonal antibody for G protein-coupled receptor crystallography. Nat Methods 11:927–929

Derewenda ZS, Vekilov PG (2006) Entropy and surface engineering in protein crystallization. Acta Crystallogr D Biol Crystallogr 62:116–124

Faham S, Watanabe A, Besserer GM, Cascio D, Specht A, Hirayama BA, Wright EM, Abramson J (2008) The crystal structure of a sodium galactose transporter reveals mechanistic insights into Na+/sugar symport. Science 321:810–814

Farwick M, Siewe RM, Krämer R (1995) Glycine betaine uptake after hyperosmotic shift in Corynebacterium glutamicum. J Bacteriol 177:4690–4695

Forrest LR, Krämer R, Ziegler C (2011) The structural basis of secondary active transport mechanisms. Biochim Biophys Acta 1807:167–88

Ge L, Perez C, Waclawska I, Ziegler C, Muller DJ (2011) Locating an extracellular K^+-dependent interaction site that modulates betaine-binding of the Na^+-coupled betaine symporter BetP. Proc Natl Acad Sci U S A 108:E8908

Hauptman H (1997) Phasing methods for protein crystallography. Curr Opin Struct Biol 7:672–680

Huber T, Steiner D, Röthlisberger D, Plückthun A (2007) In vitro selection and characterization of DARPins and Fab fragments for the co-crystallization of membrane proteins: The Na(+)-citrate symporter CitS as an example. J Struct Biol 159:206–221

Hunter MS, Fromme P (2011) Toward structure determination using membrane-protein nanocrystals and microcrystals. Methods 55:387–404

Khafizov K, Perez C, Koshy C, Quick M, Fendler K, Ziegler C, Forrest LR (2012) Investigation of the sodium-binding sites in the sodium-coupled betaine transporter BetP. Proc Natl Acad Sci USA 109(44):E3035–E3044. doi:10.1073/pnas.1209039109

Khelashvili G, Levine MV, Shi L, Quick M, Javitch JA, Weinstein H (2013) The membrane protein LeuT in micellar systems: aggregation dynamics and detergent binding to the S2 site. J Am Chem Soc 135(38):14266–14275

Kniazeff J, Shi L, Loland CJ, Javitch JA, Weinstein H, Gether U (2008) An intracellular interaction network regulates conformational transitions in the dopamine transporter. J Biol Chem 283:17691–17701

Korkmaz F, Ressl S, Ziegler C, Mäntele W (2013) K(+)-induced conformational changes in the trimeric betaine transporter BetP monitored by ATR-FTIR spectroscopy. Biochim Biophys Acta 1828:1181–1191

Koshy C, Schweikhard E, Gärtner R, Perez C, Yildiz O, Ziegler C (2013) Structural evidence for functional lipid interactions in the betaine transporter BetP. EMBO J 32:3096–3105

Krämer R, Ziegler C (2009) Regulative interactions of the osmosensing C-terminal domain in the trimeric glycine betaine transporter BetP from Corynebacterium glutamicum. Biol Chem 390:685–691

Krishnamurthy H, Gouaux E (2012) X-ray structures of LeuT in substrate-free outward-open and apo inward-open states. Nature 481:469–474

Kühlbrandt W (2013) Introduction to electron crystallography. Methods Mol Biol 955:1–16

Lehner I, Basting D, Meyer B, Haase W, Manolikas T, Kaiser C, Karas M, Glaubitz C (2008) The key residue for substrate transport (Glu14) in the EmrE dimer is asymmetric. J Biol Chem 283:3281–3288

Liu W, Cherezov V (2011) Crystallization of membrane proteins in lipidic mesophases. J Vis Exp. doi:pii:2501.10.3791/2501

Mittal A, Böhm S, Grütter MG, Bordignon E, Seeger MA (2012) Asymmetry in the homodimeric ABC transporter MsbA recognized by a DARPin. J Biol Chem 287:20395–20406

Möker N, Brocker M, Schaffer S, Krämer R, Morbach S, Bott M (2004) Deletion of the genes encoding the MtrA-MtrB two-component system of Corynebacterium glutamicum has a strong influence on cell morphology, antibiotics susceptibility and expression of genes involved in osmoprotection. Mol Microbiol 54:420–438

Ott V, Koch J, Spate K, Morbach S, Kramer R (2008) Regulatory properties and interaction of the C- and N-terminal domains of BetP, an osmoregulated betaine transporter from Corynebacterium glutamicum. Biochemistry 47:12208–12218

Otzen DE, Andersen KK (2013) Folding of outer membrane proteins. Arch Biochem Biophys 531 (1–2):34–43

Ozcan N, Ejsing CS, Shevchenko A, Lipski A, Morbach S, Krämer R (2007) Osmolality, temperature, and membrane lipid composition modulate the activity of betaine transporter BetP in Corynebacterium glutamicum. J Bacteriol 189:7485–7496

Padan E, Kozachkov L, Herz K, Rimon A (2009) NhaA crystal structure: functional-structural insights. J Exp Biol 212(Pt 11):1593–1603

Park SH, Das BB, Casagrande F, Tian Y, Nothnagel HJ, Chu M, Kiefer H, Maier K, De Angelis AA, Marassi FM, Opella SJ (2012) Structure of the chemokine receptor CXCR1 in phospholipid bilayers. Nature 491:779–83

Penmatsa A, Wang KH, Gouaux E (2013) X-ray structure of dopamine transporter elucidates antidepressant mechanism. Nature. doi:10.1038/nature12533

Perez C, Ziegler C (2013) Mechanistic aspects of sodium-binding sites in LeuT-like fold symporters. Biol Chem 394:641–648

Perez C, Khafizov K, Forrest LR, Krämer R, Ziegler C (2011a) The role of trimerization in the osmoregulated betaine transporter BetP. EMBO Rep 12:804–810

Perez C, Koshy C, Ressl S, Nicklisch S, Krämer R, Ziegler C (2011b) Substrate specificity and ion coupling in the Na^+/betaine symporter BetP. EMBO J 30:1221–1229

Perez C, Koshy C, Yildiz O, Ziegler C (2012) Alternating-access mechanism in conformationally asymmetric trimers of the betaine transporter BetP. Nature 490:126–130

Peter H, Burkovski A, Kramer R (1998a) Osmo-sensing by N- and C-terminal extensions of the glycine betaine uptake system BetP of Corynebacterium glutamicum. J Biol Chem 273:2567–2574

Peter H, Weil B, Burkovski A, Krämer R, Morbach S (1998b) Corynebacterium glutamicum is equipped with four secondary carriers for compatible solutes: identification, sequencing, and characterization of the proline/ectoine uptake system, ProP, and the ectoine/proline/glycine betaine carrier, EctP. J Bacteriol 180:6005–6012

Pope CR, Unger VM (2012) Electron crystallography–the waking beauty of structural biology. Curr Opin Struct Biol 22(4):514–519

Ressl S, Terwisscha van Scheltinga AC, Vonrhein C, Ott V, Ziegler C (2009) Molecular basis of transport and regulation in the Na(+)/betaine symporter BetP. Nature 458:47–52

Rübenhagen R, Rönsch H, Jung H, Krämer R, Morbach S (2000) Osmosensor and osmoregulator properties of the betaine carrier BetP from Corynebacterium glutamicum in proteoliposomes. J Biol Chem 275:735–741

Rübenhagen R, Morbach S, Krämer R (2001) The osmoreactive betaine carrier BetP from Corynebacterium glutamicum is a sensor for cytoplasmic K+. EMBO J 20:5412–5420

Schiller D, Krämer R, Morbach S (2004a) Cation specificity of osmosensing by the betaine carrier BetP of Corynebacterium glutamicum. FEBS Lett 563:108–112

Schiller D, Rübenhagen R, Krämer R, Morbach S (2004b) The C-terminal domain of the betaine carrier BetP of Corynebacterium glutamicum is directly involved in sensing K + as an osmotic stimulus. Biochemistry 43:5583–5591

Schiller D, Ott V, Kramer R, Morbach S (2006) Influence of membrane composition on osmosensing by the betaine carrier BetP from Corynebacterium glutamicum. J Biol Chem 281:7737–7746

Schulze S, Köster S, Geldmacher U, Terwisscha van Scheltinga AC, Kühlbrandt W (2010) Structural basis of Na(+)-independent and cooperative substrate/product antiport in CaiT. Nature 467:233–236

Seeger MA, Schiefner A, Eicher T, Verrey F, Diederichs K, Pos KM (2006) Structural asymmetry of AcrB trimer suggests a peristaltic pump mechanism. Science 313:1295–1298

Seeger MA, Zbinden R, Flütsch A, Gutte PG, Engeler S, Roschitzki-Voser H, Grütter MG (2013) Design, construction, and characterization of a second-generation DARPin library with reduced hydrophobicity. Protein Sci 22:1239–1257

Steiner-Mordoch S, Soskine M, Solomon D, Rotem D, Gold A, Yechieli M, Adam Y, Schuldiner S (2008) Parallel topology of genetically fused EmrE homodimers. EMBO J 27:17–26

Tsai CJ, Ziegler C (2010) Coupling electron cryomicroscopy and X-ray crystallography to understand secondary active transport. Curr Opin Struct Biol 20:448–55

Tsai CJ, Ejsing CS, Shevchenko A, Ziegler C (2007) The role of lipids in two-dimensional crystallization of BetP, a glycine-betaine transporter from Corynebacterium glutamicum. J Struct Biol 160:275–286

Tsai CJ, Khafizov K, Hakulinen J, Forrest LR, Krämer R, Kühlbrandt W, Ziegler C (2011) Structural asymmetry in a trimeric Na^+/betaine symporter, BetP, from Corynebacterium glutamicum. J Mol Biol 407:368–81

Tsatskis Y, Khambati J, Dobson M, Bogdanov M, Dowhan W, Wood JM (2005) The osmotic activation of transporter ProP is tuned by both its C-terminal coiled-coil and osmotically induced changes in phospholipid composition. J Biol Chem 280:41387–41394

Ubarretxena-Belandia I, Stokes DL (2012) Membrane protein structure determination by electron crystallography. Curr Opin Struct Biol 22(4):520–528 van der Heide T, Poolman B (2000) Osmoregulated ABC-transport system of Lactococcus lactis senses water stress via changes in the physical state of the membrane. Proc Natl Acad Sci U S A 97:7102–7106

Verdon G, Boudker O (2012) Crystal structure of an asymmetric trimer of a bacterial glutamate transporter homolog. Nat Struct Mol Biol 19:355–357

Wang H, Elferich J, Gouaux E (2012) Structures of LeuT in bicelles define conformation and substrate binding in a membrane-like context. Nat Struct Mol Biol 19:212–219

Weyand S, Shimamura T, Yajima S, Suzuki S, Mirza O, Krusong K, Carpenter EP, Rutherford NG, Hadden JM, O'Reilly J, Ma P, Saidijam M, Patching SG, Hope RJ, Norbertczak HT, Roach PC, Iwata S, Henderson PJ, Cameron AD (2008) Structure and molecular mechanism of a nucleobase-cation-symport-1 family transporter. Science 322:709–713

Wood JM (1999) Osmosensing by bacteria: signals and membrane-based sensors. Microbiol Mol Biol Rev 63:230–262

Wood JM (2011) Bacterial osmoregulation: a paradigm for the study of cellular homeostasis. Annu Rev Microbiol 65:215–238

Yamashita A, Singh SK, Kawate T, Jin Y, Gouaux E (2005) Crystal structure of a bacterial homologue of Na+/Cl–dependent neurotransmitter transporters. Nature 437:215–223

Zdravkovic I, Zhao C, Lev B, Cuervo JE, Noskov SY (2012) Atomistic models of ion and solute transport by the sodium-dependent secondary active transporters. Biochim Biophys Acta 1818:337–347

Zhao Y, Terry D, Shi L, Weinstein H, Blanchard SC, Javitch JA (2010) Single-molecule dynamics of gating in a neurotransmitter transporter homologue. Nature 465:188–193

Zhao Y, Terry DS, Shi L, Quick M, Weinstein H, Blanchard SC, Javitch JA (2011) Substrate-modulated gating dynamics in a Na + -coupled neurotransmitter transporter homologue. Nature 474:109–113

Zhao C, Stolzenberg S, Gracia L, Weinstein H, Noskov S, Shi L (2012) Ion-controlled conformational dynamics in the outward-open transition from an occluded state of LeuT. Biophys J 103:878–888

Ziegler C, Bremer E, Kramer R (2010) The BCCT family of carriers: from physiology to crystal structure. Mol Microbiol 78:13–34

Zomot E, Bahar I (2012) A conformational switch in a partially unwound helix selectively determines the pathway for substrate release from the carnitine/γ-butyrobetaine antiporter CaiT. J Biol Chem 287:31823–31832

Zou Y, Weis WI, Kobilka BK (2012) N-terminal T4 lysozyme fusion facilitates crystallization of a G protein coupled receptor. PLoS One 7(10):e46039

Chapter 5
Development of Refined Homology Models: Adding the Missing Information to the Medically Relevant Neurotransmitter Transporters

Thomas Stockner, Andreas Jurik, René Weissensteiner, Michael Freissmuth, Gerhard F. Ecker, and Harald H. Sitte

Abstract Neurotransmitter:sodium symporters are located on presynaptic neurons and terminate neurotransmission by removing the monoamine substrates from the synaptic cleft. Until very recently, only several conformational snapshots/structures of a bacterial homolog of neurotransmitter:sodium symporters, namely, the leucine/alanine transporter LeuT from Aquifex aeolicus, were available. However, this transporter shares only 21 % overall sequence identity with its human homologs. In this chapter, we describe how a model can be developed from a template with such low identity. The effort of model building will strongly depend on the purpose. We discuss this process and focus on the important steps that allowed us to obtain a model which can be used for molecular dynamics simulations. Furthermore, we also highlight the inherent limitations of the proposed approaches. Prediction of ligand binding brings in additional complexity. Therefore, experimental scrutiny of the resulting models is a key component to successful validation. We describe two specific examples: model building of the dopamine transporter and ligand docking to the serotonin transporter. We evaluate our modeling approach by direct comparison of our models to the recently published first eukaryotic neurotransmitter: sodium symporter, the drosophila melanogaster DAT transporter.

Keywords Dopamine transporter • Homology modelling • Homology model optimization • Molecular dynamics simulations • Neurotransmitter:sodium symporter models • Protein ligand docking • Serotonin transporter • Tricyclic antidepressants

---

T. Stockner (✉) • M. Freissmuth • H.H. Sitte
Center of Physiology and Pharmacology, Institute of Pharmacology, Medical University Vienna, Waehringerstr. 13A, 1090, Vienna, Austria
e-mail: thomas.stockner@meduniwien.ac.at; michael.freissmuth@meduniwien.ac.at; harald.sitte@meduniwien.ac.at

A. Jurik • R. Weissensteiner • G.F. Ecker
Department of Pharmaceutical Chemistry, University of Vienna, Althanstr. 14, 1090, Vienna, Austria
e-mail: andreas.jurik@univie.ac.at; rene.weissensteiner@univie.ac.at; gerhard.f.ecker@univie.ac.at

R. Krämer and C. Ziegler (eds.), *Membrane Transport Mechanism*, Springer Series in Biophysics 17, DOI 10.1007/978-3-642-53839-1_5,

5.1 Protein Structure Prediction

The three-dimensional structure is determined by its primary sequence. This fundamental insight by Christian Anfinsen was awarded the Nobel Prize in 1972. X-ray crystallography and NMR spectroscopy allow for determining protein structures and to gain mechanistic insights into their function. The Protein Data Bank (PDB) (http://www.pdb.org), which collects all solved structures, contains now 95,000 entries (October 2013). The human genome contains approximately 25,000 genes and encodes for 6,000–7,000 membrane proteins. Structural knowledge of membrane proteins is still limited, with the total number of solved membrane protein structures approaching 1,000. Protein folds are much more conserved as compared to their sequences. The number of structural folds or domains is limited and has been estimated to be less than 5,000 for compact globular protein (Brenner et al. 1997), while the number of protein families has been predicted to be in the range of 10,000–30,000 (Liu et al. 2004). Paralog proteins of the same species have evolved by gene duplication and orthologous proteins developed after speciation events. These homologous proteins, while divergent in sequence, most likely share the same fold (see Chaps. 3 and 4). If the structure of one of the homologous proteins has been determined, homology or comparative modeling can be used to infer the structure of the protein of interest. In those instances where no template is available, de novo structure prediction paradigms offer a solution: recent developments do potentially allow for building models at medium or better resolution. This ought to suffice if the modeling exercise was aimed at supporting experimental projects that make use of site-directed mutagenesis.

5.1.1 The Membrane Environment

The similarity of the template–target pairs declines with increasing sequence divergence. The divergence grows slower for membrane proteins as compared to water-soluble proteins, because the two-dimensional layer of the membrane environment provides additional restraints. This environmental effect allows for building models for template–target pairs with lower sequence identity than possible for water-soluble proteins (Punta et al. 2007). The membrane consists of a roughly 3-nm-thick hydrophobic center and a highly polar region comprising the head groups. Side chains with strong polarity are not well tolerated within the hydrophobic membrane core, while the energetic cost for charges is prohibitively high: moving an arginine side chain from the water environment into the hydrophobic interior of a membrane has been estimated to require 58.1 kJ/mol and an asparagine side chain required 23.9 kJ/mol. In contrast, this is an energetically favorable process for the side chains of hydrophobic residues; the partitioning of a leucine side chain from water into the membrane, for instance, releases -15.2 kJ/mol (MacCallum et al. 2007). Transmembrane helices are typically not interrupted, if they are in direct contact with the lipid environment. Indeed, the backbone peptide bond contains the polar and hydrogen bonding amide and carbonyl groups. These functional groups participate in secondary structure interactions, typically in α-helices or β-sheet structures. The surface of a membrane protein is typically smoother and more compact than that of a water-soluble counterpart. The driving force is the large entropic cost of confining cholesterol or lipid tails in the small volume of protein pockets or crevasses (Tieleman et al. 1997).

Transport of substrates requires that conformational changes can be accommodated that switch between at least two states. The same is true for receptors that transmit signals across the membrane. The membrane environment imposes restraints to possible movements that are absent in aqueous environment: typically, charges cannot be moved into the membrane and exposure of hydrophobic helices to the water phase is energetically expensive. The lateral pressure profile exerts strong forces onto membrane proteins and thereby limits the range of possible asymmetric conformational changes (Cantor 1997; Jerabek et al. 2010). Mechanosensitive channels use this effect to detect forces acting on the membranes (Perozo et al. 2002; Ollila et al. 2011).

5.1.2 Protein Folding

Accurate ab initio protein folding by in silico methods describing the protein folding process has been considered a holy grail of computational biology. The folding velocity of the protein is an important limiting factor. It has been recognized some 50 years ago that the number of possible conformations is infinitely large and thus folding of a protein takes infinitely long, if all possible conformations are visited. This thought experiment was introduced by Cyrus Levinthal ("Levinthal paradox") to highlight the fact that proteins must find their minimum energy conformation by progressing through a confined search space represented by a folding funnel. Because of the inherent limitations of molecular dynamics simulations (system size and timescale limitation), only fast folding proteins of small sizes are accessible to ab initio folding simulations, while water is often described only implicitly (Lane et al. 2013; Noé et al. 2009; Voelz et al. 2010; Shirts and Pande 2000; Lindorff-Larsen et al. 2011; Bowman et al. 2011; Freddolino and Schulten 2009; Duan and Kollman 1998). Unbiased folding of polytopic membrane proteins has not yet been addressed by using ab initio approaches. The viscosity of the membrane, the heterogeneous environment with the hydrophobic membrane core and the polar head group region, as well as the much longer folding timescales represent a major challenge. Moving a charge group from one side of the membrane to the other comes with a large energetic cost, but is necessary for folding of polytopic membrane proteins. Unbiased direct folding simulation of larger polytopic membrane protein might therefore not become feasible. Hence, in the foreseeable future, ab initio folding will remain one of—if not—the major challenges in computational approaches to understanding protein structures.

5.1.3 Protein Structure Prediction by Threading

De novo structure prediction methods (Levitt 1992) have rapidly evolved in the last years and have been shown to reach the accuracy of comparative modeling in favorable case (Kinch et al. 2011; MacCallum et al. 2011). State-of-the-art programs like ROSETTA (Rohl et al. 2004) FILM3 (Nugent and Jones 2012) or TASSER (Zhang et al. 2005; Liu et al. 2012) are based on identifying segments with similar sequence in known protein structures followed by their three-dimensional assembly. Membrane constraints as implemented in ROSETTA_membrane (Yarov-Yarovoy et al. 2006) help to reduce the enormous search space. This de novo structure prediction has become a useful alternative if no homologous protein is available for comparative modeling.

5.1.4 Comparative Modeling

Comparative modeling or homology modeling can be used to infer the structure of a protein based on the known structure of a homologous protein. The model building process consists of four steps:

5.1.4.1 Template Identification

Identification of homologous proteins with known structure is the first step for comparative modeling. This task can become a challenge for distantly related proteins with diverging sequences, while close homologs can be easily detected. A fast approach for identification of homologous structures is a BLAST search against the sequences of all protein entries in the PDB database. Metaserver approaches that use consensus scoring as implemented in Phyre (Bennett-Lovsey et al. 2008) or Pcons (Lundström et al. 2001) have shown good performance to identify distant homologs (Moult 2005). Templates from remotely related proteins must be selected and their suitability verified by using a set of criteria (e.g., an extensive phylogenetic tree, comparing protein family annotations, and biochemical evidence supportive of related functions). There is always the danger of generating a model with good local geometric properties (Ramachandran plot, atom–atom contacts, bonded parameters,. . .), which nevertheless relies on using an inappropriate template and thus carries the risk of generating misleading hypotheses. A thorough description of template identification can be found in Chap. 2 by Schlessinger.

5.1.4.2 Alignment

The alignment of template and target sequence is crucial for comparative modeling, because the subsequent steps are critically dependent on the alignment. Any mistake at this point will be propagated to the final model and can hardly be corrected for at any of the later stages. Sequence alignments with template–target identity above 50 % can typically be obtained with high precision. Alignments with a sequence identity below 30 % become increasingly prone to errors (Kryshtafovych et al. 2011a; Forrest et al. 2006; Martí-Renom et al. 2000) and results from alignment programs must be carefully evaluated. The situation has been improved by the continuous progress in sequence alignment programs and techniques, such as structure-based alignment, multiple sequence alignment, profile–profile alignments, and the inclusion of secondary structure information (Liu et al. 2012; Moretti et al. 2007; Fernandez-Fuentes et al. 2007; Söding 2005). Manual adjustment of alignments by including expert knowledge can prove essential in difficult cases with low sequences identity. Candidate transmembrane helices of the target, for instance, should remain intact (i.e., they must not contain gaps), if they are so in the template. This also implies that along a helix all corresponding residues are oriented in a similar way (i.e., membrane-exposed residues must be confined to the hydrophobic membrane section in both template and target). Conversely, a single residue within a transmembrane helix suffices to correctly align an entire helix. Helix S6 of ion channels is a good example, where Ca^{2+}, Na^{+}, and TRP channels can be aligned to each other and to the recently solved $Na_{V}AB$ structure (Payandeh et al. 2011) by virtue of a single conserved asparagine residue (Shafrir et al. 2008).

In a typical modeling project, there is a high probability that experimental data are available from site-directed mutagenesis. In the case of a transporter protein, these mutational data typically cluster in the translocation path. Similar data may not be available for helices at the protein–lipid interface. The alignment may nevertheless be optimized by selecting the most probable from a limited number of possibilities. A useful strategy consists of combining bioinformatics analyses with model building for a large set of proteins that are known to share the same fold. All paralogs present in the human genome or all orthologs in different species qualify in this respect. A negative selection strategy based on biophysical considerations can be successful in rejecting the incorrect alignments. The position of polar and charged residues with respect to the hydrophobic membrane environment (Chamberlain et al. 2004) is often a good discriminatory parameter.

5.1.4.3 Model Building

If a template is identified and an alignment generated, different model building methods can be used. None of the various approaches clearly outperform any other, although some implementations are superior in the CASP (critical assessment of structure prediction) competition (MacCallum et al. 2011; Mariani et al. 2011). It was observed that models built by the different programs did not come closer to the native structure as the template itself (Wallner and Elofsson 2005), though recent advances in the model building programs seem to have improved the situation (MacCallum et al. 2011). Molecular dynamics simulations can further improve the conformation of structures over the starting model (MacCallum et al. 2011). Simulations started from both the homology model and the target structure have been shown to converge to an overlapping region in phase space (Raval et al. 2012). This region was not always overlapping with the experimentally determined structure, pointing towards effects from crystal contacts and/or limitations of the employed force field on long timescales. The choice of a model building program will depend on the actual problem, preferences, and the availability of correct templates. MODELLER (Sali and Blundell 1993) is a widely used comparative modeling program of general applicability. Modeling programs are typically not tailored to membrane proteins, although the properties of the membrane environment call for special treatment. MEDELLER (Kelm et al. 2010) is one of the modeling program specifically tailored to tackle membrane proteins. Its model building strategy comprises a two-step process: in the first step, the protein membrane core is built, starting from the membrane center. The loops are constructed only after the transmembrane core has been built by using the template-based loop building method FREAD (Choi and Deane 2010).

5.1.4.4 Model Building of Side Chains

Side chain conformations of the protein core can usually be predicted with higher accuracy than at the periphery of the protein. Packing quality of the core increases with sequence identity and, thus, the similarity of the backbone between template and target. Accuracy can be further improved by the use of tailored rotamer library-based approaches (Xiang et al. 2007), which take the backbone conformation into account, and by considering solvent accessibility and atom–atom contacts. Programs to predict side chain conformations for a given backbone include SCWRL (Krivov et al. 2009; Canutescu et al. 2003), SMOL (Liang and Grishin 2002), and SCAP (Faure et al. 2008). An advantage of these rotamer library-based methods is their high computational speed: they are considerably faster than methods, which use a physics-based force field in combination with a conformational search algorithm. Predictions of side chain orientations can be very successful if residue identity is above 50 %. The average prediction accuracy can reach ~75 % (Xiang 2006; Huang et al. 1998). Side chains at the protein surface have typically larger conformational freedom and might not be restrained at all. Conformational distributions can be best assessed by molecular dynamics simulations: here, the accessible phase space is probed directly while explicitly taking the contribution of the environment into account. The conformational space of surface residues is readily accessible. Exhaustive sampling of different conformations or assembly of core side chain may require prohibitively large computational resources. Fortunately, sampling of the full phase space is not always essential for biological questions. A lower resolution is sufficient if the aim of a modeling project is to provide a basis for developing hypothesis for site-directed mutagenesis studies. Conversely, detailed insights will be necessary and therefore a full sampling of the conformational state (s) is mandatory, if the project focuses on properties like the conformation of side chains in the substrate-binding site and the dependence on ligand chemistry, the knowledge of the free energy of states, or the energy barrier separating these.

5.1.4.5 Model Building of Loops

The biophysical properties of the membrane constrain the variability in the transmembrane protein core. This is not the case for the loops, which connect the transmembrane helices, because they typically reside outside the membrane. These loops can be quite diverse in both length and conformation. The extracellular loop 2 of the family of G-protein-coupled receptors (GPCRs) is a representative example: The structure of several GPCRs has now been solved. The seven-transmembrane helix core is remarkably similar between homologous GPCRs, while the extracellular loop 2 is variable in length and structure. In fact, it can fold into an α-helical, β-sheet, or loop structure, depending on the primary sequence (Cherezov et al. 2007; Warne et al. 2011; Jaakola et al. 2008; Makino et al. 2010; Wu et al. 2010; Chien et al. 2010; Haga et al. 2012).

Missing loops are normally built using either template-based or de novo loop construction methods. Short loops of less than 8–9 residues can be modeled de novo with reasonable accuracy (Jaroszewski 2009; Deane and Blundell 2001; Jacobson et al. 2004). Larger loops, however, become increasingly difficult to model. Recent development in the inclusion of implicit solvent model into PLOP showed that much larger loops of up to 20 residues can be predicted accurately (Friesner et al. 2013). In contrast, loop modeling using structural database searches for and ranks candidate loops (Choi and Deane 2010). This approach depends on sequence similarity and availability of template loops (Deane and Blundell 2001; Peng and Yang 2007; Tosatto et al. 2002). Even very long loops can be modeled with high accuracy, if an appropriate template is available and can be identified.

5.1.4.6 Assessment and Refining

The quality of models has been evaluated in template–target pairs for water-soluble (Martí-Renom et al. 2000) and membrane proteins (Forrest et al. 2006; Wallner and Elofsson 2005). It has been observed that model quality decreases with increasing sequence diversity, diminishing faster for water-soluble proteins as compared to membrane proteins. The membrane environment provides an additional restraint on protein structure that limits the available conformational space. If a structure of a protein is solved by NMR or by X-ray crystallography, the quality of the resulting model is assessed by a set of parameters such as the Ramachandran plot and the lengths of the bonds, their angles and torsions, as well as atom–atom overlap detection. Well standard in structure determination, these parameters are not well suited to assess the quality of homology models, because most programs developed for homology modeling already optimize for these parameters by design. Scores like the DOPE score (Shen and Sali 2006), Verify3D (Eisenberg et al. 1997), and ProQ (Wallner and Elofsson 2003) or energy scores can be used to identify the better models (Kryshtafovych and Fidelis 2009; Kryshtafovych et al. 2011b), but they have limited predictive power for determining, if the best model has the native conformation. Protein model quality estimates on absolute scale would be necessary for identification of the native conformation. In theory, this should be possible. A large enthalpic component must exist for stably folded proteins, because folding of a long peptide chain into a well-structured protein includes a steep drop in conformational entropy that is among other contributions counteracted by the enthalpic term. Prediction of native like protein structure is attempted by QMEAN (Benkert et al. 2008), ProQ (Wallner and Elofsson 2003), MolProbity (Davis et al. 2007), ModFOLD (McGuffin 2007), and ProSA (Sippl 1995). These measures are predictors for the physical correctness of the model and capable of distinguishing plausible from non-plausible models, but they have not been tailored to membrane proteins, for which the driving forces of folding differ from those of water-soluble proteins. Further improvements in this field will help to improve both the identification of the best structure within a set of model and a quantitative measure of how distant a model is from the native structure.

The quality of a model can currently be estimated best by independent methods: first of all, a model must be in line with all available data. A model is obviously only useful, if it has predictive power: helix orientation or loop structures can be probed by accessibility studies or constraints imposed by appropriate mutations, e.g., reduction or increases in transport rates, ion fluxes, or changes in ligand binding. Finally, molecular dynamics simulations can be used as reality check: if side chains are incorrectly oriented or packed, molecular dynamics simulations can probably detect these defects, because the structure is likely to become unstable: typically, local secondary structures unfold and/or large fluctuations can be observed. Thus, long simulations in the native environment have a good change to identify (1) incorrectly modeled loops, (2) mistakes in the alignment, (3) local differences between template and target, or (4) the selection of the wrong templates.

5.2 Application: Development of the Homology Model of the Dopamine Transporter

5.2.1 Sequence Alignments

A correct sequence alignment is an essential requirement for comparative modeling. This step becomes more and more challenging with increasing diversity between template and target. The identity of the transmembrane domain between LeuT and the dopamine transporter DAT is 21 %. Sequence alignments of human neurotransmitter transporter with LeuT, including DAT, have already been published (Beuming et al. 2006; Yamashita et al. 2005). We developed a multiple sequences alignment starting from the alignment published by Beuming et al. 2006. The alignment of each helix was individually analyzed: residues in the substrate- and sodium-binding site show much higher sequence identity and were used to align the respective helices. The position of the lipid-exposed tryptophan residues and that of the charged lysine and arginine residues were checked for their compatibility with the membrane environment.

Helix capping motifs are important for stabilizing helices (Aurora and Rose 1998). The best known and most frequent motif is the Schellman motif at the C-terminus of helices (Aurora et al. 1994; Schellman 1980), where a glycine residue next to the helix cap allows for a sharp change in backbone propagation with a backbone φ and ψ combination only possible for glycine residues. Thereby one additional carbonyl functional group of the helix can be hydrogen bonded by a backbone amide function, stabilizing the helix terminus. We identified a Schellman motif in every second transmembrane helix of DAT.

5.2.2 Model Building, Optimization, and Scrutiny

Once a correct alignment has been obtained, comparative modeling can be used to build the model. The quality of the model and therefore the effort depends on the purpose as described above. We created a model for the DAT with the intention to use it for both ligand binding studies and MD simulations (Stockner et al. 2013). The orientation of side chains can only be assumed to be unaltered given the high template target diversity. With such a low identity even conserved residues will not necessarily always share the same side chain orientation because of changes in their environment. Side chain prediction programs like SCWRL (Krivov et al. 2009; Canutescu et al. 2003) or SMOL (Liang and Grishin 2002) have been shown to achieve above 70 % correct prediction. Although impressive, this means that on average every fourth side chain might be incorrectly predicted, necessitating further optimization. Short molecular dynamics simulations of 2 ns have shown that the RMSD from the starting crystal structure correlates with its resolution (Law et al. 2005). Analysis of the structural parameters of molecular dynamics simulations can therefore identify regions of contentious side chain packing. The limited space available in the protein interior is often prohibitive for flipping or inversion of relative side chains packing, but leading to local unfolding of the secondary structure. Alteration of side chain packing requires local unfolding or a large increase of helix–helix distances to allow free rearrangement. This process is normally too slow to represent a useful strategy. A solution to this limitation is the addition of restraints aimed at maintaining secondary structure during molecular dynamics simulation-based structure optimization, but weak enough to allow for distortions. Side chains with unfavorable geometry and packing can potentially disentangle and repack if optimization is performed by relatively long simulation. Residue W84 is a good example. This residue is conserved in the human SLC6 family members, while the LeuT template carries a leucine at this position. If the orientation of this side chain would be fully conserved, then W84 would be confined in a space which is probably too small for this large side chain and the hydrophilic ε-NH group on the side chain would be buried within the protein. We observed frequent reorientations of the W84 side chain during optimization or model challenging, sometimes accompanied by destabilization of the EL4 loop. Comparison of the optimized side chain orientation to the new drosophila DAT structure (dDAT) (Penmatsa et al. 2013) revealed a similar position of this side chain. Our intention was to build a model that could be used to carry out long MD simulations. It has been shown that mutations in regions of the DAT far from the substrate-binding site can affect cocaine binding. These observations can be rationalized by assuming allosteric effects that modify the conformational equilibrium and/or energy levels between different states, potentially stabilizing a geometry that is not a stable state of the wild-type transporter (Uhl 2003). Therefore, correct modeling of all side chains is mandatory, because incorrect orientation of a single non-surface-exposed side chain can potentially lead to a similar phenomenon (of a shift in the free energy hypersurface) as experimentally observed for point mutations. We therefore built and tested models iteratively addressing these challenges. The best ten models from homology modeling were first selected and inserted into a membrane that was pre-shaped to harbor the DAT (see Fig. 5.1a). It emerged that first gradually removing restraints from the side chains and then slowly releasing the backbone worked well. The secondary structure (not the atom position) was further maintained by distance restraints for 20 ns. This allows for relative changes between helices and for side chain rearrangement, while maintaining secondary structure integrity. Finally, the models were challenged for additional 20 ns. Remaining structural imperfections can thereby develop into structural instabilities. Analysis of these instabilities was then converted into restraints for model building in the next iteration of model building/optimization/model scrutiny. Given large enough numbers of tested models, very stably structures can be obtained and identified (see Fig. 5.1b).

A comparison of the full set of our models with both the LeuT template and the new dDAT structure showed structural instabilities in the membrane inserted part of the transporter. These instabilities appeared most frequently in regions, where LeuT differs from dDAT including the ends of H2 and H7. The beginning of H3 appeared shortened by one helical turn, while the EL5 loop could only be stabilized after introduction of an additional helical turn into H10. Transmembrane helix 12 showed a distinct behavior: its position appeared uncertain, while helices 1–11 of hundred of MD optimized structures can be readily superposition. In addition, a kink at the conserved proline within H12 was frequently observed. This pointed to the assumption that H12 should be part of the oligomer interface and only stabilized in oligomeric interface. The new dDAT structure shows the same kink in H12 and revealed a different position of this helix as compared to LeuT.

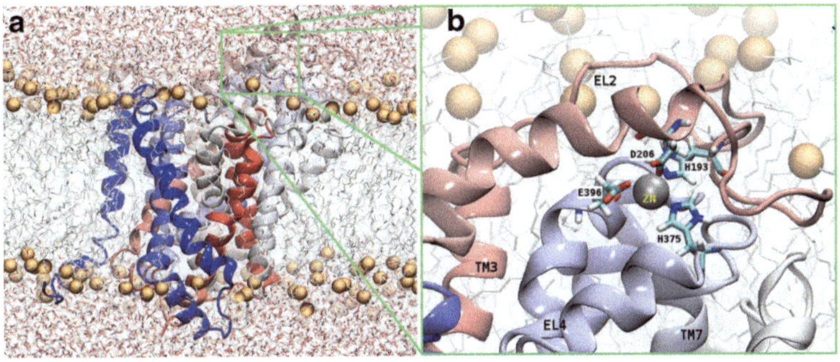

Fig. 5.1 Panel **a** shows the membrane inserted DAT system as used during model optimization. Panel **b** highlights the residues constituting the zinc binding site

The potential and the limitation of homology modeling paired with extensive MD optimization can be directly assessed in light of the new dDAT structure (Penmatsa et al. 2013). In our experience, long MD simulation and a large number of tested models have the potential to thoroughly optimize the homology model. Regions with structural deficits can often be recognized by instabilities: the task of optimizing the model remains a challenge. MD simulations appear to be key to select between different possibilities and obtain the optimal results.

5.2.3 The EL2 Loop of the DAT

Building of the extracellular loop of the DAT constituted an especially challenging task (Stockner et al. 2013). The EL2 loop is 21 residues longer than the corresponding loop in LeuT. Two residues were not resolved in the LeuT template (PDB ID: 2A65) (Yamashita et al. 2005). The residues preceding and following the insertion are not necessarily in the same orientation. Before this gap, the LeuT sequence has 3 proline residues, which do not have a counterpart in the eukaryotic homologs (see Fig. 5.2). The number of residues in the loop to be modeled is therefore 26. If we assume that each residue could be in one of three conformations (loop, α-helical, or β-sheet), the number of possibilities is 3^{26}. This is prohibitively large. What strategy can be used? A template-based strategy is not recommended, because a disulfide bond is essential for its stability and the loop is located at the protein membrane interface. In zero-order approximation the loop could be ignored. This is acceptable, if the model is to be used for generating hypotheses on the substrate permeation path. Similarly, this loop is of limited importance for ligand binding studies, because it is very distant from the binding site (above 3 nm) and beyond the typically cutoff used in docking studies. The conformational equilibrium is influenced by this loop, as it has been shown that one residue, H193, within this loop is part of an endogenous Zn^{2+}-binding site and binding of Zn^{2+} stabilizes the outward-facing state (Loland et al. 1999; Norregaard et al. 1998). The zero-order approximation is therefore not applicable.

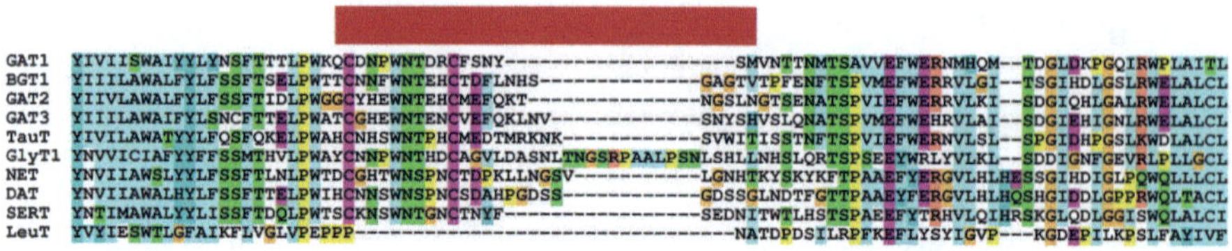

Fig. 5.2 A multiple sequence alignment showing the EL2 loop region. The *red* bar indicates the extended loop of the human NSS homolog structures, which does not have any counterpart in the bacterial LeuT homolog structure. The conservation in length and residues type of the first part of the loop and the variability of the second part is readily evident

Brute force model building without further restraints is not suitable for such a large loop. Analysis of the sequence indicated that the loop can be divided into two parts, split at residue H193 (see Fig. 5.2). The second part is highly variable in length between protein paralogs. It contains sequences which are associated with unstructured flexible and water-exposed loops. It can be assumed that it constitutes a flexible linker that only weakly interacts with the rest of the protein. The first part of the loop shows conservation of several properties, which could be used as restraints to reduce the search space to an extent that modeling becomes possible.

The final model (see Fig. 5.1) was subjected to long molecular dynamics simulations of 200 ns (Stockner et al. 2013) and resulted stable. We then developed an experiment to independently test the model. The model indicated that the zinc-binding site includes a fourth residue (D206), which had not yet been described in the literature. Binding of Zn^{2+} is known to reduce the transport rates by ~50 %. Mutation of H193, H375, or E396 has been shown to completely abolish the zinc sensitivity of the high-affinity binding site (Loland et al. 1999; Norregaard et al. 1998). Mutation of D206 to lysine or cysteine showed a significant change in the zinc-dependent inhibition of substrate uptake. In contrast, substitution by alanine did not have any effect on transport rates. These phenotypes are in line with the structural/functional role of residue 206 and support the developed model.

The crystal structure of the dDAT (Penmatsa et al. 2013) includes the conserved part of the EL2 loop, while the flexible second part has been removed to allow crystallization. The observed orientation is very different as compared to our model, burying polar and charged non-conserved region of the EL2 helix and exposing hydrophobic residues on the dDAT surface that are covered in our human DAT model. An analysis of the crystal structure revealed that the same part of the EL2 loop shows an extensive interaction interface with the co-crystallized antibody of the adjacent asymmetric unit. The conformation observed in the crystal might therefore be strongly influenced by crystal contacts and comparison of the loop must be carried out with caution.

5.3 Protein–Ligand Homology Modeling

Development of a valid structural model is not a straightforward task, if sequence identity is low, as outlined above. Investigation of potential protein–ligand interactions requires highest model quality in the substrate-binding site, while remote regions outside the cutoff play a minor role during docking. The prediction of the protein–ligand complex adds additional degrees of freedom to the modeling process, as mutual adaptation processes, or "induced-fit" mechanisms, have to be taken into account. Large-scale conformational changes require the use of a different template structure. Small-scale induced-fit effects by ligand binding can be accounted for by the inclusion of the small ligand molecule into the modeling process. The accuracy of ligand orientation and binding site geometries can be increased, but this strongly depends on ligand similarity and sequence identity (Evers et al. 2003). Dalton et al. suggested using a Tanimoto similarity measure of 0.5 as a threshold for ligands and 50 % for template–target sequence identity (Dalton and Jackson 2010). Only the natural substrate of the GABA transporters meets the requirements for the ligand (GABA versus leucine), when using LeuT as a template for modeling of the NSS family. None of the mammalian transporters shows a sufficient sequence identity. Thus, putative ligand orientations ought to be elaborated by combining placement techniques ("docking") with supporting information derived from experimental data.

5.3.1 Docking Methods

The accuracy of docking experiments correlates with the computational effort. Calculation times increase with the degree of allowed flexibility from rigid docking, over flexible treatment of selected side chains to computationally more intensive docking algorithms like Induced-Fit Docking protocols that allow for local backbone rearrangements. It has been shown that generation of correct bioactive conformations can often be achieved by docking programs (Warren et al. 2006). The Achilles' heel of docking remains the identification of that pose. Scoring functions are approximate and tailored for fast evaluation, thereby cutting back on the accuracy. This limitation of proper ranking can be alleviated by first reducing the number of solutions to a set of hypotheses using clustering approaches and then by using independent measures systematically eliminating those that are not plausible. This requires experimental testing, if published data are not available.

5.3.2 Modeling Strategies in Published Models

The number of NSS models has been growing continuously in the recent past, summarized by Kristensen et al. (2011). Most of the reported models so far rely on or partly further enhance the comprehensive sequence alignment of Beuming et al. (2006). The central binding site termed S1-binding site is rather conserved between all NSS members, sharing ~50 % sequence identity with the LeuT template. Alignments do not differ in this structural region. Modeling and docking approaches in the inner S1-binding site were constrained to comparably small molecules, as the first published LeuT structure was in an outward-occluded state that did not leave enough room for large ligands. The difference between the occluded and the outward-open conformation consists of a rotation of the bundle domain by almost 20°. Studies of binding of larger inhibitors like tiagabine to the GABA transporter 1 (GAT-1) (Skovstrup et al. 2010; Palló et al. 2007) or cocaine to the DAT (Beuming et al. 2008) became possible after the outward-facing LeuT template structure was solved (Singh et al. 2008). The recently published structure of the inward-facing conformation of LeuT can serve as a template for modeling approaches trying to identify binding pose for inhibitors like ibogaine (Bulling et al. 2012). Residues important for ligand binding in the main S1-binding site have been identified in several NSS and have mainly been tested by site-directed mutagenesis of substrate competition assays (Celik et al. 2008; Sarker et al. 2010; Andersen et al. 2009; Huang and Zhan 2007; Indarte et al. 2008; Gedeon et al. 2010; Koldsø et al. 2013; Xhaard et al. 2008; Paczkowski et al. 2007; Schlessinger et al. 2011). The putative secondary substrate-binding site (S2 site) is much less conserved and not straightforward to model. Nevertheless, docking studies indicate it as a possible interacting site (Sarker et al. 2010; Huang and Zhan 2007; Indarte et al. 2008; Gedeon et al. 2010). Model quality is typically assessed using routinely applied statistical homology modeling measures including Ramachandran plot, atom overlaps, or model quality score like the DOPE score (Shen and Sali 2006), SWISSModel (Schwede et al. 2003), or SAVES metaserver scores (http://services.mbi.ucla.edu/SAVES/). The quality of the binding site was further assessed in a few studies by either correlating docking results to measured activity values (Kardos et al. 2010; Wein and Wanner 2010), enriching known actives in virtual screening experiments (Schlessinger et al. 2012), or probing using molecular dynamics simulations (Skovstrup et al. 2010).

5.3.3 Experimental Data-Guided Docking and Scoring

As outlined above, most docking algorithms explore the conformational space sufficiently well to generate correctly docked poses. However, most often the scoring functions employed fail to rank the correct poses as the top candidates (Warren et al. 2006). Transporters are considered a difficult target, because substrate binding does often induce a large conformational change as part of the transport cycle. Furthermore, it is quite often not clear at exactly which state inhibitors are exerting their biological effect. This adds an additional layer of uncertainty to the homology modeling/docking workflow. Thus, besides thorough validation of the homology models and the docking poses by means of standard statistical tools, it is highly recommended to include as many experimental data as possible into the model validation and pose prioritization step. One interesting study in this context is the work of Barker et al. (1999), who investigated complementarity between mutant transporters and "mutant" ligands: the interaction of residue D98 in (rat) SERT with the amino group in dimethyltryptamine (DMT) was disrupted by shortening the DMT to the methylamine analog. Binding affinity was restored by lengthening side chain via mutation of D98–E98 (rSERT D98E) to reposition the carboxylate group. This was an elegant approach to support the hypothesis that the substrate and D98 interacted. Thus, these experiments lay the ground for setting constraints in subsequent docking studies. Ten years later, Celik et al. (2008) used this approach to prioritize between two binding modes of 5-HT, obtained when docking it into a homology model of hSERT. Both binding modes show a salt bridge between D98 and the charged primary amine of 5-HT. The difference between the two orientations of 5-HT is the positioning of the 5-OH group of 5-HT, being vicinal to either S438/T439 or A169/I172/A173. Thus the authors used an approach termed paired mutant–ligand analog complementation (PaMLAC) to prioritize the two binding hypotheses. Thus, the proposed ligand–protein interaction was systematically examined by disrupting it through site-directed mutagenesis and reestablishing another interaction via a ligand analog matching the mutated residue. The inclusion of mutations into the validation process bears the risk of introducing unexpected changes to the transporter conformation. If activity data of congeneric series of compounds are available, these could be used in a "ligand-only" based approach by identifying those docked poses from an exhaustively sampled pose space, which are able to explain the SAR (structure–activity relationship) of the ligand series (Klepsch et al. 2011). Docking of a structurally analogous set of compounds, such as tricyclic antidepressants including imipramine, desipramine, clomipramine, and amitryptilin, into a homology model of hSERT followed by rmsd-based common scaffold clustering led to the identification of three common scaffold clusters representing three different binding modes (see Fig. 5.3). Poses of the first cluster show ionic interaction between E493 and the charged nitrogen of the imipramine side chain as main interaction. In the second cluster, the di-benzazepine ring system is placed into the hydrophobic pocket of the central binding site, and the nitrogen atom shows an ionic interaction with D98. This pose is similar to the dDAT structure published in 2013 in complex with nortryptiline (Penmatsa et al. 2013). The orientations in cluster three show the ring system in the extracellular vestibule and the amino propylene side chain rising into the central binding site with an ionic interaction between D98 and the charged nitrogen.

All these docking poses are consistent with the observation of the competitive inhibition of SERT, but differ in the role of the interaction of D98 and Y95: (1) the

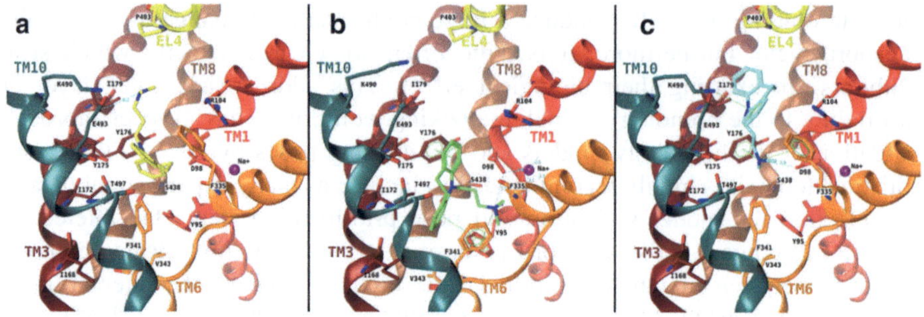

Fig. 5.3 Potential binding poses of imipramine in SERT are highlighted. Clusters 1–3 as discussed in the text are ordered from left to right (**a–c**). Figure reproduced from (Sarker et al. 2010)

positively charged nitrogen of the SERT substrates serotonin (5-HT) and DMT interact with D98 as described above. In poses 2 and 3 we observe a similar interaction of the charged nitrogen of the tricyclic antidepressants, but not in pose 1. (2) This charged nitrogen in the side chain of the tricyclic antidepressants is found within a distance (in poses 2 and 3) of the hydroxyl group of Y95, where hydrogen bonding could be possible, but not in pose 1. To prove that the charged nitrogen in the side chain of the tricyclic antidepressants adopts a similar binding mode as the SERT substrate 5-HT, Y95 was mutated to phenylalanine. The removal of the hydroxyl group of Y95 resulted in a pronounced drop in the affinity of imipramine, corresponding to a loss of 5.0 kJ/mol, which is in the expected range for the contribution of an H-bond to binding free energy (2.1–8.4 kJ/mol). These observations are consistent with the charged nitrogen interacting with the D98 and Y95, supporting the notion that charged nitrogen in the side chain of the ligands is bound to the transporter in a similar way as the endogenous substrate 5-HT.

5.4 Conclusions and Outlook

Models are most useful if they can serve to generate hypotheses and make predictions. Protein modeling crucially depends on the availability of appropriate template structures. For many membrane proteins only remote homologous structures are available, while for several protein families no template yet exists. In recent years we have seen a strong increase in new membrane protein crystal structures, including structures of new protein families. This is very promising and will provide us with good templates for most membrane proteins. Scrutiny of models becomes increasingly important with sequence divergence. Models should be independently validated by experimental evaluations before used as a structural basis for prediction.

Acknowledgments The financial support from the Austrian Science Fund (FWF) project "Transmembrane Transporters in Health and Disease" (SFB F35) and the DK + project "Ion Channels and Transporters as Molecular Drug Targets" is gratefully acknowledged.

References

Andersen J, Taboureau O, Hansen KB et al (2009) Location of the antidepressant binding site in the serotonin transporter: importance of Ser-438 in recognition of citalopram and tricyclic antidepressants. J Biol Chem 284:10276–10284. doi:10.1074/jbc.M806907200

Aurora R, Rose GD (1998) Helix capping. Protein Sci 7:21–38. doi:10.1002/pro.5560070103

Aurora R, Srinivasan R, Rose GD (1994) Rules for alpha-helix termination by glycine. Science 264:1126–1130

Barker EL, Moore KR, Rakhshan F, Blakely RD (1999) Transmembrane domain I contributes to the permeation pathway for serotonin and ions in the serotonin transporter. J Neurosci 19:4705–4717

Benkert P, Tosatto SCE, Schomburg D (2008) QMEAN: a comprehensive scoring function for model quality assessment. Proteins 71:261–277. doi:10.1002/prot.21715

Bennett-Lovsey RM, Herbert AD, Sternberg MJE, Kelley LA (2008) Exploring the extremes of sequence/structure space with ensemble fold recognition in the program Phyre. Proteins 70:611–625. doi:10.1002/prot.21688

Beuming T, Shi L, Javitch JA, Weinstein H (2006) A comprehensive structure-based alignment of prokaryotic and eukaryotic neurotransmitter/Na+ symporters (NSS) aids in the use of the LeuT structure to probe NSS structure and function. Mol Pharmacol 70:1630–1642. doi:10.1124/mol.106.026120

Beuming T, Kniazeff J, Bergmann ML et al (2008) The binding sites for cocaine and dopamine in the dopamine transporter overlap. Nat Neurosci 11:780–789. doi:10.1038/nn.2146

Bowman GR, Voelz VA, Pande VS (2011) Atomistic folding simulations of the five-helix bundle protein λ(6−85). J Am Chem Soc 133:664–667. doi:10.1021/ja106936n

Brenner SE, Chothia C, Hubbard TJ (1997) Population statistics of protein structures: lessons from structural classifications. Curr Opin Struct Biol 7:369–376

Bulling S, Schicker K, Zhang Y-W et al (2012) The mechanistic basis for noncompetitive ibogaine inhibition of serotonin and dopamine transporters. J Biol Chem 287:18524–18534. doi:10.1074/jbc.M112.343681

Cantor RS (1997) The lateral pressure profile in membranes: a physical mechanism of general anesthesia. Biochemistry 36:2339–2344. doi:10.1021/bi9627323

Canutescu AA, Shelenkov AA, Dunbrack RL (2003) A graph-theory algorithm for rapid protein side-chain prediction. Protein Sci 12:2001–2014. doi:10.1110/ps.03154503

Celik L, Sinning S, Severinsen K et al (2008) Binding of serotonin to the human serotonin transporter. Molecular modeling and experimental validation. J Am Chem Soc 130:3853–3865. doi:10.1021/ja076403h

Chamberlain AK, Lee Y, Kim S, Bowie JU (2004) Snorkeling preferences foster an amino acid composition bias in transmembrane helices. J Mol Biol 339:471–479. doi:10.1016/j.jmb.2004.03.072

Cherezov V, Rosenbaum DM, Hanson MA et al (2007) High-resolution crystal structure of an engineered human beta2-adrenergic G protein-coupled receptor. Science 318:1258–1265. doi:10.1126/science.1150577

Chien EYT, Liu W, Zhao Q et al (2010) Structure of the human dopamine D3 receptor in complex with a D2/D3 selective antagonist. Science 330:1091–1095. doi:10.1126/science.1197410

Choi Y, Deane CM (2010) FREAD revisited: accurate loop structure prediction using a database search algorithm. Proteins 78:1431–1440. doi:10.1002/prot.22658

Dalton JAR, Jackson RM (2010) Homology-modelling protein–ligand interactions: allowing for ligand-induced conformational change. J Mol Biol 399:645–661. doi:10.1016/j.jmb.2010.04.047

Davis IW, Leaver-Fay A, Chen VB et al (2007) MolProbity: all-atom contacts and structure validation for proteins and nucleic acids. Nucleic Acids Res 35:W375–W383. doi:10.1093/nar/gkm216

Deane CM, Blundell TL (2001) CODA: a combined algorithm for predicting the structurally variable regions of protein models. Protein Sci 10:599–612. doi:10.1110/ps.37601

Duan Y, Kollman PA (1998) Pathways to a protein folding intermediate observed in a 1-microsecond simulation in aqueous solution. Science 282:740–744

Eisenberg D, Lüthy R, Bowie JU (1997) VERIFY3D: assessment of protein models with three-dimensional profiles. Methods Enzymol 277:396–404

Evers A, Gohlke H, Klebe G (2003) Ligand-supported homology modelling of protein binding-sites using knowledge-based potentials. J Mol Biol 334:327–345. doi:10.1016/j.jmb.2003.09.032

Faure G, Bornot A, de Brevern AG (2008) Protein contacts, inter-residue interactions and side-chain modelling. Biochimie 90:626–639. doi:10.1016/j.biochi.2007.11.007

Fernandez-Fuentes N, Madrid-Aliste CJ, Rai BK et al (2007) M4T: a comparative protein structure modeling server. Nucleic Acids Res 35:W363–W368. doi:10.1093/nar/gkm341

Forrest LR, Tang CL, Honig B (2006) On the accuracy of homology modeling and sequence alignment methods applied to membrane proteins. Biophys J 91:508–517. doi:10.1529/biophysj.106.082313

Freddolino PL, Schulten K (2009) Common structural transitions in explicit-solvent simulations of villin headpiece folding. Biophys J 97:2338–2347. doi:10.1016/j.bpj.2009.08.012

Friesner RA, Abel R, Goldfeld DA et al (2013) Computational methods for high resolution prediction and refinement of protein structures. Curr Opin Struct Biol 23:177–184. doi:10.1016/j.sbi.2013.01.010

Gedeon PC, Indarte M, Surratt CK, Madura JD (2010) Molecular dynamics of leucine and dopamine transporter proteins in a model cell membrane lipid bilayer. Proteins 78:797–811. doi:10.1002/prot.22601

Haga K, Kruse AC, Asada H et al (2012) Structure of the human M2 muscarinic acetylcholine receptor bound to an antagonist. Nature 482:547–551. doi:10.1038/nature10753

Huang X, Zhan C-G (2007) How dopamine transporter interacts with dopamine: insights from molecular modeling and simulation. Biophys J 93:3627–3639. doi:10.1529/biophysj.107.110924

Huang ES, Koehl P, Levitt M et al (1998) Accuracy of side-chain prediction upon near-native protein backbones generated by Ab initio folding methods. Proteins 33:204–217

Indarte M, Madura JD, Surratt CK (2008) Dopamine transporter comparative molecular modeling and binding site prediction using the LeuT(Aa) leucine transporter as a template. Proteins 70:1033–1046. doi:10.1002/prot.21598

Jaakola V-P, Griffith MT, Hanson MA et al (2008) The 2.6 angstrom crystal structure of a human A2A adenosine receptor bound to an antagonist. Science 322:1211–1217. doi:10.1126/science.1164772

Jacobson MP, Pincus DL, Rapp CS et al (2004) A hierarchical approach to all-atom protein loop prediction. Proteins 55:351–367. doi:10.1002/prot.10613

Jaroszewski L (2009) Protein structure prediction based on sequence similarity. Methods Mol Biol 569:129–156. doi:10.1007/978-1-59745-524-4_7

Jerabek H, Pabst G, Rappolt M, Stockner T (2010) Membrane-mediated effect on ion channels induced by the anesthetic drug ketamine. J Am Chem Soc 132:7990–7997. doi:10.1021/ja910843d

Kardos J, Palló A, Bencsura A, Simon A (2010) Assessing structure, function and druggability of major inhibitory neurotransmitter gamma-aminobutyrate symporter subtypes. Curr Med Chem 17:2203–2213

Kelm S, Shi J, Deane CM (2010) MEDELLER: homology-based coordinate generation for membrane proteins. Bioinformatics 26:2833–2840. doi:10.1093/bioinformatics/btq554

Kinch L, Yong Shi S, Cong Q et al (2011) CASP9 assessment of free modeling target predictions. Proteins 79(Suppl 1):59–73. doi:10.1002/prot.23181

Klepsch F, Chiba P, Ecker GF (2011) Exhaustive sampling of docking poses reveals binding hypotheses for propafenone type inhibitors of P-glycoprotein. PLoS Comput Biol 7(5): e1002036. doi:10.1371/journal.pcbi.1002036

Koldsø H, Christiansen AB, Sinning S, Schiøtt B (2013) Comparative modeling of the human monoamine transporters: similarities in substrate binding. ACS Chem Neurosci 4:295–309. doi:10.1021/cn300148r

Kristensen AS, Andersen J, Jørgensen TN et al (2011) SLC6 neurotransmitter transporters: structure, function, and regulation. Pharmacol Rev 63:585–640. doi:10.1124/pr.108.000869

Krivov GG, Shapovalov MV, Dunbrack RL (2009) Improved prediction of protein side-chain conformations with SCWRL4. Proteins 77:778–795. doi:10.1002/prot.22488

Kryshtafovych A, Fidelis K (2009) Protein structure prediction and model quality assessment. Drug Discov Today 14:386–393. doi:10.1016/j.drudis.2008.11.010

Kryshtafovych A, Fidelis K, Moult J (2011a) CASP9 results compared to those of previous CASP experiments. Proteins 79(Suppl 1):196–207. doi:10.1002/prot.23182

Kryshtafovych A, Fidelis K, Tramontano A (2011b) Evaluation of model quality predictions in CASP9. Proteins 79(Suppl 1):91–106. doi:10.1002/prot.23180

Lane TJ, Shukla D, Beauchamp KA, Pande VS (2013) To milliseconds and beyond: challenges in the simulation of protein folding. Curr Opin Struct Biol 23:58–65. doi:10.1016/j.sbi.2012.11.002

Law RJ, Capener C, Baaden M et al (2005) Membrane protein structure quality in molecular dynamics simulation. J Mol Graph Model 24:157–165. doi:10.1016/j.jmgm.2005.05.006

Levitt M (1992) Accurate modeling of protein conformation by automatic segment matching. J Mol Biol 226:507–533

Liang S, Grishin NV (2002) Side-chain modeling with an optimized scoring function. Protein Sci 11:322–331. doi:10.1110/ps.24902

Lindorff-Larsen K, Piana S, Dror RO, Shaw DE (2011) How fast-folding proteins fold. Science 334:517–520. doi:10.1126/science.1208351

Liu X, Fan K, Wang W (2004) The number of protein folds and their distribution over families in nature. Proteins 54:491–499. doi:10.1002/prot.10514

Liu T, Geng X, Zheng X et al (2012) Accurate prediction of protein structural class using auto covariance transformation of PSI-BLAST profiles. Amino Acids 42:2243–2249. doi:10.1007/s00726-011-0964-5

Loland CJ, Norregaard L, Gether U (1999) Defining proximity relationships in the tertiary structure of the dopamine transporter. J Biol Chem 274:36928–36934

Lundström J, Rychlewski L, Bujnicki J, Elofsson A (2001) Pcons: a neural-network-based consensus predictor that improves fold recognition. Protein Sci 10:2354–2362

MacCallum JL, Bennett WFD, Tieleman DP (2007) Partitioning of amino acid side chains into lipid bilayers: results from computer simulations and comparison to experiment. J Gen Physiol 129:371–377. doi:10.1085/jgp.200709745

MacCallum JL, Pérez A, Schnieders MJ et al (2011) Assessment of protein structure refinement in CASP9. Proteins 79(Suppl 1):74–90. doi:10.1002/prot.23131

Makino CL, Riley CK, Looney J et al (2010) Binding of more than one retinoid to visual opsins. Biophys J 99:2366–2373. doi:10.1016/j.bpj.2010.08.003

Mariani V, Kiefer F, Schmidt T et al (2011) Assessment of template based protein structure predictions in CASP9. Proteins 79(Suppl 1):37–58. doi:10.1002/prot.23177

Martí-Renom MA, Stuart AC, Fiser A et al (2000) Comparative protein structure modeling of genes and genomes. Annu Rev Biophys Biomol Struct 29:291–325. doi:10.1146/annurev.biophys.29.1.291

McGuffin LJ (2007) Benchmarking consensus model quality assessment for protein fold recognition. BMC Bioinforma 8:345. doi:10.1186/1471-2105-8-345
Moretti S, Armougom F, Wallace IM et al (2007) The M-Coffee web server: a meta-method for computing multiple sequence alignments by combining alternative alignment methods. Nucleic Acids Res 35:W645–W648. doi:10.1093/nar/gkm333
Moult J (2005) A decade of CASP: progress, bottlenecks and prognosis in protein structure prediction. Curr Opin Struct Biol 15:285–289. doi:10.1016/j.sbi.2005.05.011
Noé F, Schütte C, Vanden-Eijnden E et al (2009) Constructing the equilibrium ensemble of folding pathways from short off-equilibrium simulations. Proc Natl Acad Sci U S A 106:19011–19016. doi:10.1073/pnas.0905466106
Norregaard L, Frederiksen D, Nielsen EO, Gether U (1998) Delineation of an endogenous zinc-binding site in the human dopamine transporter. EMBO J 17:4266–4273. doi:10.1093/emboj/17.15.4266
Nugent T, Jones DT (2012) Accurate de novo structure prediction of large transmembrane protein domains using fragment-assembly and correlated mutation analysis. Proc Natl Acad Sci U S A 109:E1540–E1547. doi:10.1073/pnas.1120036109
Paczkowski FA, Sharpe IA, Dutertre S, Lewis RJ (2007) χ-Conotoxin and tricyclic antidepressant interactions at the norepinephrine transporter define a new transporter model. J Biol Chem 282:17837–17844. doi:10.1074/jbc.M610813200
Palló A, Bencsura A, Héja L et al (2007) Major human gamma-aminobutyrate transporter: in silico prediction of substrate efficacy. Biochem Biophys Res Commun 364:952–958. doi:10.1016/j.bbrc.2007.10.108
Payandeh J, Scheuer T, Zheng N, Catterall WA (2011) The crystal structure of a voltage-gated sodium channel. Nature 475:353–358. doi:10.1038/nature10238
Peng H-P, Yang A-S (2007) Modeling protein loops with knowledge-based prediction of sequence-structure alignment. Bioinformatics 23:2836–2842. doi:10.1093/bioinformatics/btm456
Penmatsa A, Wang KH, Gouaux E (2013) X-ray structure of dopamine transporter elucidates antidepressant mechanism. Nature 503(7474):85–90. doi:10.1038/nature12533
Perozo E, Kloda A, Cortes DM, Martinac B (2002) Physical principles underlying the transduction of bilayer deformation forces during mechanosensitive channel gating. Nat Struct Biol 9:696–703. doi:10.1038/nsb827
Punta M, Forrest LR, Bigelow H et al (2007) Membrane protein prediction methods. Methods 41:460–474. doi:10.1016/j.ymeth.2006.07.026
Raval A, Piana S, Eastwood MP et al (2012) Refinement of protein structure homology models via long, all-atom molecular dynamics simulations. Proteins 80:2071–2079. doi:10.1002/prot.24098
Rohl CA, Strauss CE, Misura KM, Baker D (2004) Protein structure prediction using Rosetta. Methods Enzymol 383:66–93
Sali A, Blundell TL (1993) Comparative protein modelling by satisfaction of spatial restraints. J Mol Biol 234:779–815. doi:10.1006/jmbi.1993.1626
Samuli Ollila OH, Louhivuori M, Marrink SJ, Vattulainen I (2011) Protein shape change has a major effect on the gating energy of a mechanosensitive channel. Biophys J 100:1651–1659. doi:10.1016/j.bpj.2011.02.027
Sarker S, Weissensteiner R, Steiner I et al (2010) The high-affinity binding site for tricyclic antidepressants resides in the outer vestibule of the serotonin transporter. Mol Pharmacol 78:1026–1035. doi:10.1124/mol.110.067538
Schellman C (1980) The alpha-L conformation at the ends of helices. In: Jaenicke R (ed) Protein fold. Elsevier, New York, NY, pp 53–56
Schlessinger A, Geier E, Fan H et al (2011) Structure-based discovery of prescription drugs that interact with the norepinephrine transporter, NET. Proc Natl Acad Sci U S A 108:15810–15815. doi:10.1073/pnas.1106030108

Schlessinger A, Wittwer MB, Dahlin A et al (2012) High selectivity of the γ-aminobutyric acid transporter 2 (GAT-2, SLC6A13) revealed by structure-based approach. J Biol Chem 287:37745–37756. doi:10.1074/jbc.M112.388157

Schwede T, Kopp J, Guex N, Peitsch MC (2003) SWISS-MODEL: an automated protein homology-modeling server. Nucleic Acids Res 31:3381–3385

Shafrir Y, Durell SR, Guy HR (2008) Models of the structure and gating mechanisms of the pore domain of the NaChBac ion channel. Biophys J 95:3650–3662. doi:10.1529/biophysj.108.135327

Shen M-Y, Sali A (2006) Statistical potential for assessment and prediction of protein structures. Protein Sci 15:2507–2524. doi:10.1110/ps.062416606

Shirts M, Pande VS (2000) COMPUTING: screen savers of the world unite! Science 290:1903–1904. doi:10.1126/science.290.5498.1903

Singh SK, Piscitelli CL, Yamashita A, Gouaux E (2008) A competitive inhibitor traps LeuT in an open-to-out conformation. Science 322:1655–1661. doi:10.1126/science.1166777

Sippl MJ (1995) Knowledge-based potentials for proteins. Curr Opin Struct Biol 5:229–235

Skovstrup S, Taboureau O, Bräuner-Osborne H, Jørgensen FS (2010) Homology modelling of the GABA transporter and analysis of tiagabine binding. ChemMedChem 5:986–1000. doi:10.1002/cmdc.201000100

Söding J (2005) Protein homology detection by HMM–HMM comparison. Bioinformatics 21:951–960. doi:10.1093/bioinformatics/bti125

Stockner T, Montgomery TR, Kudlacek O et al (2013) Mutational analysis of the high-affinity zinc binding site validates a refined human dopamine transporter homology model. PLoS Comput Biol 9:e1002909. doi:10.1371/journal.pcbi.1002909

Tieleman DP, Marrink SJ, Berendsen HJ (1997) A computer perspective of membranes: molecular dynamics studies of lipid bilayer systems. Biochim Biophys Acta 1331:235–270

Tosatto SCE, Bindewald E, Hesser J, Männer R (2002) A divide and conquer approach to fast loop modeling. Protein Eng 15:279–286

Uhl G (2003) The top 20 dopamine transporter mutants: structure–function relationships and cocaine actions. Eur J Pharmacol 479:71–82. doi:10.1016/j.ejphar.2003.08.058

Voelz VA, Bowman GR, Beauchamp K, Pande VS (2010) Molecular simulation of ab initio protein folding for a millisecond folder NTL9(1-39). J Am Chem Soc 132:1526–1528. doi:10.1021/ja9090353

Wallner B, Elofsson A (2003) Can correct protein models be identified? Protein Sci 12:1073–1086. doi:10.1110/ps.0236803.a

Wallner B, Elofsson A (2005) All are not equal: a benchmark of different homology modeling programs. Protein Sci 14:1315–1327. doi:10.1110/ps.041253405

Warne T, Moukhametzianov R, Baker JG et al (2011) The structural basis for agonist and partial agonist action on a β(1)-adrenergic receptor. Nature 469:241–244. doi:10.1038/nature09746

Warren GL, Andrews CW, Capelli A-M et al (2006) A critical assessment of docking programs and scoring functions. J Med Chem 49:5912–5931. doi:10.1021/jm050362n

Wein T, Wanner KT (2010) Generation of a 3D model for human GABA transporter hGAT-1 using molecular modeling and investigation of the binding of GABA. J Mol Model 16:155–161. doi:10.1007/s00894-009-0520-3

Wu B, Chien EYT, Mol CD et al (2010) Structures of the CXCR4 chemokine GPCR with small-molecule and cyclic peptide antagonists. Science 330:1066–1071. doi:10.1126/science.1194396

Xhaard H, Backström V, Denessiouk K, Johnson MS (2008) Coordination of Na+ by monoamine ligands in dopamine, norepinephrine, and serotonin transporters. J Chem Inf Model 48:1423–1437. doi:10.1021/ci700255d

Xiang Z (2006) Advances in homology protein structure modeling. Curr Protein Pept Sci 7:217–227

Xiang Z, Steinbach PJ, Jacobson MP et al (2007) Prediction of side-chain conformations on protein surfaces. Proteins 66:814–823. doi:10.1002/prot.21099

Yamashita A, Singh SK, Kawate T et al (2005) Crystal structure of a bacterial homologue of Na^+/Cl^--dependent neurotransmitter transporters. Nature 437:215–223. doi:10.1038/nature03978

Yarov-Yarovoy V, Schonbrun J, Baker D (2006) Multipass membrane protein structure prediction using Rosetta. Proteins 62:1010–1025. doi:10.1002/prot.20817

Zhang Y, Arakaki AK, Skolnick J (2005) TASSER: an automated method for the prediction of protein tertiary structures in CASP6. Proteins 61(Suppl 7):91–98. doi:10.1002/prot.20724

Chapter 6
The Life and Times of Lac Permease: Crystals Ain't Everything, but They Certainly Do Help

M. Gregor Madej and H. Ronald Kaback

Abstract This chapter focuses on the lactose permease of *Escherichia coli* (LacY), a galactoside/H^+ symporter, as this membrane transport protein is the grandparent for the major facilitator superfamily (MFS) and arguably the most intensively studied secondary transporter known at present.

LacY couples the free energy released from downhill translocation of H^+ in response to an H^+ electrochemical gradient to drive the stoichiometric accumulation of galactopyranosides against a concentration gradient under physiological conditions. X-ray structures of an inward-facing conformation and most recently, an almost occluded conformation with a narrow periplasmic opening have been solved, which confirm many conclusions from biochemical and biophysical studies. Although structure models are critical, they are not sufficient to explain the catalysis of transport. The clues to understanding transport mechanisms are based on the principles of enzyme kinetics. Secondary transport is a dynamic process that can be described only partially by the static snapshots provided by X-ray crystallography. However, without structural information it is virtually impossible to deduce the chemistry underlying ion-coupled transport. By combining a large body of biochemical/biophysical data derived from systematic studies of site-directed mutants in LacY, residues involved in substrate binding and H^+ translocation have been identified. On the basis of the functional properties of the mutants

M.G. Madej
Department of Physiology, David Geffen School of Medicine, University of California Los Angeles, Los Angeles, CA 90095, USA

H.R. Kaback (✉)
Department of Physiology, David Geffen School of Medicine, University of California Los Angeles, Los Angeles, CA 90095, USA

Department of Microbiology, Immunology and Molecular Genetics, David Geffen School of Medicine, University of California Los Angeles, Los Angeles, CA 90095, USA

Molecular Biology Institute, David Geffen School of Medicine, University of California Los Angeles, Los Angeles, CA 90095, USA
e-mail: rkaback@mednet.ucla.ed

R. Krämer and C. Ziegler (eds.), *Membrane Transport Mechanism*, Springer Series in Biophysics 17, DOI 10.1007/978-3-642-53839-1_6,

and the X-ray structures, a working model for the symport mechanism that involves alternating access of the binding site is presented. The use of molecular biology to engineer LacY for dynamic studies combined with computational modeling has led to the postulate that *the transport reaction is driven by thermodynamics, but is controlled kinetically*.

Keywords Membrane transport • Symporters • Permeases • Protons • Galactopyranosides • Bioenergetics

6.1 A Little History

In 1955, when the existence of cell membranes themselves was being questioned, let alone the existence of proteins that transport solutes specifically across them, Cohen and Rickenberg (1955) found an inducible transport system for lactose in *Escherichia coli*, and it was subsequently found to be part of the famous lac operon (Müller-Hill 1996). This discovery was the first time that a transport function was associated with genetics and indicated that a protein might be involved. In addition to regulatory loci, it is now known widely that the lac operon contains three structural genes: (i) the Z gene encoding β-galactosidase, a cytosolic enzyme that catalyzes cleavage of lactose into galactose and glucose once it enters the cell; (ii) the Y gene encoding LacY (*aka*, lactose or lac permease, lac carrier, and later the galactoside/H^+ symporter); and (iii) the A gene encoding galactoside transacetylase, a cytosolic enzyme that catalyzes acetylation of mainly thio-β-D-galactopyranosides with acetyl-CoA as the acetyl donor and has an unknown physiological function. With the advent of molecular biology, *lacY* became the first gene encoding a membrane protein to be cloned and sequenced (Buchel et al. 1980) leading to overexpression of LacY (Teather et al. 1978). This success in the early days of molecular biology opened the study of secondary active transport at the molecular level. Overexpression of LacY combined with the use of a highly specific photoaffinity label (Kaczorowski et al. 1980) led to the functional reconstitution of LacY into proteoliposomes (Foster et al. 1982; Newman et al. 1981; Newman and Wilson 1980; Viitanen et al. 1984). It was then shown that purified LacY catalyzes all the translocation reactions typical of the transport system in vivo with comparable turnover numbers (Viitanen et al. 1984; Matsushita et al. 1983). It was also demonstrated that reconstitution of proteolipsomes with only two purified proteins, LacY and cytochrome *o*, leads to respiration-driven active lactose transport in a tightly coupled, highly efficient manner (Matsushita et al. 1983).

6.2 Bioenergetics and Kinetics

Although it was initially postulated (Kaback 1971) that LacY might be a respiratory intermediate undergoing sulfhydryl/disulfide interchange during turnover, with the development of methodology to determine and quantitate membrane potentials

($\Delta\psi$) and pH gradients (ΔpH) in microscopic systems, it became readily apparent that the chemiosmotic concept of Peter Mitchell (Mitchell 1963, 1967, 1968) unequivocally provides the best explanation for the active transport of galactosides, as well as many other solutes (Ramos et al. 1976; Ramos and Kaback 1977a, b). Accordingly, the steady-state level of accumulation of a wide variety of solutes in bacteria is driven by an electrochemical H^+ gradient ($\Delta\widetilde{\mu}_{H^+}$) composed of an electrical component ($\Delta\psi$; interior negative) and/or a pH gradient (ΔpH; interior alkaline) according to the following relationship:

$$\Delta\widetilde{\mu}_{H^+} = \Delta\psi - 2.3\ RT/F\ \Delta pH,$$

where 2.3 RT/F is equal to 58.8 at 20 °C.

A comic-book representation of chemiosmotic lactose transport is shown in Fig. 6.1. At the top (Fig. 6.1a) is an idealized right-side-out (RSO) membrane vesicle (*aka* "Kabackosome") or cell pumping out H^+ by means of the membrane-embedded respiratory chain or hydrolysis of ATP by F_1/F_0 ATPase with the generation of a $\Delta\widetilde{\mu}_{H^+}$ (interior negative and/or alkaline), thereby creating an internal sink for H^+. LacY catalyzes the obligatorily coupled stoichiometric transport of a galactoside and an H^+, utilizing the free energy released from the energetically downhill movement of H^+ to drive uphill accumulation of lactose or other galactosides against a concentration gradient. This means that a $\Delta\widetilde{\mu}_{H^+}$ of −60 mV in the form of a $\Delta\psi$ and/or a ΔpH will theoretically generate a tenfold concentration gradient of sugar, −120 mV a 100-fold gradient, −180 mV a 1,000-fold gradient, and so forth.

It is particularly important to realize that coupling between sugar and H^+ translocation is obligatory. Thus, in the absence of $\Delta\widetilde{\mu}_{H^+}$, downhill sugar influx (Fig. 6.1b) or efflux (Fig. 6.1c) will drive uphill H^+ transport with the generation of a $\Delta\widetilde{\mu}_{H^+}$, the polarity of which depends on the direction of the lactose gradient (Fig. 6.1b, c).

Another important mode of transport is equilibrium exchange or counterflow (Fig. 6.1d, e). For the former, RSO vesicles are pre-equilibrated with radiolabeled lactose and diluted into medium containing an equimolar concentration of unlabeled lactose. Under saturating conditions, rapid exchange of internal radiolabeled sugar with external cold sugar leads to rapid loss of radiolabeled sugar from the vesicles (Fig. 6.1d). In the converse experiment (counterflow), vesicles pre-equilibrated with a high concentration of unlabeled lactose (e.g., 10 mM) are diluted into medium containing a lower concentration of radiolabeled lactose (Fig. 6.1e). Exchange of internal cold with external hot lactose leads to rapid accumulation of radioactivity by the vesicles that terminates abruptly when all of the internal cold lactose has been exchanged with external hot lactose. At this point, the radioactivity in the vesicles decreases for two reasons: (i) the concentration gradient (in > out) dissipates; and (ii) exchange continues, but the specific activity of the radiolabeled lactose is continuously reduced by the cold lactose that is exchanging out of the vesicles.

Notably, neither exchange nor counterflow is influenced to any extent whatsoever by imposition of a $\Delta\psi$ or a ΔpH or by a wide range of pH values. Moreover,

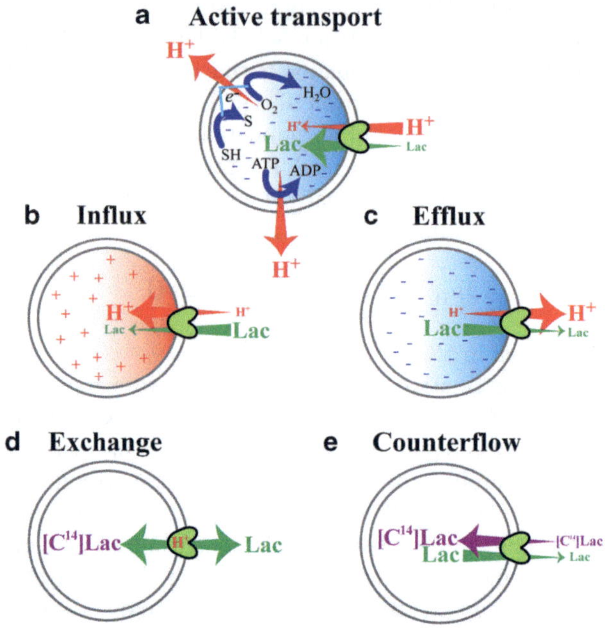

Fig. 6.1 Transport reactions catalyzed by LacY. (**a**) Uphill lactose accumulation in response to $\Delta\widetilde{\mu}_{H^+}$ generated either by H^+ respiration or ATP hydrolysis by F_1F_0-ATPase. (**b**) Uphill H^+ transport in response to an inwardly directed downhill lactose gradient. (**c**) Uphill H^+ transport in response to an outwardly directed downhill lactose gradient. Substrate gradients generate $\Delta\widetilde{\mu}_{H^+}$, the polarity of which depends upon the direction of the substrate concentration gradient. (**d**) Equilibrium exchange: concentrated suspensions of vesicles are equilibrated passively with radioactive lactose. Aliquots are then diluted into buffer with cold lactose (equimolar concentration). (**e**) Counterflow experiment: concentrated suspensions of vesicles are equilibrated passively with substrate, and aliquots are diluted into buffer containing radioactively labeled substrate at a much lower concentration

the apparent K_m for counterflow is essentially the same as that obtained for $\Delta\widetilde{\mu}_{H^+}$-driven uphill transport. *Therefore, the global conformational change involved in symport is driven by sugar binding and dissociation, not by* $\Delta\widetilde{\mu}_{H^+}$.

Figure 6.2 depicts a simplified kinetic scheme for lactose/H^+ symport. The cycle represents the overall symport process, while the orange area represents exchange and counterflow. As will be discussed in more detail later, symport starts with protonation of LacY (step 1), which is required for high-affinity binding of lactose. Sugar binding to protonated LacY (step 2) causes a conformational change to an occluded state (step 3), which can relax to the inside where sugar dissociates first (step 5), followed by deprotonation (step 6), and return of unloaded LacY via an apo occluded intermediate (steps 7 and 8). As shown, exchange or counterflow involves only steps 2–5. The most convincing evidence for this ordered mechanism is provided by the behavior of mutants with neutral replacements for Glu325 that catalyze exchange and counterflow at least as well as the wild type (WT), but cannot do any reaction that involves net H^+ transport (Carrasco et al. 1989).

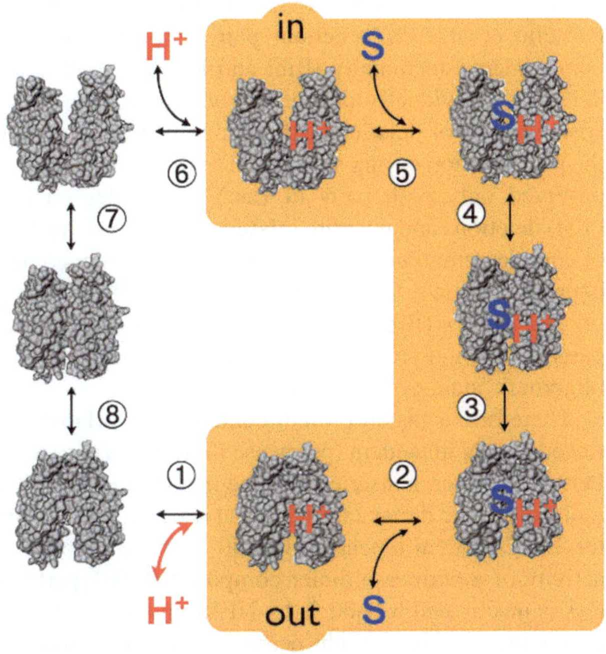

Fig. 6.2 Kinetic scheme for galactoside/H^+ symport, exchange, and counterflow. Symport starts with protonation of LacY (step 1), which is required for high-affinity binding of lactose. Sugar binding to protonated LacY (step 2) causes a conformational change to an occluded state (step 3), which can relax to the inside where sugar dissociates first (step 5), followed by deprotonation (step 6), and return of unloaded LacY via an apo-occluded intermediate (steps 7 and 8). As shown, exchange or counterflow involves only steps 2–5 (*orange shaded area*)

Since LacY catalyzes an electrogenic reaction, it would be expected that the effect of $\Delta\psi$ or ΔpH on the kinetics of the transport reaction would be an increase in V_{max}. Surprisingly, however, either component of $\Delta\tilde{\mu}_{H^+}$ causes the K_m to decrease 50–100-fold with little or no change in V_{max} (Robertson et al. 1980). Even more surprisingly, the K_D for sugar binding is the same from both sides of the membrane and is unaffected by imposition of $\Delta\tilde{\mu}_{H^+}$ (Guan and Kaback 2004). *Therefore, the widely held notion that active transport involves a change in affinity on opposite sides of the membrane is not valid, at least for LacY*. Theoretically, from the bacterium's point of view, this makes perfect sense, as a free-living organism with a $\Delta\tilde{\mu}_{H^+}$-coupled LacY that finds itself in an environment with a low lactose concentration will rapidly outgrow a bug that does not.

Many of these points will come up again later when the mechanism of coupling is discussed in detail.

6.3 The Middle Ages

6.3.1 Oligomeric State of LacY

One of the most difficult problems to resolve for a membrane protein that does not catalyze a covalent reaction is its functional oligomeric state. Notwithstanding strong evidence that LacY is functionally a monomer (Dornmair et al. 1985; Costello et al. 1987), certain paired in-frame deletion mutants were found to complement functionally (Bibi and Kaback 1992). Although cells expressing the deletions individually do not catalyze transport, cells simultaneously expressing specific pairs of deletions exhibit significant transport activity, and it is clear that the phenomenon occurs at the level of the protein. Remarkably, complementation is observed only with pairs of LacY molecules containing large transmembrane (TM) deletions and not with missense mutations or point deletions. The mechanism of complementation is probably related to the phenomenon whereby independently expressed N- and C-terminal fragments of the permease interact to form a functional complex (Bibi and Kaback 1990). In any case, the observation that certain pairs of deletion mutants can interact revived the concern regarding the functional oligomeric state of WT LacY.

Therefore, a fusion protein containing two lac permease molecules covalently linked C to N in tandem (permease dimer) was engineered (Sahin-Tóth et al. 1994). Permease dimer is inserted into the membrane in a functional state, and each LacY molecule in the dimer exhibits equal activity. Thus, point mutations in either half of the LacY tandem repeat lead to 50 % inactivation of transport. Furthermore, the activity of a permease dimer composed of WT permease and a mutant devoid of Cys is inactivated by ~60 % by NEM. In order to test the caveat that oligomerization between dimers might occur in trans, a permease dimer was constructed that contains two different deletion mutants that complement when expressed as untethered molecules. This construct does not exhibit any activity whatsoever indicating that dimers do not undergo oligomerization in trans. The experiments are clearly consistent with the conclusion that WT LacY is functional as a monomer.

Finally, intermolecular thiol cross-linking was utilized to determine surface-exposed positions containing single-Cys replacements in each transmembrane helix or loop in RSO vesicles (Guan et al. 2002a; Ermolova et al. 2003). Single-Cys mutants located on opposite sides of the molecule exhibit similar rates of cross-linking with sigmoid kinetics. Furthermore, cross-linking is markedly decreased at 0 °C, suggesting that lateral diffusion of the permease within the plane of the membrane is important for intermolecular cross-linking. The findings indicate that intermolecular cross-linking is a stochastic process resulting from random collisions and support the other lines of evidence that lactose permease is structurally, as well as functionally, a monomer.

6.3.2 Hydrogen/Deuterium (H/D) Exchange

Although X-ray crystal structures of secondary transport proteins are essential, the transport mechanism(s) is a dynamic process by nature that cannot be fully described by static snapshots. H/D exchange of ~85 % of the backbone amide protons in LacY occurs within 10–15 min at ~20 °C (le Coutre et al. 1998; Patzlaff et al. 1998; Sayeed and Baenziger 2009) with 100 % exchange at elevated, non-denaturing temperatures (Sayeed and Baenziger 2009). In contrast, with the K^+ channel KcsA at room temperature, it takes 3 h to exchange ~45 % of its amide protons for deuterium (le Coutre et al. 1998). This amazing property of LacY clearly indicates that this very hydrophobic protein (~65 % unequivocally hydrophobic residues) is in a highly dynamic state.

6.3.3 Structure Without Crystals

Because LacY was apparently so resistant to traditional means of structural analysis, alternative approaches were developed to approximate the overall 3D fold, as well as local interactions. LacY has been the test bed for the development of most of these methodologies with transport proteins, and they include mapping transmembrane segments by phoA fusions (Calamia and Manoil 1990), protein insertion into loops and deletion analysis (Prive and Kaback 1996; Weinglass and Kaback 2000; Wolin and Kaback 1999), accessibility/reactivity of natural or uniquely engineered Cys residues to membrane-permeant and impermeant reagents (Guan and Kaback 2007), determination of secondary structure by circular dichroism (CD) (Foster et al. 1983a), laser Raman (Vogel et al. 1985), or Fourier transform infrared spectroscopy (FTIR) (le Coutre et al. 1997), identifying long-range contacts by suppressor analysis, thiol cross-linking (Wu and Kaback 1996; Zhou et al. 2008), excimer fluorescence (Jung et al. 1993), engineered Mn(II) binding sites (Jung et al. 1995), site-directed electron paramagnetic resonance (Zhao et al. 1999a; Smirnova et al. 2007), or discontinuous mAb epitope mapping (Sun et al. 1996, 1997).

On the one hand, data obtained with these methods provided an overview of the organization of the protein as well as insight into local interactions between essential side chains. For example, a combination of hydropathy profiling and CD led to the correct indication that LacY contains 12 transmembrane helices (Foster et al. 1983b). This type of information together with functional observations led to a proposed mechanism that was consistent with the available evidence (Kaback et al. 2001). On the other hand, a structure model based on thiol cross-linking data was incorrect because thiol cross-linking yields the closest distance between Cys residues, which leads to underestimation of distances on the cytoplasmic side of LacY (Sorgen et al. 2002). Therefore, the large central cavity was missed due to the inherent flexibility of LacY. However, proximities derived from excimer fluorescence (Jung et al. 1993, 1994a) or engineered Mn(II) binding sites (Jung et al. 1995; He et al. 1995a, b) were largely confirmed by the X-ray structures (Abramson et al. 2003; Kumar et al. 2014), and the secondary structure content calculated from CD was remarkably accurate (85 % helix by CD versus 86 % from the X-ray structure). In contrast, helix content was significantly underestimated by laser Raman or FTIR.

6.3.4 Cys-Scanning Mutagenesis

The notion behind Cys-scanning mutagenesis began with the identification of native Cys148 as the residue in LacY that is protected from alkylation and inactivation by *N*-ethylmaleimide (NEM) (Fox and Kennedy 1965) and the development of site-directed mutagenesis in the early 1980s. Despite the interpretation that Cys148 was close to or part of the sugar-binding site, initial site-directed mutagenesis (Trumble et al. 1984; Viitanen et al. 1985) showed that it is not essential for activity. But since Cys was thought to play a major role in the mechanism of LacY (Kaback and Barnes 1971), each of the other 7 native Cys residues was mutated, and none was found to be essential. And then the unthinkable (at the time) was carried out—all 8 native Cys residues were mutated simultaneously, which led to a Cys-less mutant with ~60 % of WT activity (van Iwaarden et al. 1991). Since no one had successfully crystallized a permease at this time, a *lacY* gene encoding Cys-less LacY with unique restriction sites approximately every 100 base pairs was constructed, and over the next decade or so, each residue was changed initially to a Cys [with the exception of the C-terminal 17 residues, which can be deleted with no effect on expression or activity (McKenna et al. 1992)]. By this means, a library was built with the intention of readily carrying out multiple site-directed cross-linking studies in order to obtain a model for helix packing. Although such a model was constructed (Sorgen et al. 2002), once a crystal structure was obtained, it became clear that no amount of cross-linking would suffice to yield a realistic structure because of the inherent flexibility of LacY and because thiol cross-linking gives the closest distance between Cys residues.

However, Cys-scanning mutagenesis (Frillingos et al. 1998) has been particularly valuable for determining which residues are irreplaceable for stability, for the symport mechanism, and for all sorts of biochemical/biophysical studies. Cys is average in bulk, relatively hydrophobic, and amenable to highly specific modification with biochemical and biophysical probes. As evidenced from the widespread use of Cys mutagenesis with all manner of membrane proteins, the technique has gained great favor in general.

LacY comprises 417 amino acid residues. In order to determine which residues play an obligatory role in the mechanism and to create a library of mutants with a single-Cys residue at each position of the molecule for structure/function studies, each residue was replaced individually with Cys. The great majority of the 401 single-Cys mutants are expressed normally in the membrane and catalyze accumulation of lactose against a significant concentration gradient, thereby demonstrating that Cys replacement at most positions does not induce severe perturbations in the structure of the permease or in the symport mechanism. Mutation of only 92 residues has a significant effect on activity, inhibiting the steady-state level of accumulation by >50 % (Frillingos et al. 1998). Moreover, it is striking that less than ten residues are absolutely irreplaceable for symport: Glu126 (helix IV) and Arg144 (helix V), which are critical for substrate binding, as well as Trp151 (helix V) where an aromatic is essential; Glu269 (helix VIII), His322 (helix X), and

Tyr236 (helix VII), which are involved in sugar binding and may be involved in H^+ translocation as well; and Arg302 (helix IX) and Glu325 (helix X), which are exclusively involved in H^+ translocation.

6.4 The Modern Era

6.4.1 X-ray Crystal Structures

After more than two decades of crystallization attempts with WT LacY, a structure was solved for a conformationally restricted mutant in 2003 (Abramson et al. 2003), and remarkably, the structure of GlpT, the P_i/glycerol-3-phosphate antiporter, another MFS member, was published in the same issue of *Science* (Huang et al. 2003). The GlpT structure was solved by using a so-called funnel approach in which several homologues were screened in order to find one that crystallized (Lemieux et al. 2003). In contrast, the LacY structure was determined by using a conformationally restricted mutant that favors a single conformation. The C154G mutant binds substrate with high affinity, but catalyzes little or no transport (Smirnova and Kaback 2003; Ermolova et al. 2005; Nie et al. 2008). In addition, the mutant is thermostable and has a reduced tendency to aggregate relative to the WT, properties that favor successful crystallization.

The overall structure of an inward-open conformer of LacY is shown in Fig. 6.3. The asymmetric unit of the LacY crystal is composed of an artificial dimer, with two molecules oriented in opposite directions, consistent with the conclusion that the monomer is the functional unit as discussed. The structures of the two monomers are almost identical with an rmsd of 0.04 Å for 417 Cα atoms. Viewed from the side (Fig. 6.3a), the monomer is heart shaped with an internal cavity open on the cytoplasmic side (Fig. 6.3a, b, blue surface), and the largest dimensions of the molecule are 60 Å (along the membrane) by 60 Å (perpendicular to the membrane). Viewed from the cytoplasmic face (Fig. 6.3b), the molecule has a distorted oval shape with dimensions of 35 Å × 60 Å. The molecule exhibits a large interior hydrophilic cavity, which is open only on the cytoplasmic side with dimensions of 25 Å × 15 Å. As established by biochemical studies [reviewed in Kaback et al. (2001)], the side chains involved in galactoside binding are at the apex of the cavity in the approximate middle of the membrane (Fig. 6.3, colored in orange).

The monomer is composed of 12 transmembrane helices as predicted (Abramson et al. 2003; Guan et al. 2007), many of which are very distorted. The N- and C-terminal six helices form two distinct helical bundles connected by a long loop between helices VI and VII (loop VI-VII; Fig. 6.3a, b). Although the loop has two short helical segments at the N- and C-termini, it is an extended, flexible structure. The N- and C-terminal 6-helix bundles have the same topology and are related by an approximate twofold symmetry, as proposed for OxlT and other members of the MFS (Abramson et al. 2003; Guan et al. 2007; Chaptal et al. 2011;

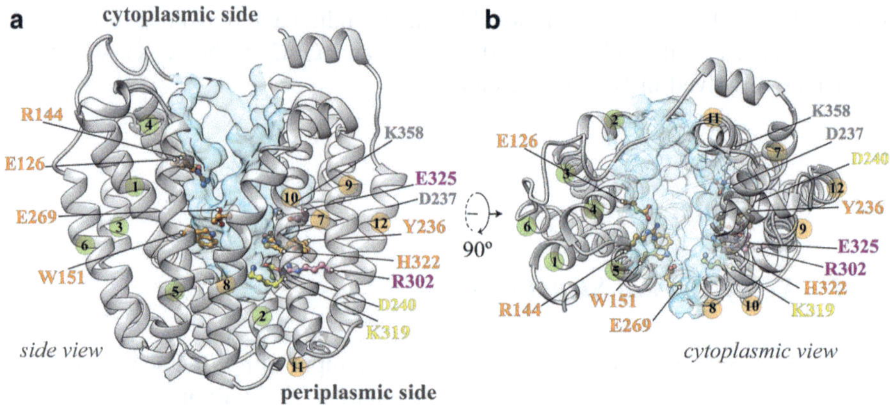

Fig. 6.3 LacY structure. (**a**) Side view of the inward-facing conformation of LacY. Helices are shown as *ribbons*; the numbers of helices in the N-terminal six-helix bundle are colored *green*, and the numbers of C-terminal six-helix bundle are colored *orange*. Critical residues are indicated (*orange*, residues involved in sugar binding; *magenta*, residues involved in H^+ translocation; *yellow*, residues with unknown function; *red*, oxygen; *blue*, nitrogen; *yellow*, sulfur; details are provided in the main text). The water-accessible surface of the cavity is shown in *light blue* (surface was calculated using the Computed Atlas of Surface Topography of proteins (CASTp) Web tool with a probe size of 1.4 Å). (**b**) Cytoplasmic view of the inward-open conformation; the color coding is as in (**a**)

Huang et al. 2003; Mirza et al. 2006; Dang et al. 2010; Yin et al. 2006; Newstead et al. 2011; Solcan et al. 2012; Sun et al. 2012; Pedersen et al. 2013; Yan et al. 2013; Zheng et al. 2013). The central hydrophilic cavity is formed between helices I, II, IV, and V of the N-terminal domain and helices VII, VIII, X, and XI of the C-terminal domain, while helices III, VI, IX, and XII are largely embedded in the bilayer.

Subsequently, WT LacY was crystallized by ensuring that the purified protein is not depleted of bound phospholipid (Guan et al. 2006, 2007). Remarkably, the structures of the C154G mutant and the WT appear to be identical, although biochemical studies indicate strongly that the periplasmic aspect of the mutant is paralyzed in a partially open conformation (Nie et al. 2008; Jiang et al. 2012). It is also notable that the transport activity of the C154G mutant (helix V) is largely restored by replacing Gly24 in neighboring helix I with a Cys residue (Ermolova et al. 2005). This and other results suggest that neighboring Gly residues in helices I and V, which cross in the middle of the membrane, allow the helices to come into closer contact, thereby compromising conformational flexibility and inhibiting transport. However, this explanation should be taken with a grain of salt, as Cys is the only replacement for Gly24 that effectively rescues the C154G mutant (Ermolova et al. 2005).

In addition to two structures of the C154G mutant, one at ~3.5 Å (Abramson et al. 2003) and one at ~2.9 Å resolution (Mirza et al. 2006), the WT LacY structure

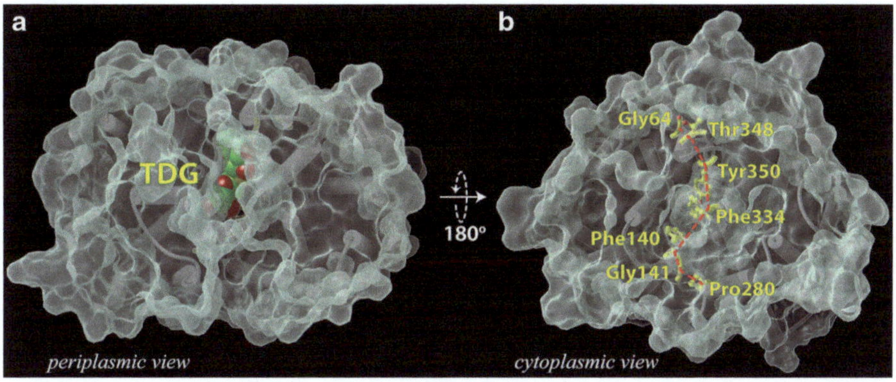

Fig. 6.4 Space-filling models of LacY G46W/G262W (pdb-ID 4OAA). (**a**) Viewed from the periplasmic side and showing TDG just visible within the molecule. (**b**) Viewed from the cytoplasmic side and showing the residues that form a zipper-like motif to seal that side

(Guan et al. 2007) and the structure of single-Cys122 LacY with a covalently bound inactivator (Chaptal et al. 2011) all exhibit the same inward-facing conformation.

However, very recently a new X-ray crystal structure was obtained with the LacY double-Trp mutant G46W/G262W (Kumar et al. 2014). Transport activity is abrogated in the mutant with little or no effect on galactoside binding (Smirnova et al. 2013). Although thought to be arrested in an open-outward conformation (Smirnova et al. 2013), the structure is almost occluded and partially open to the periplasmic side (Fig. 6.4a); the cytoplasmic side is tightly sealed (Fig. 6.4b). Surprisingly, the opening on the periplasmic side is sufficiently narrow that sugar cannot get in or out of the binding site and a clearly defined density for a bound sugar at the apex of the cavity in the middle of the protein is observed (Kumar et al. 2014).

6.5 The Active Site

6.5.1 Side Chains Involved in Sugar Binding

Galactoside transport is inactivated by *N*-ethylmaleimide (NEM), and LacY can be selectively labeled with radioactive NEM by substrate protection against alkylation (Fox and Kennedy 1965). The substrate-protected residue was shown to be Cys148 (Bieseler et al. 1985), thus providing initial evidence that the substrate-binding site may involve helix V. However, when Cys148 is replaced with various side chains, neither binding nor active transport is abolished (Trumble et al. 1984; Jung et al. 1994b). As shown in the new LacY structure (Kumar et al. 2014), Cys148 is near the bound sugar, but does not make direct contact with the galactopyranosyl ring (Fig. 6.5, top panel). Thus, the inactivating effect of

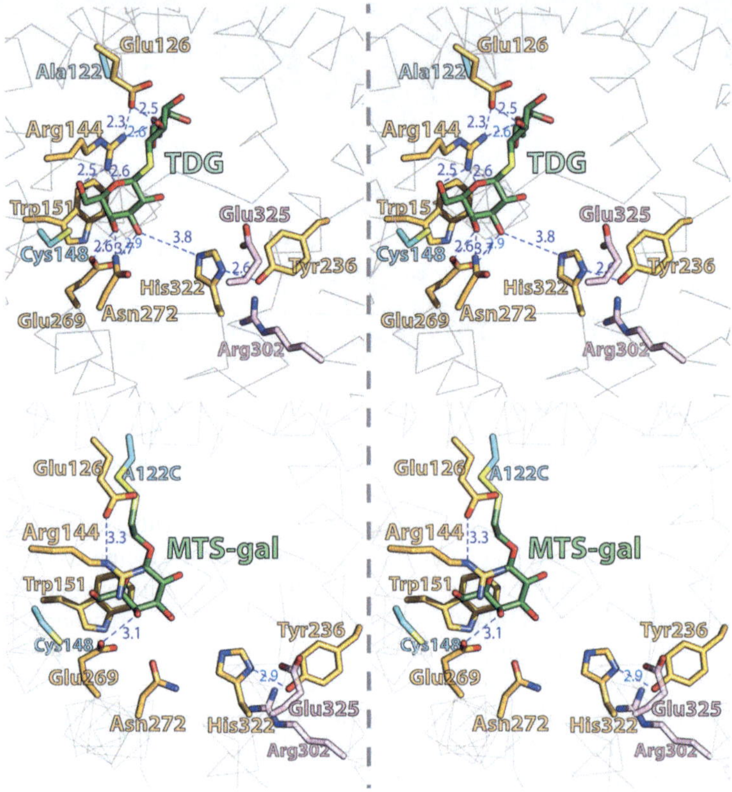

Fig. 6.5 Stereo view of the sugar-binding site. TDG is shown as *green stick* model and side chains forming hydrogen bonds with TDG are in *gold*. Ala122 and Cys148, which are close to TDG, but do not make direct contact, are shown in *cyan* (heteroatoms are colored *red*, oxygen; *blue*, nitrogen; *yellow*, sulfur). *Top panel*: (pdb-ID 4OAA) molecule A with TDG. *Lower panel*: LacY with covalently bound MTS-Gal (pdb-ID 2Y5Y)

alkylation is likely due to steric blockade of galactoside binding. A similar steric effect is observed when Ala122 (helix IV) is replaced with Cys (Kwaw et al. 2001). However, it is remarkable that alkylation of mutant A122C or replacement with a bulky side chain causes LacY to become specific for binding and transport of the monosaccharide galactose (Guan et al. 2002b). Thus, anomeric substituents on β-D-galactopyranosides lie in close proximity to Ala122 (helix IV) (Fig. 6.5, top panel).

Cys-scanning mutagenesis reveals that replacement of Glu126 (helix IV) or Arg144 (helix V) with neutral amino acyl side chains abolishes transport, and activity is not observed with double-neutral substitutions or when the residues are interchanged (Frillingos et al. 1997). Only mutants E126D or R144K exhibit any activity whatsoever. Mutant E126D has six to tenfold lower affinity for sugar and accumulates lactose at a lower rate to a normal steady state, whereas mutant R144K displays little or no galactoside affinity and transports lactose at a negligible rate to a very low steady state (Frillingos et al. 1997).

Glu126 and Arg144 are charge paired. Replacement of either residue with Ala in a LacY mutant containing a single-Cys residue at position 148 markedly decreases NEM labeling of Cys148, but the double-Ala mutant labels normally. Thus, an unpaired charge causes a conformational perturbation that decreases the reactivity of Cys148, while double-neutral replacement with Ala, replacement of Arg144 with Lys, or when Glu126 and Arg144 are interchanged has no such effect (Venkatesan and Kaback 1998). Further evidence that the two positions are close physically is provided by spontaneous disulfide cross-linking (Wolin and Kaback 2000) and site-directed spin labeling (Zhao et al. 1999b). Direct binding assays show that none of the Arg144 mutants, including R144K, bind lactose or the high-affinity substrate analogue *p*-nitrophenyl-α-D-galactopyranoside (NPG), while neutral replacements for Glu126 do not bind NPG or β-D-galactopyranosyl 1-thio-β-D-galactopyranoside (TDG), but mutant E126D exhibits significant binding of TDG, although affinity is decreased (Sahin-Tóth et al. 1999; Smirnova et al. 2009a). In addition, there is no substrate protection against NEM labeling when Glu126 and Arg144 are interchanged. As a whole, the results demonstrate that a carboxyl group at position 126 and a guanidinium group at position 144 are absolute requirements for substrate binding and indicate that the two residues are likely charge paired. Electrospray Ionization Mass Spectrometry (ESI-MS) studies also support the interaction of Arg144 and Glu126 by covalent modification of the guanidinium group with the Arg-specific reagent butane 2,3-dione (BD) (Weinglass et al. 2004). The reactivity of Arg144 with BD is low and reduced in the presence of ligand. But replacement of Glu126 with Ala results in an increase in the BD reactivity of Arg144, consistent with the presence of a charge pair between Arg144 and Glu126 in the absence of sugar.

Trp151, two turns of helix V from Arg144, also plays an important role in substrate binding (Guan et al. 2003), although mutants W151Y and W151F catalyze active lactose transport with time courses similar to the WT. Mutant W151F or W151Y binds NPG and TDG relatively poorly, although there is relatively little change in the K_m of lactose transport. In addition, amino acid replacements with an alkyl side chain exhibit little or no transport and no significant binding affinity. Thus, an aromatic side chain at position 151 is obligatory for galactoside binding and transport.

The fluorescent properties of Trp151 from a fully functional mutant devoid of all other Trp residues were also investigated (Vazquez-Ibar et al. 2003). The steady-state fluorescence spectrum of Trp151 and fluorescence quenching experiments with water-soluble quenchers demonstrate that Trp151 is in a hydrophilic environment. Furthermore, substrate binding leads to a blue shift in the fluorescence spectrum and reduction in accessibility to polar quenchers, indicating that Trp151 becomes less exposed to aqueous solvent. In addition, the phosphorescence spectrum of Trp151 is red shifted in the presence of substrate, indicating a direct stacking interaction between the galactopyranosyl and indole rings (Fig. 6.5).

Glu269 (helix VIII), another irreplaceable residue, is also critical for substrate recognition and it was believed that it might be involved in H^+ translocation as well (see below) (Venkatesan and Kaback 1998; Ujwal et al. 1994; Franco and Brooker 1994; He and Kaback 1997; Sahin-Tóth et al. 2000; Weinglass et al. 2002). Neutral replacements abolish lactose transport, while replacement with Asp leads to a marked decrease in affinity for TDG or NPG. Mutant E269D exhibits significant transport of TDG, and there is about a threefold increase in H^+/TDG stoichiometry (Ujwal et al. 1994; Franco and Brooker 1994). Furthermore, Gly-scanning mutagenesis of mutant E269D shows that mutant E269D can be rescued with respect to binding and all translocation reactions, suggesting that positioning of the carboxyl group at position 269 in the binding site is critical (Weinglass et al. 2002). As a final point, analysis of the CNBr fragment containing Glu269 by ES-MSI demonstrates that this carboxyl group is protected by substrate against reaction with hydrophobic carbodiimides.

Importantly, galactoside binding probably involves induced fit (see below). Initially, there may be a nonspecific hydrophobic interaction between the hydrophobic face of the galactopyranosyl moiety and Trp151. When proper geometry is achieved, specific interactions with the galactopyranosyl ring occur with Arg144 and Glu126 and other side chains that cause protonated LacY to assume an occluded conformation (Fig. 6.5).

6.5.2 Side Chains Involved in H^+ Binding and Coupling

One fundamentally important problem is the identification of the residues involved in H^+ binding and the mechanism of coupling between sugar and H^+ translocation. Current evidence indicates strongly that LacY is protonated prior to ligand binding (Sahin-Tóth et al. 2000; Smirnova et al. 2008). Thus, in addition to the evidence for the kinetic scheme discussed above (Fig. 6.2), galactoside binding to LacY as a function of pH exhibits a pK_a of ~10.5 (Smirnova et al. 2008). Thus, in the physiological pH range, LacY clearly has a very high probability of being protonated. Remarkably, neutral replacement of Glu325 abolishes the pK_a, and high-affinity binding is observed up to pH 11 (Smirnova et al. 2009a). Thus, it is apparent that a negative charge at position 325 (i.e., unprotonated Glu325) abrogates binding. Furthermore, it has been shown directly (Smirnova et al. 2012) that addition of TDG to a concentrated solution of purified, detergent solubilized LacY induces no change in pH, while a positive control with the antiporter EmrE (provided by S. Schuldiner, Hebrew University, Jerusalem) under identical conditions releases 1 H^+/mol EmrE upon addition of tetraphenylphosphonium (Soskine et al. 2004).

Although LacY must be protonated prior to galactoside binding, which represents a part of the coupling mechanism, it is difficult to study the mechanism of H^+ translocation. However, in addition to $\Delta\widetilde{\mu}_{H^+}$-driven active transport, LacY catalyzes other modes of translocation that are important for studying coupled H^+ transport. Because individual steps in the overall transport cycle cannot be delineated by studying $\Delta\widetilde{\mu}_{H^+}$-driven active transport, carrier-mediated efflux down a concentration gradient, equilibrium exchange, and entrance counterflow are used to probe the mechanism (Kaczorowski and Kaback 1979; Kaczorowski et al. 1979). Efflux, exchange, and counterflow with WT LacY are explained by the kinetic scheme shown in Fig. 6.2 (see Sect. 6.2).

Many enzyme reactions involve H^+ transfer in the rate-limiting step, and as a consequence, these reactions may exhibit a solvent isotope effect when studied in deuterium oxide (D_2O). In brief, such reactions proceed slower in D_2O because of differences in the zero-point stretch vibrations of bonds to protium relative to deuterium (Jenks 1969; Schowen 1977). With right-side-out (RSO) vesicles or proteoliposomes reconstituted with purified LacY, three to fourfold inhibition of the rate of H^+-coupled downhill lactose efflux is observed in D_2O from pH 5.5 to 7.5, with no effect on $\Delta\tilde{\mu}_{H^+}$-driven active transport, exchange, counterflow, or affinity for sugar (Kaczorowski et al. 1979; Viitanen et al. 1983; Carrasco et al. 1984). These observations and others indicate that reactions involved in protonation or deprotonation are not rate determining for $\Delta\tilde{\mu}_{H^+}$-driven active transport, exchange, or counterflow, while protonation or deprotonation is rate limiting when a lactose concentration gradient drives uphill H^+ translocation (Fig. 6.1b, c). This conclusion also receives strong support from solid supported membrane electrophysiology with proteoliposomes reconstituted with purified LacY, as well as some of the mutants described above (Garcia-Celma et al. 2009, 2010; Gaiko et al. 2013).

Efflux, exchange, and counterflow are blocked in His322 mutants, and replacement with Asn or Gln results in a marked decrease in affinity for sugar (Smirnova et al. 2009a; Sahin-Tóth et al. 2000). However, the mutants catalyze lactose influx down a concentration gradient at a slow rate (Padan et al. 1985; Puttner et al. 1986; Puttner and Kaback 1988) without H^+ translocation. In contrast, neutral Glu325 mutants are specifically defective in all steps involving net H^+ translocation but catalyze lactose exchange and counterflow as well or better than WT LacY [reviewed in Kaback (1987)]. Thus, Glu325 is required for protonation and deprotonation (Fig. 6.2, steps 1 and 6). Glu325 mutations mimic D_2O or mAb 4B1 (Carrasco et al. 1982, 1984), but affinity is unaffected by neutral replacements, D_2O or mAb 4B1. Similarly, replacement of Arg302 with Ala or Ser maintains affinity and exchange activity, although active transport is completely inhibited (Sahin-Tóth and Kaback 2001). Extensive mutagenesis and functional characterization reveal that neutral replacements for Glu269 cause LacY to become defective in all translocation reactions. Even replacement of Glu269 with Asp yields LacY that hardly catalyzes active lactose transport, efflux down a concentration gradient, or equilibrium exchange. However, as discussed above, mutant E269D accumulates TDG with an increase in H^+/TDG stoichiometry (Ujwal et al. 1994; Franco and Brooker 1994) and also exhibits markedly decreased galactoside affinity (Venkatesan and Kaback 1998; Smirnova et al. 2009a; He and Kaback 1997; Sahin-Tóth et al. 2000; Weinglass et al. 2002). Therefore, Glu269 was also believed to be involved in H^+ translocation and sugar binding, as well as coupling between sugar and H^+ translocation (see below).

Interestingly, when Tyr236 (helix VII) is replaced with Phe, active transport and efflux are abolished, and the mutant catalyzes limited equilibrium exchange (Roepe and Kaback 1989). In contrast, mutants with Cys or Ala in place of Tyr236 catalyze active transport lactose 30–50 % as well as WT (Vadyvaloo et al. 2006).

In the inward-facing structures (Fig. 6.3) (Abramson et al. 2003; Guan et al. 2007; Chaptal et al. 2011; Mirza et al. 2006), a complex salt bridge/H-bond network is observed, which is composed of residues from helix VII (Tyr236 and Asp240), helix X (Lys319, His322 and Glu325), and helix IX (Arg302). It is noteworthy that Glu325 is embedded in a hydrophobic milieu formed by Met299 and Ala295 (helix IX), Leu329 (helix X), and Tyr236 (helix VII), which is consistent with the notion that Glu325 is protonated in this conformation (Kaback et al. 2001). Therefore, the structures represent the protonated inward-facing conformation. It has been suggested (Sahin-Tóth and Kaback 2001) that Arg302 could interact with Glu325 to drive deprotonation. However, in the structures shown, the side chain of Arg302 is ~7 Å removed from Glu325, suggesting that a conformational rearrangement may occur. On the other hand, LacY with two Cys residues at position 302 and 325 exhibits excimer fluorescence (Jung et al. 1993), and with two His residues, a Mn(II) binding site is observed (He et al. 1995a). The structural data combined with biochemical/molecular biological studies discussed above indicate that His322 might be the immediate H^+ donor to Glu325 (see below). Since mutants with simultaneous neutral replacements for Asp240 and Lys319 maintain low but significant transport activity (Sahin-Toth and Kaback 1993), it is unlikely that this salt bridge is directly involved in H^+ translocation; however, the two residues could be involved in stabilization of the salt bridge/H-bond network. Interestingly, when Asp240 and Lys319 are replaced with Cys residues and then cross-linked, transport is abolished (Zhang et al. 2002).

6.6 Induced Fit

In a recent study (Zhou et al. 2012), seven of the irreplaceable residues were mutated individually, and galactoside-induced opening or closing of periplasmic or cytoplasmic cavities was probed by site-directed alkylation (SDA). As expected, mutation of Glu126 (helix IV) or Arg144 (helix V), which are critically involved in sugar binding, completely blocks sugar-induced periplasmic opening. Remarkably, however, replacement of Glu269 (helix VIII), His322 (helix X), or Tyr236 (helix VII) causes spontaneous opening of the periplasmic cavity in the absence of sugar and decreased closing of the cytoplasmic cavity in the presence of galactoside. Moreover, mutations in the triad exhibit poor affinity for sugar, and galactoside/H^+ symport is abolished as well. In contrast, mutation of Arg302 (helix IX) or Glu325 (helix X) has no such effect, and sugar binding induces normal opening and closing of periplasmic and cytoplasmic cavities. It has been postulated that Glu269, His322, and Tyr236 act in concert to coordinate opening and closing of the cavities.

The recently solved structure of the LacY G46W/G269W mutant provides support for the induced-fit mechanism for sugar binding in LacY (Mirza et al. 2006; Vazquez-Ibar et al. 2004). This double-Trp mutant behaves as if it is arrested in an outward-open conformation with diffusion-limited access to the binding site based on biochemical and spectroscopic measurements (Smirnova et al. 2013). The X-ray structure (Kumar et al. 2014) exhibits an almost occluded bound TDG molecule since the periplasmic opening is too narrow for the bound TDG to exit from the binding site without concomitant and concerted structure change, and the cytoplasmic side is tightly sealed (Fig. 6.4a, b). A simple explanation is that in the absence of sugar, the periplasmic pathway in the mutant is sufficiently open to allow access of TDG to the binding site. But when binding occurs and the mutant attempts to transition into an occluded state, it cannot do so completely because the bulky Trp residues block complete closure. Thus, the mutant binds sugar and can initiate but cannot complete the transition into a fully occluded state. Therefore, the mutant is completely unable to catalyze transport of any type across the membrane.

Furthermore, the density observed in the cavity in the middle of the protein clearly defines the conformation of TDG and the amino acid side chains ligating the sugar in the structure are entirely consistent with extensive biochemical/spectroscopic studies. A notable implication of the present structure is that Glu269, His322, and Tyr236, long thought to be involved in coupled H^+ translocation by implication [reviewed in Guan and Kaback (2006)], are clearly ligands to the galactopyranosyl moiety of the substrate. Although mutation of these residues causes the periplasmic side of LacY to open spontaneously (Zhou et al. 2012), there is little or no direct evidence that they are involved in H^+ translocation, although it is conceivable that they may play a dual role in the transport mechanism.

Remarkably, as opposed to the almost occluded, open-outward conformation of TDG-bound G46W/G262W, LacY with covalently bound MTS-Gal in the binding site exhibits an inward-open conformation (66) (Fig. 6.5, lower panel). Furthermore, although MTS-Gal is covalently bound to a Cys in place of Ala122, the galactopyranoside interacts with Trp151 and Glu269, but not with the other side chains that interact in the double-Trp structure. This difference clearly suggests that the transition into the occluded state requires fully ligated sugar. Given this possibility and observations indicating that the alternating access mechanism of LacY is driven by galactoside binding and dissociation, and not by $\Delta\tilde{\mu}_{H^+}$, it seems highly likely that sugar binding involves induced fit. By these means, the N- and C-terminal bundles converge as given side chains from both the N- and C-terminal helix bundles ligate the galactoside. The energetic cost of binding and the resultant conformational change are regained upon sugar dissociation and provide the energy for a further structural change that allows deprotonation. In brief, the mechanism of LacY resembles that of an enzyme, the difference being that the intermediate is a conformer(s) of the protein rather than a transition-state intermediate of the substrate.

6.7 Alternating Access

As stated by Mitchell (1968), "The protein moves around the substrate, allowing alternating access of binding sites to either side of the membrane." In this section of the chapter, the experimental approaches that provide a strong case for the alternating access mechanism in LacY are reviewed.

6.7.1 Site-Directed Alkylation

Alkylation of Cys side chains by radiolabeled NEM or fluorescent tetramethylrhodamine-5-maleimide (TMRM), which are membrane-permeant reagents, has been used extensively to study reactivity/accessibility of single-Cys residues introduced into Cys-less LacY. This approach, which is based on alkylation of single-Cys LacY in RSO membrane vesicles containing single-Cys mutants, provides important information about the structure, function, and dynamics of LacY [reviewed in Kaback et al. (2007) and Nie et al. (2007)]. The reactivity/accessibility of Cys residues depends on the surrounding environment and is limited by close contacts between transmembrane helices and/or the low dielectric of the environment. Almost every amino acid in LacY was replaced individually with Cys and tested by SDA for reactivity with NEM, which provided initial support for an alternating access mechanism. Engineered Cys replacements located near or within the inward-facing hydrophilic cavity react well with this alkylating agent. TDG binding *increases* NEM reactivity of single-Cys replacements located predominantly on the periplasmic side of LacY (Fig. 6.6a) and *decreases* reactivity of replacements located predominantly on the cytoplasmic side (Fig. 6.6b). Furthermore, both sets of Cys replacements in the putative cavities are located at the periplasmic (increased reactivity) and cytoplasmic (decreased reactivity) ends of the same helices and distributed in a pseudo-symmetrical manner. The pattern is consistent with a model in which a the single sugar-binding site in the approximate middle of LacY is alternatively exposed to either side of the membrane due to opening and closing of cytoplasmic and periplasmic hydrophilic pathways.

More recently, a simple, more facile SDA method with TMRM (Nie et al. 2007, 2008, 2009; Jiang et al. 2011) was developed to examine the effect of sugar binding on alkylation of single-Cys LacY mutants either in RSO membrane vesicles or in dodecyl-β,D-maltopyranoside (DDM) micelles (Nie and Kaback 2010). The experiments were carried out at 0 °C, where thermal motion is restricted (Venkatesan and Kaback 1998; Venkatesan et al. 2000a, b), and linear rates of labeling are readily obtained (Fig. 6.6c). TMRM labeling is almost negligible with LacY containing each of five single-Cys residues at positions on the periplasmic side in RSO membrane vesicles or with purified protein solubilized in DDM. Binding of TDG markedly *increases* the rate of labeling (Fig. 6.6c, right panel). Consistently, on the cytoplasmic side, each of four single-Cys replacement mutants labels at a rapid rate in the absence of sugar both in RSO membrane vesicles and with purified protein in DDM. Binding of TDG markedly *decreases* the rate of TMRM labeling either in the membrane or with purified protein in DDM (Fig. 6.6c, left panel).

The observations are certainly consistent with the interpretation that WT LacY in the native bacterial membrane is in a conformation similar to that of the X-ray crystal structures in the absence of ligand (Abramson et al. 2003; Guan et al. 2007; Chaptal et al. 2011; Mirza et al. 2006). The periplasmic side is tightly closed, and an open cavity is present facing the cytoplasm (the inward-facing conformation). Sugar binding leads to closure of the cytoplasmic-facing cavity with opening of a

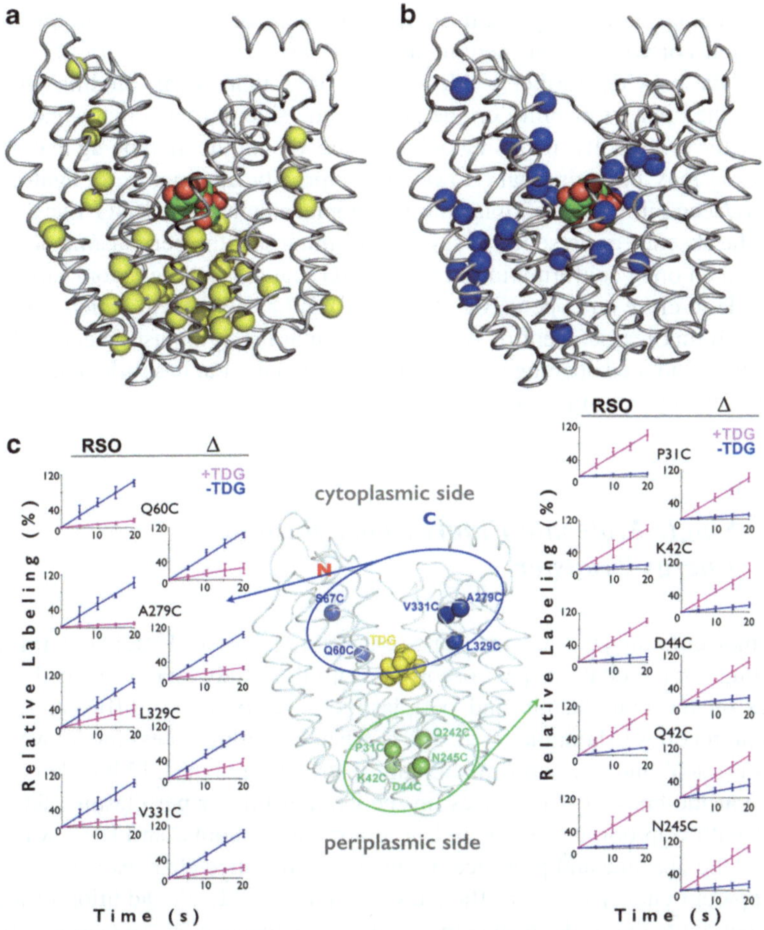

Fig. 6.6 Distribution of Cys replacements that exhibit changes in reactivity with NEM upon galactoside binding. *Spheres* represent Cα atoms of single-Cys replacements on the backbone of the LacY structure in an inward-facing conformation (pdb-ID 1PV7). (**a**) Positions of Cys residues that exhibit a clearly significant increase in reactivity. (**b**) Positions of Cys residues that exhibit a clearly significant decrease in reactivity. (**c**) TMRM labeling of cytoplasmic (*left*) or periplasmic (*right*) single-Cys mutants in RSO membrane vesicles or as purified proteins in DDM (indicated as *triangle*). Labeling of cytoplasmic single-Cys LacY mutants, or periplasmic single-Cys LacY mutants, was performed with TMRM for a given time at 0 °C in the absence of TDG (−TDG, in *blue*) or preincubated for 10 min with TDG prior to addition of TMRM (+TDG, in *pink*). Relative TMRM labeling rates are plotted relative to maximum labeling at 20 s

cavity on the periplasmic side. The average increase in periplasmic TMRM labeling observed in the presence of TDG in RSO vesicles is approximately tenfold, and the average cytoplasmic decrease in the presence of TDG is very similar (approximately ninefold). With purified single-Cys proteins in DDM, comparable averages are approximately sixfold and approximately fivefold. Therefore, the change in

TMRM labeling induced by sugar on opposite faces of LacY is about the same in RSO vesicles or with the purified mutants in DDM.

The data provide further evidence not only that sugar binding markedly increases the open probability on the periplasmic side, but also that sugar binding increases the probability of closing on the inside, the implication being that opening and closing may be reciprocal. However, reciprocity may not be obligatory. The reactivity/accessibility of periplasmic Cys replacements in C154G LacY is very high in the absence or presence of sugar confirming that the periplasmic pathway is arrested in an open conformation in the C154G mutant, but the cytoplasmic cavity is still able to close and open (Nie et al. 2008). It has also been demonstrated (Liu et al. 2010) that replacement of Asp68 with Glu at the cytoplasmic end of helix II blocks sugar-induced opening of the periplasmic cavity, but has little or no effect on closing of the cytoplasmic cavity.

6.7.2 Single Molecule Fluorescence Resonance Energy Transfer

Single molecule fluorescence resonance energy transfer (sm-FRET) determined by alternating laser excitation spectroscopy was used to study ligand-induced distance changes on the cytoplasmic and periplasmic sides of LacY diffusing freely in detergent micelles (Majumdar et al. 2007). The alternating access model was tested with WT LacY and the conformationally restricted mutant C154G. Pairs of Cys residues at the ends of two helices on the cytoplasmic or periplasmic sides were labeled with appropriate donor and acceptor fluorophores, smFRET was determined in the absence and presence of sugar, and distance changes were estimated from apparent energy transfer efficiency E*. With WT LacY, addition of a galactoside, but not a glucoside, results in a decrease in distance on the cytoplasmic side and an increase in distance and in distance distribution on the periplasmic side. In contrast, with the C154G mutant, a more pronounced decrease in distance and in distance distribution is observed on the cytoplasmic side, but there is no change on the periplasmic side. The results are consistent with the alternating access model and indicate that the functional defect in the mutant is due to impaired ligand-induced flexibility on the periplasmic side.

6.7.3 Double Electron–Electron Resonance

Four-pulse double electron–electron resonance (DEER) (Pannier et al. 2000; Jeschke 2002) combined with site-directed spin labeling is highly sensitive to distance change within 20–60 Å range in proteins, which is compatible with the size of LacY. The nitroxide probes are much smaller than the Alexa dyes used in smFRET and quantify distance changes more accurately. Nitroxide-labeled paired-Cys replacements at the ends of transmembrane helices on the cytoplasmic or periplasmic sides of WT LacY and C154G mutant were used for distance measurements (Smirnova et al. 2007). Seven cytoplasmic and fifteen periplasmic pairs were individually tested for distance changes in the presence of galactosidic or non-galactosidic sugars (Smirnova et al. 2007; Madej et al. 2012). Remarkably, binding of galactosidic sugars causes conformational rearrangements on both sides of WT LacY (Fig. 6.7). On the cytoplasmic side, each nitroxide-labeled pair exhibits decreased interspin distances. Distance distributions shifted toward shorter distances (from 4 to 21 Å for different pairs) in the presence of galactoside (compare upper and lower distribution plots for the 73–340 pair in Fig. 6.7, right panel). Conversely, on the periplasmic side, each of three spin-labeled pairs shows increased distances ranging from 4 to 14 Å for WT-based protein (upper and lower distribution plots for the 164–310 pair; Fig. 6.7, left panel). Thus, in the presence of the galactoside specifically, the inward-facing cytoplasmic cavity closes and a cavity opens on the tightly sealed periplasmic side. In the conformationally restricted C154G mutant, sugar-induced closure is observed on the cytoplasmic side, but little or no change occurs on periplasmic side, which is partially open in the absence of sugar. The DEER measurements in conjunction with molecular modeling based on the X-ray structures (Madej et al. 2012) provide strong support for the alternating access model and suggest a structure for the outward-facing conformer of LacY, which agrees remarkably well with the model proposed by Radestock and Forrest (Radestock and Forrest 2011). In addition, the measurements are consistent with the presence of intermediate conformation(s) in LacY (see Fig. 6.7, top panel), since multiple distance distributions have been typically observed for the WT symporter with or without bound sugar (Madej et al. 2012).

6.7.4 Site-Directed Cross-Linking

As discussed above, the side chains essential for sugar recognition and H^+ binding are located at the apex of the cytoplasmic cavity and inaccessible from the outside. On the periplasmic side, helices I/II and VII from the N- and C-terminal six-helix bundles, respectively, participate in sealing the cavity from the outside. Three double-Cys mutants (I40C/N245C, T45C/N245C, and I32C/N245C) located in the interface between helices I/II and VII on the periplasmic side of LacY were constructed with a factor Xa protease cleavage site in the loop between helices IV and V for detection of cross-linking (Zhou et al. 2008). All three pairs cross-link quantitatively and reversibly with flexible homo-bifunctional thiol reagents. Strikingly, relatively short reagents (less than 10 Å) block lactose transport in all three mutants, whereas full or partial activity is observed when cross-linking is mediated by flexible reagents greater than about 10 Å in length. The rigid cross-linking reagent naphthyl dimaleimide (NDM) practically eliminates transport. Therefore,

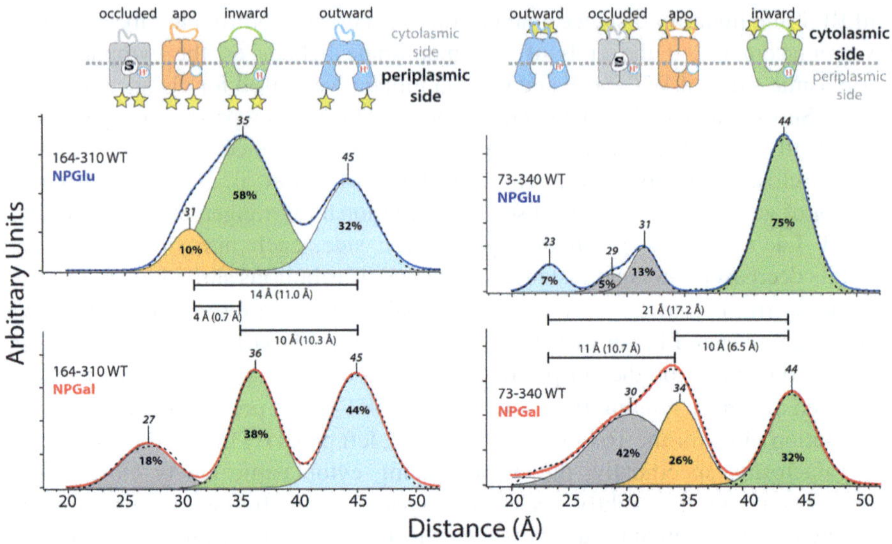

Fig. 6.7 Interspin distance distributions detected with nitroxide-labeled periplasmic Cys pair 164/310 (on *left*) and cytoplasmic Cys pair 73/340 (on *right*; indicated by *yellow stars*) on WT LacY. Distance distributions obtained by Tikhonov regularization of dipolar spectra are shown (*solid blue line* with no bound sugar; *solid red line* with bound sugar). Relative distributions of conformational populations are obtained by multi-Gaussian deconvolution [details are provided in the study by Smirnova et al. (2007) and Madej et al. (2012)]. Peak centers, indicated on *top* of the Gaussian peaks, represent the distance between the nitroxides at these positions (interspin distance). The interspin distances are attributed to a distinct conformation (shown on *top*: *blue*, outward-facing; *gray*, occluded intermediate; *orange*, apo intermediate; *green*, inward-facing; as in Fig. 6.8). The relative area of the respective Gaussian peaks is indicated (%) and represents the fraction of each conformational population in the multi-Gaussian fit (*broken black line*). The distance differences between the peak centers represent distance changes between these positions in different conformations and are indicated as *horizontal bars*. The *numbers in parentheses* are the distance differences between Cα atoms in the structure models of the respective conformations

the transport mechanism of LacY must involve closing, as well as opening, of a large periplasmic cavity. Satisfyingly, 17 Å is the minimum length of cross-linker required for maximum transport activity, a distance similar to that obtained from DEER measurements for opening of periplasmic side in the presence of sugar.

6.7.5 Trp Fluorescence

Fluorescence of intrinsic Trp residues in LacY provides a sensitive tool for functional studies using recently developed approaches that allow direct measurement of sugar binding, as well as global conformational changes in LacY (Smirnova et al. 2006, 2009b, 2011). As a typical membrane transporter, LacY contains multiple Trp residues located predominantly at the interface of the phospholipid bilayer that are important for insertion and stability (Yau et al. 1998; White 2007). The exception is Trp151, which is a component of the sugar-binding site in LacY and the only Trp residue out of six in close proximity to bound galactopyranoside (Figs. 6.5 and 6.8a). The short distance between Trp151 and the sugar is favorable for the formation of a donor–acceptor pair between Trp and nitrophenyl or dansyl derivatives at the anomeric position of the galactosyl moiety resulting in fluorescence resonance energy transfer (FRET). Placing of NPG in the binding site of LacY positions the nitrophenyl moiety in close distance to Trp151, commensurate with the Förster distance for a Trp-nitrophenyl pair (Fig. 6.8a). Indeed, NPG binding to LacY containing all six native Trp residues exhibits FRET from Trp151 (Trp151 $\rightarrow$ NPG FRET) specific for the galactopyranoside, which can be measured in steady-state or stopped-flow experiments as a decrease of Trp fluorescence (Smirnova et al. 2006). Importantly, binding of galactopyranosides that do not absorb UV light (e.g., lactose, TDG, melibiose) has a negligible effect on the fluorescence of WT LacY, and can be used to displace NPG from the sugar-binding site (Fig. 6.8c). Addition of saturating concentrations of TDG in the absence of NPG causes little or no change in the emission spectrum of Trp. However, when TDG is added after incubation with NPG, a significant increase in Trp fluorescence is observed because of the displacement of NPG from the binding site. The replacement of Trp151 with Tyr completely abolishes FRET as a result of NPG binding.

According to the alternating access mechanism, the tightly sealed periplasmic side of LacY observed in the X-ray structures, as well as in the bacterial membrane, must open in order to allow access of sugar to the binding site in the middle of the protein. Ergo the question arises as to whether or not opening of the periplasmic cavity is limiting for substrate binding and possibly transport (Fig. 6.8b). In an effort to address this question, rates of NPG binding were measured by Trp151 $\rightarrow$ NPG FRET using stopped flow with purified LacY in DDM and reconstituted into proteoliposomes (Smirnova et al. 2011). A decrease of Trp fluorescence resulting from Trp151 $\rightarrow$ NPG FRET was observed with LacY in DDM as well as in proteoliposomes, although pre-steady-state kinetics of sugar binding to solubilized LacY is completely different from that observed with the reconstituted protein. In DDM, the rate (k_{obs}) increases linearly with increasing substrate concentration (Fig. 6.8d, filled circles). Thus, in an inward-facing conformation—as is the case with LacY in DDM—the sugar-binding site in LacY is readily accessible from cytoplasmic side, and binding can be described as single-step reversible process:

$$[\mathrm{E}] + [\mathrm{S}] \underset{k_{\mathrm{off}}}{\overset{k_{\mathrm{on}}}{\leftrightarrows}} [\mathrm{ES}]$$

The k_{obs} (reciprocal relaxation time $1/\tau$) for this reaction depends linearly on ligand concentration ($k_{obs} = k_{off} + k_{on}[S]$).

In marked contrast, LacY reconstituted into proteoliposomes binds NPG at a slow rate that is independent of NPG concentration (Fig. 6.8d, filled diamonds). Addition

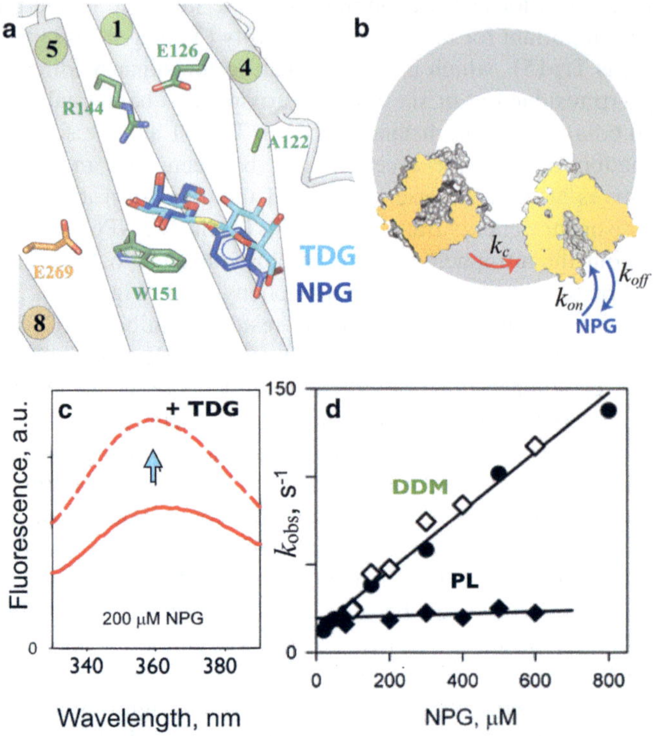

Fig. 6.8 Sugar binding to LacY detected by FRET. (**a**) NPG and TDG are shown in the active site; Trp151 and other significant residues are indicated (color coded as in Fig. 6.3). (**b**) *Cartoon* representing the general postulate that the opening of the periplasmic cavity (k_c) rate-limiting step in sugar binding with reconstituted LacY. (**c**) Trp151-NPG FRET measured as increase of Trp fluorescence measured after displacement of bound NPG with excess TDG. Trp emission spectra of LacY recorded in the presence of NPG before (*red solid line*) and after addition of 10 mM TDG (*red dashed line*). (**d**) Concentration dependence of the rates (k_{obs}) of NPG binding measured by stopped flow in DDM (*filled circle*), reconstituted into proteoliposomes at LPR 10 (*filled diamond*), and after addition of DDM to proteoliposomes at LPR 10 (*open diamond*)

of DDM to the same proteoliposomes dissolves the membrane and restores linear concentration dependence of NPG binding rates identical to that observed with purified protein in DDM (Fig. 6.8d, open diamonds). This finding suggests that in LacY reconstituted into proteoliposomes sugar-binding site is not readily accessible and that the binding reaction includes a limiting step that is likely represented by opening of the periplasmic cavity. LacY reconstituted into proteoliposomes is oriented with the periplasmic side facing the external medium (Herzlinger et al. 1984), as in the native bacterial membrane. Therefore, as shown in the X-ray structures (Abramson et al. 2003; Guan et al. 2007; Mirza et al. 2006), as well as with RSO membrane vesicles (Nie and Kaback 2010), reconstituted LacY exists predominantly in an inward-facing conformation with a sealed periplasmic cavity (Fig. 6.8b), since sugar-binding rates do not increase even at super-saturating NPG concentrations. Evidently, the periplasmic cavity leading to the sugar-binding site must open first, and this slow opening appears to be limiting for substrate binding.

Trp fluorescence is an invaluable probe for detection of conformational changes in LacY (Smirnova et al. 2009b, 2011). Trp fluorescence in proteins is quenched by certain amino acyl side chains such as a protonated His, an amino group, or a Cys residue (Chen and Barkley 1998). Measurements of quenching/unquenching of fluorescence of strategically introduced Trp residues allow differentiation between inward- and outward-facing conformations of LacY and provide strong support for the alternating access mechanism (Smirnova et al. 2009b). Thus, the sugar-binding effect has been tested with mutants, where an additional Trp residue was introduced on either side of WT LacY far from sugar binding site and predicted to be in close proximity to side chains of natural quenchers in either the inward- or outward-facing conformers (Fig. 6.9a). In mutant N245W, Trp is located on the periplasmic side of LacY (helix VII) where fluorescence is quenched by native His35 (helix I) in the inward-facing conformation. The fluorescence intensity of mutant N245W increases as a result of sugar binding (Fig. 6.9b) due to *unquenching* caused by an increase in distance between Trp245 and quencher His35. The opposite effect is observed in mutant F140W/F334H where Trp140 (helix V) and His334 (helix X) are located across the wide-open cytoplasmic cavity. Sugar binding leads to *quenching* of Trp fluorescence (Fig. 6.9d) due to direct collision between Trp140 and His334, which results from the closing of cytoplasmic cavity. The pH dependency of Trp245 *unquenching* (Fig. 6.9c) and Trp140 *quenching* (Fig. 6.9e) exhibits a pK_a of ~8, typical for a His side chain interacting with an aromatic group (Loewenthal et al. 1992), thereby confirming His as the quencher in both mutants. The results provide yet another strong, independent line of evidence for the alternating access mechanism and demonstrate that the methodology described provides a sensitive probe to measure conformational changes.

The N245W LacY mutant with Trp introduced on the periplasmic side has been used for a direct comparison between rates of opening of the periplasmic pathway and sugar-binding rates with LacY reconstituted into proteoliposomes by following either unquenching of Trp fluorescence, or Trp151 → NPG FRET (Kaback et al. 2011). The kinetics of sugar binding to N245W mutant is similar to that described previously for WT LacY. In DDM, the rate of sugar binding exhibits a linear dependence on sugar concentration. However, the mutant reconstituted into proteoliposomes binds sugar with k_{obs} of ~60 s^{-1}, and the rate is independent of NPG concentration. In order to measure rates of opening of the periplasmic cavity with other sugars, four galactosidic sugars that do not change Trp fluorescence—TDG, melibiose, octyl-α-D-galactoside, or methyl-α-D-galactoside—were used (Fig. 6.9f). Rates of unquenching of Trp fluorescence in mutant N245W due to binding of these galactosides to LacY in DDM are very similar to the NPG binding rates observed in proteoliposomes. Moreover, the rates of both processes are essentially independent of sugar concentration (Fig. 6.9f, compare filled diamonds with open symbols fitted with dashed line). The similarity of the kinetics of two processes suggests that opening of the periplasmic cavity is likely the limiting step for sugar access to the binding site in LacY (Fig. 6.8b).

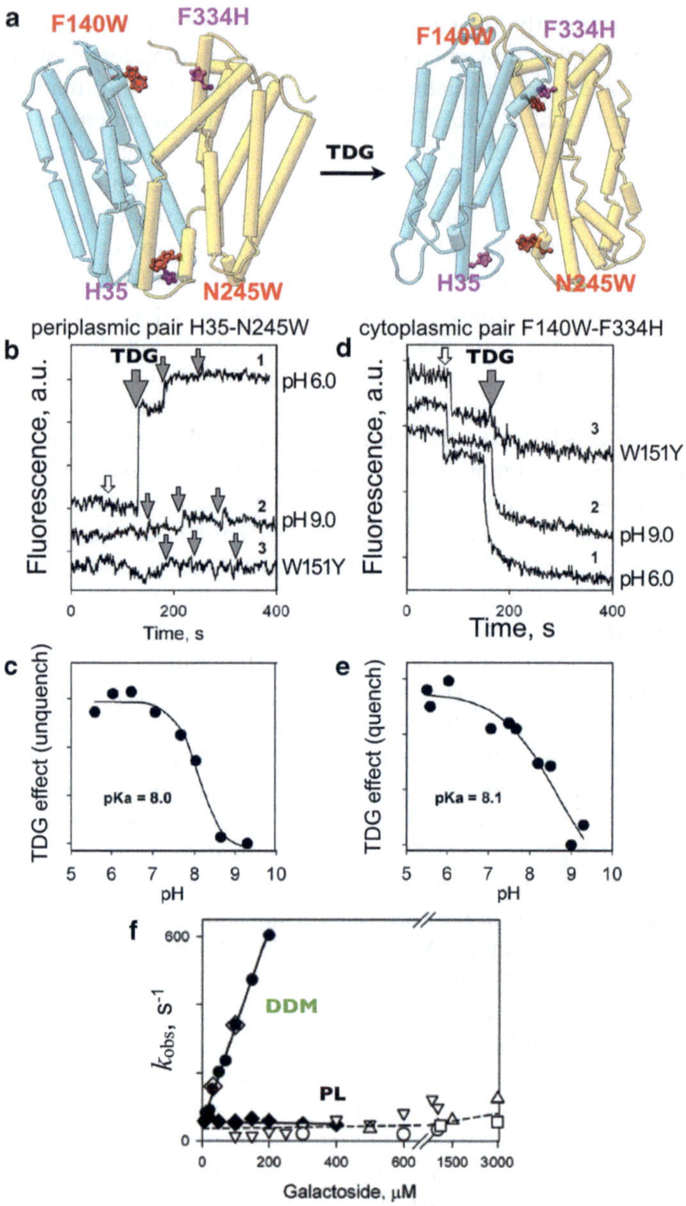

Fig. 6.9 Intrinsic fluorescence quenching. (**a**) Side view of the structure model in the inward-facing conformation is shown on the *left*, and a model in the outward-facing conformation is shown on the *right*. Pairs of residues selected for Trp substitutions are shown on the *rod model* of LacY. The H35–N245W pair (periplasmic) and F140W–F334H (cytoplasmic) pairs are shown as *red* (Trp) and *magenta* (His) balls-and-stick model. The *arrow* indicates the conformational change resulting from sugar binding. (**b**) TDG effect on fluorescence of Trp introduced on the periplasmic side. Sequential additions of 6 μL of buffer (*open arrow*) and 6 μL aliquots of 1.8 M

A turnover number of 16–20 s^{-1} was estimated for $\Delta\tilde{\mu}_{H^+}$-driven, lactose/H^+ symport by WT LacY in RSO membrane vesicles and with the purified protein reconstituted into proteoliposomes (Viitanen et al. 1984). Very much the same rate (21 ± 4 s^{-1}) is observed for sugar binding to WT LacY in proteoliposomes (Fig. 6.8d). Therefore, opening of the periplasmic pathway may also be a limiting step in the symport mechanism catalyzed by WT LacY.

6.8 An Overall Mechanism for Coupling in LacY

The following description provides an overall symport mechanism for LacY:

1. Lactose/H^+ symport in the uphill or downhill mode is precisely the same reaction. The difference is in the rate-limiting step. For downhill symport, it is deprotonation; for uphill transport, where there is a driving force on the H^+, deprotonation is no longer limiting, and either dissociation of sugar or a conformational change becomes limiting.
2. Sugar binding and dissociation—not $\Delta\tilde{\mu}_{H^+}$—are the driving force for alternating access.
3. LacY must be protonated to bind sugar (the pK_a for sugar binding is ~10.5).
4. Sugar binding involves induced fit causing transition to an occluded state, which undergoes alternating access.
5. Sugar dissociates first, releasing the energy consumed by induced-fit binding.
6. Upon sugar dissociation, there is a conformational change that causes Arg302 (helix IX) to approximate Glu325 leading to deprotonation of LacY.

Fig. 6.9 (continued) TDG (*black arrows*) to the mutant W151Y/N245W at pH 6.0 (trace 1), or at pH 9.0 (trace 2), and to the control mutant W151Y at pH 6.0 (trace 3). (**c**) Trp fluorescence unquench for W151Y/N245W mutant as a function of pH. TDG (15 mM) was added to protein solution at the pH indicated fluorescence change was corrected for dilution and plotted versus pH. *Solid line* is data fit with an estimated pK_a of ~8.0. (**d**) TDG effect on Trp fluorescence of cytoplasmic pair F140W/F334H at pH 5.5 (trace 1), pH 8.5 (trace 2), and pH 9.0 (trace 3). Additions of 20 μL of buffer (*open arrow*) or 20 μL of 1.5 M TDG (*black arrow*). (**e**) Trp fluorescence quench for mutant F140W/F334H as a function of pH. Effect of 15 mM TDG at indicated pH is plotted as in (**b**). *Solid line* is a data fit with a pK_a of 8.1. (**f**) Concentration dependence of sugar-binding rates and rates of periplasmic pathway opening. Binding rates (k_{obs}) estimated for the purified mutant in DDM (*filled circle*), reconstituted in proteoliposomes (*filled diamond*), and after proteoliposomes had been dissolved in DDM (*open diamond*). The reconstituted mutant binds NPG with a k_{obs} of 56 ± 7 s^{-1}. Rates of unquenching of Trp245 fluorescence resulting from opening of the periplasmic cavity after sugar binding were measured in a DDM solution and are presented: TDG (*inverted triangle*), melibiose (*open triangle*), octyl α-D-galactoside (*open circle*), and methyl α-D-galactoside (*open square*). The rate of opening of the periplasmic cavity in DDM is 50–100 s^{-1} at the saturating concentrations of all four galactosides tested

6.8.1 Transport Intermediates

Based on interspin-distance measurements from DEER and SDA, the presence of occluded intermediates was suggested (Smirnova et al. 2007; Jiang et al. 2012). Further analysis of interspin distances combined with comparative modeling suggests the existence of two structurally distinct intermediates in which the central cavity is occluded (Madej et al. 2012)—a sugar-free apo intermediate and a protonated, sugar-bound occluded intermediate (Fig. 6.10a). Moreover, the analysis of the conformational distributions of the LacY molecule in the presence and absence of substrate sheds light on the relative energy levels of the respective conformations. A hypothetical energy landscape for the transport cycle is suggested in which the occluded intermediates accommodate a higher energy level relative to a relatively facile energy profile of the open conformers (Fig. 6.10b). Thus, LacY operates like an enzyme, except that the intermediate is a conformer(s) of the symporter rather than a transition state intermediate of the substrate (Mitchell 1967). In this view, the intermediate conformation is induced by sugar binding and binding leads to lowering of the activation energy barriers for the transition between the inward-facing and outward-facing conformers. Such energetic disposition, where high-energy barriers separate the inward- and outward-facing conformers, indicates that the transport reaction is thermodynamically driven, but controlled kinetically.

6.9 Mix-and-Match Evolution in the MFS

Homology of primary amino acid sequences reflects the evolutionary process and therefore provides a guide to structure, mechanism, and function. Proteins that are related by common descent are expected to exhibit homologous structures and functions to a degree proportional to the degree of their sequence similarity. This principle provides the motivation to define protein phylogenetic relationships and interrelate families when possible.

A principal difficulty with MFS proteins is their low sequence homology. Despite the conserved fold motifs and in some cases overlapping function, the sequence identity ranges around 12–18 %. However, taking into account that most MFS members have a hydrophobic amino acid content of ~60–70 %, it is likely that the detected sequence homology results from overlapping hydrophobic residues, which are not directly involved in the mechanism in most cases. Moreover, the inherent symmetry of the proteins with regard to the helix kinks and bends provides further promiscuous overlaps of residues promoting irregularities of helices (e.g., Pro or Gly) (Madej et al. 2012). However, regions of functional similarity (e.g., substrate- and/or H^+-binding sites) align only for very closely related MFS transporters (Kasho et al. 2006; Sugihara et al. 2011); in less

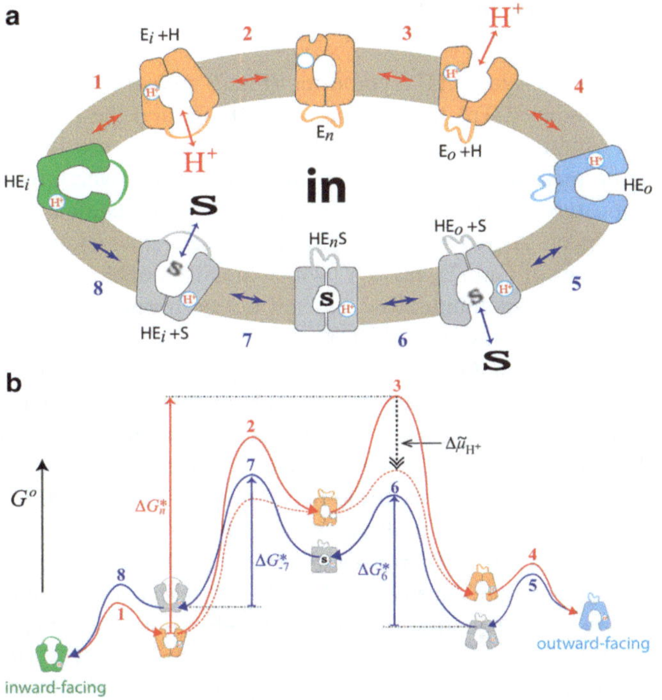

Fig. 6.10 Transport cycle of LacY. (**a**) Overview of the postulated steps in the transport model. Inward-facing (*green*) and outward-facing (*blue*) conformations are separated by the apo intermediate conformational cluster (*orange*) or by the occluded intermediate conformational cluster (*gray*). Steps are numbered consecutively: (1) conformational transition allowing the deprotonation of Glu325; (2) formation of the deprotonated apo intermediate; (3) reprotonation of LacY and opening of the periplasmic cavity; (4 and 5) reorientation of helices, where LacY passes through a low-energy conformation [see for details Madej et al. (2012)]; (6) binding of sugar and induced fit to the occluded intermediate; (7) opening of the cytoplasmic cavity and release of sugar; and (8) conformational transition to inward-facing conformation. All steps are reversible (indicated by *double-headed arrows*). (**b**) Hypothetical energy profile for the transport cycle. Conformational states in (**a**) are translated into relative energy states (indicated by the *icons* of conformations defined in (**a**) with the cytoplasmic side of LacY facing up) based on their occurrence as a conformational fraction in interspin distance distribution. The scheme can be read in cycle following the *arrowheads*. The *red line* represents the transition between steps 1 and 4 for opening of the empty carrier to the outward-facing conformation. The *blue line* corresponds to steps 5–8 for sugar transport from outside to inside (and to the exchange reaction when steps 5–8 operate in a reverse manner). The free energy of the putative rate-determining step in the absence of $\Delta\tilde{\mu}_{H^+}$ (opening of the periplasmic cavity) is indicated by the *vertical red arrow* (ΔG_n*). The hypothetical effect of an imposed $\Delta\tilde{\mu}_{H^+}$ is shown as a *dashed black vertical arrow*, and the resulting energy profile is shown by the *dashed red lines*. The energy of sugar binding on the periplasmic (ΔG_6*) and cytoplasmic (ΔG_{-7}*) sides is suggested to be equivalent from either side of the membrane (indicated by the *vertical blue arrows*)

homologous transporters the functional counterparts are often located in distant parts of the protein. This section addresses the specific challenges with the MFS to compare or detect functionally homologous positions.

6.9.1 Comparative Sequence–Function Analysis

Mutagenesis of the L-fucose/H^+ symporter (FucP) reveals that Asp46 mutants exhibit L-fucose-induced Trp fluorescence quenching, and the mutants bind with essentially WT affinity (Sugihara et al. 2012), a conclusion consistent with isothermal calorimetry measurements (Dang et al. 2010). Moreover, although Asp46 mutants do not catalyze L-fucose/H^+ symport, the D46A mutant catalyzes counterflow. In this respect, Asp46 mutants exhibit the main characteristics of neutral replacement mutants of Glu325 in LacY, which are also unable to catalyze lactose/H^+ symport, but catalyze counterflow, as well as equilibrium exchange. This and additional evidence (Kaback et al. 2001; Garcia-Celma et al. 2009, 2010; Guan and Kaback 2006) indicate that Glu325 is an essential player in the symport mechanism of LacY and directly involved in deprotonation. However, Glu325 in the C-terminal six-helix bundle of LacY is located at a different position from Asp46 in FucP, which is in the N-terminal six-helix bundle. This observation elicited the feasibility of alternative arrangements of the triple-helix motifs, since the MFS transporters may have arisen by intragenic multiplication and fusion of the structural motifs (Radestock and Forrest 2011; Hvorup and Saier 2002). To shed light on the location of functionally significant residues in LacY and FucP with respect to each other, the four subdomains corresponding to the triple-helix inverted symmetry motifs (Radestock and Forrest 2011) (repeat-A, H1–H3; repeat-B, H4–H6; repeat-C, H7–H9; and repeat-D, H10–H12) were superposed. As a result of the strong symmetry within each transporter, all superimpositions display similar qualities with root mean square deviations (rmsd) ranging from ~1.8 Å to ~2.6 Å. The respective sequence alignments of the triple-helix motifs are all in the same range of (~25 % similarity) and do not provide any preference for a functional parity. This is likely a result of the transporters disposition as α-helical transmembrane protein; it consists of about 65 % of hydrophobic residues, and a sequence alignment is biased by nonspecific hydrophobic overlap. However, if the conservation of residues in FucP is analyzed with respect to mutations in LacY that cause greater than 50 % inhibition of the lactose transport rate (Frillingos et al. 1998), only one alignment combination exhibits a remarkable conservation of these functionally significant residues (51 % similarity and 27 % identity). Importantly, the consecutive order and the orientation of the superimposed triple-helix motifs in LacY and FucP are inverted with respect to each other (Fig. 6.11a). This superposition was tested based on the phenotype of 34 mutants in FucP (Fig. 6.11b); the transport activity of the corresponding positions in LacY and FucP reveals equivalent susceptibility to mutations (Madej et al. 2013).

Homology of primary amino acid sequences in spatially distant motifs of the protein may be a consequence of evolutionary relationships between them. It appears that the evolutionary origins of these transporters are triple-helix bundles, which correspond to the inverted structural symmetry motifs (Radestock and Forrest 2011), fused in tandem by intragenic duplication. The observations are consistent with the notion that the MFS evolved by intragenic duplication of helix

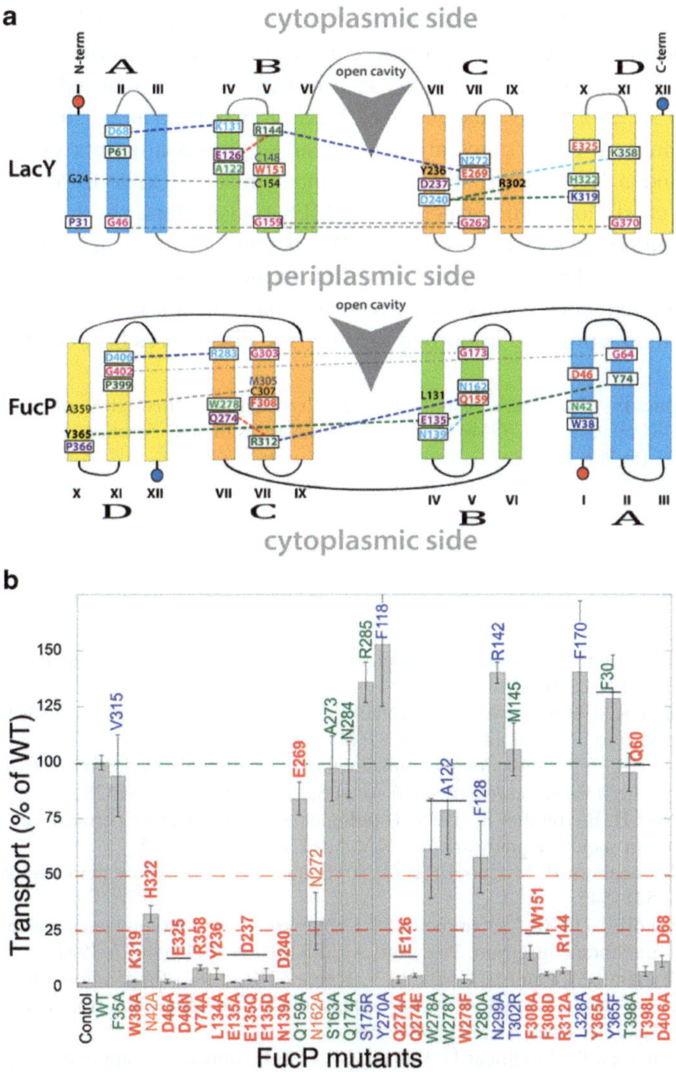

Fig. 6.11 Alignment of LacY and FucP. (**a**) Schematic representation of the superimposition of the triple-helix motifs. Helix triplets are numbered from A to D according to the sequence order. Overlapping side-chain positions are shown in the *same color* for corresponding helices in LacY and FucP. Conserved, significant side-chains are *boxed*. Contacts between side chains are indicated as *broken lines*: *blue*, conserved charged pairs; *red*, conserved intratriplet interaction of Arg in the sugar-binding site with a polar side chain; *green*, conserved coordination of an acidic side chain with two alkaline residues; *gray*, other prominent interactions (*dashed dotted line*, putative interactions between Gly residues in FucP). (**b**) In vivo transport: cell-based active transport assay. Transport activities of the FucP mutants are means of four measurements and are normalized to WT. Control refers to transport by fucP-deficient *E. coli* cells transformed with empty vector. *Error bars* represent SDs. The residues are colored according to residual activity of WT: 0–25 %, *red*; 25–50 %, *orange*; 50–100 %, *green*; >100 %, *blue*. On top of given histograms, the activities of the corresponding mutated residues in LacY are indicated for comparison

triplets (Hvorup and Saier 2002). However, fusion of the triple-helix bundles likely occurred multiple times during MFS evolution, and, simultaneously, mutations may have occurred with selective effects on transport function that led to a mix-and-match type of process. Consequently, MFS transporters may share only weak sequence similarities in conventional, linear sequence aliments despite similar functions and a likely alternating access mechanism of action. With a growing number of crystallographic structures of MFS transporters, the location of functionally important residues within the MFS becomes increasingly important in a structural context. This concept of evolution in the MFS may allow a more useful annotation of sequence motifs than a conventional, linear sequence aliment. It may also reveal interesting and important mechanistic differences between transporters.

Acknowledgments This work was supported by National Institutes of Health Grants DK51131, DK069463, and GM073210, as well as National Science Foundation Grant MCB-1129551 (to H.R.K.).

References

Abramson J et al (2003) Structure and mechanism of the lactose permease of Escherichia coli. Science 301(5633):610–615

Bibi E, Kaback HR (1990) In vivo expression of the *lacY* gene in two segments leads to functional *lac* permease. Proc Natl Acad Sci USA 87(11):4325–4329

Bibi E, Kaback HR (1992) Functional complementation of internal deletion mutants in the lactose permease of *Escherichia coli*. Proc Natl Acad Sci USA 89(5):1524–1528

Bieseler B, Prinz H, Beyreuther K (1985) Topological studies of lactose permease of *Escherichia coli* by protein sequence analysis. Ann N Y Acad Sci 456:309–325

Buchel DE, Gronenborn B, Muller-Hill B (1980) Sequence of the lactose permease gene. Nature 283(5747):541–545

Calamia J, Manoil C (1990) *Lac* permease of *Escherichia coli*: topology and sequence elements promoting membrane insertion. Proc Natl Acad Sci USA 87(13):4937–4941

Carrasco N, Tahara SM, Patel L, Goldkorn T, Kaback HR (1982) Preparation, characterization, and properties of monoclonal antibodies against the *lac* carrier protein from *Escherichia coli*. Proc Natl Acad Sci USA 79(22):6894–6898

Carrasco N, Viitanen P, Herzlinger D, Kaback HR (1984) Monoclonal antibodies against the *lac* carrier protein from *Escherichia coli*. 1. Functional studies. Biochemistry 23(16):3681–3687

Carrasco N et al (1989) Characterization of site-directed mutants in the lac permease of Escherichia coli. 2. Glutamate-325 replacements. Biochemistry 28(6):2533–2539

Chaptal V et al (2011) Crystal structure of lactose permease in complex with an affinity inactivator yields unique insight into sugar recognition. Proc Natl Acad Sci USA 108:9361–9366

Chen Y, Barkley MD (1998) Toward understanding tryptophan fluorescence in proteins. Biochemistry 37(28):9976–9982

Cohen GN, Rickenberg HV (1955) Etude directe de la fixation d'un inducteur de la b-galactosidase par les cellules d'*Escherichia coli*. Compt Rendu 240:466–468

Costello MJ et al (1987) Purified *lac* permease and cytochrome o oxidase are functional as monomers. J Biol Chem 262(35):17072–17082

Dang S et al (2010) Structure of a fucose transporter in an outward-open conformation. Nature 467 (7316):734–738

Dornmair K, Corni AF, Wright JK, Jähnig F (1985) The size of the lactose permease derived from rotational diffusion measurements. EMBO J 4(13A):3633–3638
Ermolova N, Guan L, Kaback HR (2003) Intermolecular thiol cross-linking via loops in the lactose permease of *Escherichia coli*. Proc Natl Acad Sci USA 100(18):10187–10192
Ermolova NV, Smirnova IN, Kasho VN, Kaback HR (2005) Interhelical packing modulates conformational flexibility in the lactose permease of *Escherichia coli*. Biochemistry 44 (21):7669–7677
Foster DL, Garcia ML, Newman MJ, Patel L, Kaback HR (1982) Lactose-proton symport by purified *lac* carrier protein. Biochemistry 21(22):5634–5638
Foster DL, Boublik M, Kaback HR (1983) Structure of the *lac* carrier protein of *Escherichia coli*. J Biol Chem 258(1):31–34
Fox CF, Kennedy EP (1965) Specific labeling and partial purification of the M protein, a component of the β-galactoside transport system of Escherichia coli. Proc Natl Acad Sci USA 54:891–899
Franco PJ, Brooker RJ (1994) Functional roles of Glu-269 and Glu-325 within the lactose permease of *Escherichia coli*. J Biol Chem 269(10):7379–7386
Frillingos S, Gonzalez A, Kaback HR (1997) Cysteine-scanning mutagenesis of helix IV and the adjoining loops in the lactose permease of Escherichia coli: Glu126 and Arg144 are essential off. Biochemistry 36(47):14284–14290
Frillingos S, Sahin-Toth M, Wu J, Kaback HR (1998) Cys-scanning mutagenesis: a novel approach to structure function relationships in polytopic membrane proteins. FASEB J 12 (13):1281–1299
Gaiko O, Bazzone A, Fendler K, Kaback HR (2013) Electrophysiological characterization of uncoupled mutants of LacY. Biochemistry 52(46):8261–8266
Garcia-Celma JJ, Smirnova IN, Kaback HR, Fendler K (2009) Electrophysiological characterization of LacY. Proc Natl Acad Sci USA 106(18):7373–7378
Garcia-Celma JJ, Ploch J, Smirnova I, Kaback HR, Fendler K (2010) Delineating electrogenic reactions during lactose/H + symport. Biochemistry 49(29):6115–6121
Guan L, Kaback HR (2004) Binding affinity of lactose permease is not altered by the H + electrochemical gradient. Proc Natl Acad Sci USA 101(33):12148–12152
Guan L, Kaback HR (2006) Lessons from lactose permease. Annu Rev Biophys Biomol Struct 35:67–91
Guan L, Kaback HR (2007) Site-directed alkylation of cysteine to test solvent accessibility of membrane proteins. Nat Protoc 2(8):2012–2017
Guan L, Murphy FD, Kaback HR (2002a) Surface-exposed positions in the transmembrane helices of the lactose permease of Escherichia coli determined by intermolecular thiol cross-linking. Proc Natl Acad Sci USA 99(6):3475–3480
Guan L, Sahin-Toth M, Kaback HR (2002b) Changing the lactose permease of Escherichia coli into a galactose-specific symporter. Proc Natl Acad Sci USA 99(10):6613–6618
Guan L, Hu Y, Kaback HR (2003) Aromatic stacking in the sugar binding site of the lactose permease. Biochemistry 42(6):1377–1382
Guan L, Smirnova IN, Verner G, Nagamori S, Kaback HR (2006) Manipulating phospholipids for crystallization of a membrane transport protein. Proc Natl Acad Sci USA 103(6):1723–1726
Guan L, Mirza O, Verner G, Iwata S, Kaback HR (2007) Structural determination of wild-type lactose permease. Proc Natl Acad Sci USA 104(39):15294–15298
He M, Kaback HR (1997) Interaction between residues Glu269 (Helix VIII) and His322 (Helix X) of the lactose permease of *Escherichia coli* is essential for substrate binding. Biochemistry 36:13688–13692
He MM, Voss J, Hubbell WL, Kaback HR (1995a) Use of designed metal-binding sites to study helix proximity in the lactose permease of Escherichia coli. 2. Proximity of helix IX (Arg302) with helix X (His322 and Glu325). Biochemistry 34(48):15667–15670

He MM, Voss J, Hubbell WL, Kaback HR (1995b) Use of designed metal-binding sites to study helix proximity in the lactose permease of Escherichia coli. 1. Proximity of helix VII (Asp237 and Asp240) with helices X (Lys319) and XI (Lys358). Biochemistry 34(48):15661–15666

Herzlinger D, Viitanen P, Carrasco N, Kaback HR (1984) Monoclonal antibodies against the *lac* carrier protein from *Escherichia coli.* 2. Binding studies with membrane vesicles and proteoliposomes reconstituted with purified *lac* carrier protein. Biochemistry 23 (16):3688–3693

Huang Y, Lemieux MJ, Song J, Auer M, Wang DN (2003) Structure and mechanism of the glycerol-3-phosphate transporter from Escherichia coli. Science 301(5633):616–620

Hvorup RN, Saier MH Jr (2002) Sequence similarity between the channel-forming domains of voltage-gated ion channel proteins and the C-terminal domains of secondary carriers of the major facilitator superfamily. Microbiology 148(Pt 12):3760–3762

Jenks WP (1969) Catalysis in chemistry and enzymology. McGraw-Hill, New York, NY

Jeschke G (2002) Distance measurements in the nanometer range by pulse EPR. Chemphyschem 3 (11):927–932

Jiang X, Nie Y, Kaback HR (2011) Site-directed alkylation studies with LacY provide evidence for the alternating access model of transport. Biochemistry 50(10):1634–1640

Jiang X et al (2012) Evidence for an intermediate conformational state of LacY. Proc Natl Acad Sci USA 109(12):E698–E704

Jung K, Jung H, Wu J, Privé GG, Kaback HR (1993) Use of site-directed fluorescence labeling to study proximity relationships in the lactose permease of *Escherichia coli.* Biochemistry 32:12273–12278

Jung K, Jung H, Kaback HR (1994a) Dynamics of lactose permease of Escherichia coli determined by site-directed fluorescence labeling. Biochemistry 33(13):3980–3985

Jung H, Jung K, Kaback HR (1994b) Cysteine 148 in the lactose permease of *Escherichia coli* is a component of a substrate binding site. I. Site-directed mutagenesis studies. Biochemistry 33:12160–12165

Jung K, Voss J, He M, Hubbell WL, Kaback HR (1995) Engineering a metal binding site within a polytopic membrane protein, the lactose permease of Escherichia coli. Biochemistry 34 (19):6272–6277

Kaback HR (1971) Bacterial membranes. In: Kaplan NP, Jakoby WB, Colowick NP (eds) Methods in enzymol, vol XXII. Elsevier, New York, pp 99–120

Kaback HR (1987) Permease on parade: application of site-directed mutagenesis to ion-gradient driven active transport. Bioessays 7(6):261–265

Kaback HR, Barnes EM Jr (1971) Mechanisms of active transport in isolated membrane vesicles. II. The mechanism of energy coupling between D-lactic dehydrogenase and β-galactoside transport in membrane preparations from Escherichia coli. J Biol Chem 246(17):5523–5531

Kaback HR, Sahin-Toth M, Weinglass AB (2001) The kamikaze approach to membrane transport. Nat Rev Mol Cell Biol 2(8):610–620

Kaback HR et al (2007) Site-directed alkylation and the alternating access model for LacY. Proc Natl Acad Sci USA 104(2):491–494

Kaback HR, Smirnova I, Kasho V, Nie Y, Zhou Y (2011) The alternating access transport mechanism in LacY. J Membr Biol 239(1–2):85–93

Kaczorowski GJ, Kaback HR (1979) Mechanism of lactose translocation in membrane vesicles from *Escherichia coli.* 1. Effect of pH on efflux, exchange, and counterflow. Biochemistry 18 (17):3691–3697

Kaczorowski GJ, Robertson DE, Kaback HR (1979) Mechanism of lactose translocation in membrane vesicles from *Escherichia coli.* 2. Effect of imposed delta psi, delta pH, and delta mu H+. Biochemistry 18(17):3697–3704

Kaczorowski GJ, Leblanc G, Kaback HR (1980) Specific labeling of the *lac* carrier protein in membrane vesicles of *Escherichia coli* by a photoaffinity reagent. Proc Natl Acad Sci USA 77 (11):6319–6323

Kasho VN, Smirnova IN, Kaback HR (2006) Sequence alignment and homology threading reveals prokaryotic and eukaryotic proteins similar to lactose permease. J Mol Biol 358(4):1060–1070

Kumar H et al (2014) Structure of sugar-bound LacY. Proc Natl Acad Sci USA 111(5):1784–1788

Kwaw I, Zen KC, Hu Y, Kaback HR (2001) Site-directed sulfhydryl labeling of the lactose permease of *Escherichia coli*: helices IV and V that contain the major determinants for substrate binding. Biochemistry 40(35):10491–10499

le Coutre J, Narasimhan LR, Patel CK, Kaback HR (1997) The lipid bilayer determines helical tilt angle and function in lactose permease of *Escherichia coli*. Proc Natl Acad Sci USA 94 (19):10167–10171

le Coutre J, Kaback HR, Patel CK, Heginbotham L, Miller C (1998) Fourier transform infrared spectroscopy reveals a rigid alpha-helical assembly for the tetrameric Streptomyces lividans K + channel. Proc Natl Acad Sci USA 95(11):6114–6117

Lemieux MJ et al (2003) Three-dimensional crystallization of the *Escherichia coli* glycerol-3-phosphate transporter: a member of the major facilitator superfamily. Protein Sci 12 (12):2748–2756

Liu Z, Madej MG, Kaback HR (2010) Helix dynamics in LacY: helices II and IV. J Mol Biol 396 (3):617–626

Loewenthal R, Sancho J, Fersht AR (1992) Histidine-aromatic interactions in barnase. Elevation of histidine pKa and contribution to protein stability. J Mol Biol 224(3):759–770

Madej MG, Soro SN, Kaback HR (2012) Apo-intermediate in the transport cycle of lactose permease (LacY). Proc Natl Acad Sci USA 109(44):E2970–E2978

Madej MG, Dang S, Yan N, Kaback HR (2013) Evolutionary mix-and-match with MFS transporters. Proc Natl Acad Sci USA 110(15):5870–5874

Majumdar DS et al (2007) Single-molecule FRET reveals sugar-induced conformational dynamics in LacY. Proc Natl Acad Sci USA 104(31):12640–12645

Matsushita K, Patel L, Gennis RB, Kaback HR (1983) Reconstitution of active transport in proteoliposomes containing cytochrome o oxidase and *lac* carrier protein purified from *Escherichia coli*. Proc Natl Acad Sci USA 80(16):4889–4893

McKenna E, Hardy D, Kaback HR (1992) Evidence that the final turn of the last transmembrane helix in the lactose permease is required for folding. J Biol Chem 267(10):6471–6474

Mirza O, Guan L, Verner G, Iwata S, Kaback HR (2006) Structural evidence for induced fit and a mechanism for sugar/H(+) symport in LacY. EMBO J 25:1177–1183

Mitchell P (1963) Molecule, group and electron transport through natural membranes. Biochem Soc Symp 22:142–168

Mitchell P (1967) Translocations through natural membranes. Adv Enzymol Relat Areas Mol Biol 29:33–87

Mitchell P (1968) Chemiosmotic coupling and energy transduction. Glynn Research Ltd, Bodmin, England

Müller-Hill B (1996) The lac Operon: a short history of a genetic paradigm. Walter de Gruyter, Berlin, New York

Newman MJ, Wilson TH (1980) Solubilization and reconstitution of the lactose transport system from Escherichia coli. J Biol Chem 255(22):10583–10586

Newman MJ, Foster DL, Wilson TH, Kaback HR (1981) Purification and reconstitution of functional lactose carrier from *Escherichia coli*. J Biol Chem 256(22):11804–11808

Newstead S et al (2011) Crystal structure of a prokaryotic homologue of the mammalian oligopeptide-proton symporters, PepT1 and PepT2. EMBO J 30(2):417–426

Nie Y, Kaback HR (2010) Sugar binding induces the same global conformational change in purified LacY as in the native bacterial membrane. Proc Natl Acad Sci USA 107 (21):9903–9908

Nie Y, Ermolova N, Kaback HR (2007) Site-directed alkylation of LacY: effect of the proton electrochemical gradient. J Mol Biol 374(2):356–364

Nie Y, Sabetfard FE, Kaback HR (2008) The Cys154→Gly mutation in LacY causes constitutive opening of the hydrophilic periplasmic pathway. J Mol Biol 379(4):695–703

Nie Y, Zhou Y, Kaback HR (2009) Clogging the periplasmic pathway in LacY. Biochemistry 48 (4):738–743

Padan E, Sarkar HK, Viitanen PV, Poonian MS, Kaback HR (1985) Site-specific mutagenesis of histidine residues in the *lac* permease of *Escherichia coli*. Proc Natl Acad Sci USA 82:6765–6768

Pannier M, Veit S, Godt A, Jeschke G, Spiess HW (2000) Dead-time free measurement of dipole-dipole interactions between electron spins. J Magn Reson 142(2):331–340

Patzlaff JS, Moeller JA, Barry BA, Brooker RJ (1998) Fourier transform infrared analysis of purified lactose permease: a monodisperse lactose permease preparation is stably folded, alpha-helical, and highly accessible to deuterium exchange. Biochemistry 37(44):15363–15375

Pedersen BP et al (2013) Crystal structure of a eukaryotic phosphate transporter. Nature 496 (7446):533–536

Prive GG, Kaback HR (1996) Engineering the lac permease for purification and crystallization. J Bioenerg Biomembr 28(1):29–34

Puttner IB, Kaback HR (1988) lac permease of Escherichia coli containing a single histidine residue is fully functional. Proc Natl Acad Sci USA 85(5):1467–1471

Puttner IB, Sarkar HK, Poonian MS, Kaback HR (1986) lac permease of Escherichia coli: histidine-205 and histidine-322 play different roles in lactose/H + symport. Biochemistry 25 (16):4483–4485

Radestock S, Forrest LR (2011) The alternating-access mechanism of MFS transporters arises from inverted-topology repeats. J Mol Biol 407(5):698–715

Ramos S, Kaback HR (1977a) The relationship between the electrochemical proton gradient and active transport in Escherichia coli membrane vesicles. Biochemistry 16(5):854–859

Ramos S, Kaback HR (1977b) The electrochemical proton gradient in Escherichia coli membrane vesicles. Biochemistry 16(5):848–854

Ramos S, Schuldiner S, Kaback HR (1976) The electrochemical gradient of protons and its relationship to active transport in Escherichia coli membrane vesicles. Proc Natl Acad Sci USA 73(6):1892–1896

Robertson DE, Kaczorowski GJ, Garcia ML, Kaback HR (1980) Active transport in membrane vesicles from *Escherichia coli*: the electrochemical proton gradient alters the distribution of the *lac* carrier between two different kinetic states. Biochemistry 19(25):5692–5702

Roepe PD, Kaback HR (1989) Site-directed mutagenesis of tyrosine residues in the lac permease of Escherichia coli. Biochemistry 28(14):6127–6132

Sahin-Toth M, Kaback HR (1993) Properties of interacting aspartic acid and lysine residues in the lactose permease of Escherichia coli. Biochemistry 32(38):10027–10035

Sahin-Tóth M, Kaback HR (2001) Arg-302 facilitates deprotonation of Glu-325 in the transport mechanism of the lactose permease from *Escherichia coli*. Proc Natl Acad Sci USA 98 (11):6068–6073

Sahin-Tóth M, Lawrence MC, Kaback HR (1994) Properties of permease dimer, a fusion protein containing two lactose permease molecules from *Escherichia coli*. Proc Natl Acad Sci USA 91 (12):5421–5425

Sahin-Tóth M et al (1999) Characterization of Glu126 and Arg144, two residues that are indispensable for substrate binding in the lactose permease of *Escherichia coli*. Biochemistry 38 (2):813–819

Sahin-Tóth M, Karlin A, Kaback HR (2000) Unraveling the mechanism of the lactose permease of Escherichia coli. Proc Natl Acad Sci USA 97(20):10729–10732

Sayeed WM, Baenziger JE (2009) Structural characterization of the osmosensor ProP. Biochim Biophys Acta 1788(5):1108–1115

Schowen RL (1977) Isotope effects on enzymes-catalyzed reactions. University Park Press, Baltimore, MD, pp 64–99

Smirnova IN, Kaback HR (2003) A mutation in the lactose permease of Escherichia coli that decreases conformational flexibility and increases protein stability. Biochemistry 42 (10):3025–3031

Smirnova IN, Kasho VN, Kaback HR (2006) Direct sugar binding to lacy measured by resonance energy transfer. Biochemistry 45(51):15279–15287

Smirnova I et al (2007) Sugar binding induces an outward facing conformation of LacY. Proc Natl Acad Sci USA 104:16504–16509

Smirnova IN, Kasho V, Kaback HR (2008) Protonation and sugar binding to LacY. Proc Natl Acad Sci USA 105(26):8896–8901

Smirnova IN, Kasho VN, Sugihara J, Choe JY, Kaback HR (2009a) Residues in the H + translocation site define the pKa for sugar binding to LacY. Biochemistry 48(37):8852–8860

Smirnova I, Kasho V, Sugihara J, Kaback HR (2009b) Probing of the rates of alternating access in LacY with Trp fluorescence. Proc Natl Acad Sci USA 106(51):21561–21566

Smirnova I, Kasho V, Sugihara J, Kaback HR (2011) Opening the periplasmic cavity in lactose permease is the limiting step for sugar binding. Proc Natl Acad Sci USA 108(37):15147–15151

Smirnova I, Kasho V, Sugihara J, Vazquez-Ibar JL, Kaback HR (2012) Role of protons in sugar binding to LacY. Proc Natl Acad Sci USA 109(42):16835–16840

Smirnova I, Kasho V, Sugihara J, Kaback HR (2013) Trp replacements for tightly interacting Gly-Gly pairs in LacY stabilize an outward-facing conformation. Proc Natl Acad Sci USA 110 (22):8876–8881

Solcan N et al (2012) Alternating access mechanism in the POT family of oligopeptide transporters. EMBO J 31(16):3411–3421

Sorgen PL, Hu Y, Guan L, Kaback HR, Girvin ME (2002) An approach to membrane protein structure without crystals. Proc Natl Acad Sci USA 99(22):14037–14040

Soskine M, Adam Y, Schuldiner S (2004) Direct evidence for substrate-induced proton release in detergent-solubilized EmrE, a multidrug transporter. J Biol Chem 279(11):9951–9955

Sugihara J, Smirnova I, Kasho V, Kaback HR (2011) Sugar recognition by CscB and LacY. Biochemistry 50(51):11009–11014

Sugihara J, Sun L, Yan N, Kaback HR (2012) Dynamics of the L-fucose/H + symporter revealed by fluorescence spectroscopy. Proc Natl Acad Sci USA 109(37):14847–14851

Sun J, Wu J, Carrasco N, Kaback HR (1996) Identification of the epitope for monoclonal antibody 4B1 which uncouples lactose and proton translocation in the lactose permease of *Escherichia coli*. Biochemistry 35(3):990–998

Sun J, Li J, Carrasco N, Kaback HR (1997) The last two cytoplasmic loops in the lactose permease of Escherichia coli comprise a discontinuous epitope for a monoclonal antibody. Biochemistry 36(1):274–280

Sun L et al (2012) Crystal structure of a bacterial homologue of glucose transporters GLUT1-4. Nature 490(7420):361–366

Teather RM, Müller-Hill B, Abrutsch U, Aichele G, Overath P (1978) Amplification of the lactose carrier protein in *Escherichia coli* using a plasmid vector. Mol Gen Genet 159:239–248

Trumble WR, Viitanen PV, Sarkar HK, Poonian MS, Kaback HR (1984) Site-directed mutagenesis of cys148 in the *lac* carrier protein of *Escherichia coli*. Biochem Biophys Res Commun 119(3):860–867

Ujwal ML, Sahin-Toth M, Persson B, Kaback HR (1994) Role of glutamate-269 in the lactose permease of Escherichia coli. Mol Membr Biol 11(1):9–16

Vadyvaloo V, Smirnova IN, Kasho VN, Kaback HR (2006) Conservation of residues involved in sugar/H(+) symport by the sucrose permease of Escherichia coli relative to lactose permease. J Mol Biol 358(4):1051–1059

van Iwaarden PR, Pastore JC, Konings WN, Kaback HR (1991) Construction of a functional lactose permease devoid of cysteine residues. Biochemistry 30(40):9595–9600

Vazquez-Ibar JL, Guan L, Svrakic M, Kaback HR (2003) Exploiting luminescence spectroscopy to elucidate the interaction between sugar and a tryptophan residue in the lactose permease of *Escherichia coli*. Proc Natl Acad Sci USA 100(22):12706–12711

Vazquez-Ibar JL et al (2004) Sugar Recognition by the Lactose Permease of *Escherichia coli*. J Biol Chem 279(47):49214–49221

Venkatesan P, Kaback HR (1998) The substrate-binding site in the lactose permease of *Escherichia coli*. Proc Natl Acad Sci USA 95(17):9802–9807
Venkatesan P, Kwaw I, Hu Y, Kaback HR (2000a) Site-directed sulfhydryl labeling of the lactose permease of *Escherichia coli*: helix VII. Biochemistry 39:10641–10648
Venkatesan P, Liu Z, Hu Y, Kaback HR (2000b) Site-directed sulfhydryl labeling of the lactose permease of *Escherichia coli*: helix II. Biochemistry 39:10649–10655
Viitanen P, Garcia ML, Foster DL, Kaczorowski GJ, Kaback HR (1983) Mechanism of lactose translocation in proteoliposomes reconstituted with lac carrier protein purified from *Escherichia coli*. 2. Deuterium solvent isotope effects. Biochemistry 22(10):2531–2536
Viitanen P, Garcia ML, Kaback HR (1984) Purified reconstituted lac carrier protein from *Escherichia coli* is fully functional. Proc Natl Acad Sci USA 81(6):1629–1633
Viitanen PV, Menick DR, Sarkar HK, Trumble WR, Kaback HR (1985) Site-directed mutagenesis of cysteine-148 in the *lac* permease of *Escherichia coli*: effect on transport, binding, and sulfhydryl inactivation. Biochemistry 24(26):7628–7635
Vogel H, Wright JK, Jähnig F (1985) The structure of the lactose permease derived from Raman spectroscopy and prediction methods. EMBO J 4(13A):3625–3631
Weinglass AB, Kaback HR (2000) The central cytoplasmic loop of the major facilitator superfamily of transport proteins governs efficient membrane insertion. Proc Natl Acad Sci USA 97 (16):8938–8943
Weinglass AB, Sondej M, Kaback HR (2002) Manipulating conformational equilibria in the lactose permease of Escherichia coli. J Mol Biol 315(4):561–571
Weinglass A, Whitelegge JP, Faull KF, Kaback HR (2004) Monitoring Conformational Rearrangements in the Substrate-binding Site of a Membrane Transport Protein by Mass Spectrometry. J Biol Chem 279(40):41858–41865
White SH (2007) Membrane protein insertion: the biology-physics nexus. J Gen Physiol 129 (5):363–369
Wolin C, Kaback HR (1999) Estimating loop-helix interfaces in a polytopic membrane protein by deletion analysis. Biochemistry 38(26):8590–8597
Wolin CD, Kaback HR (2000) Thiol cross-linking of transmembrane domains IV and V in the lactose permease of Escherichia coli. Biochemistry 39(20):6130–6135
Wu J, Kaback HR (1996) A general method for determining helix packing in membrane proteins in situ: helices I and II are close to helix VII in the lactose permease of *Escherichia coli*. Proc Natl Acad Sci USA 93(25):14498–14502
Yan H et al (2013) Structure and mechanism of a nitrate transporter. Cell Rep 3(3):716–723
Yau WM, Wimley WC, Gawrisch K, White SH (1998) The preference of tryptophan for membrane interfaces. Biochemistry 37(42):14713–14718
Yin Y, He X, Szewczyk P, Nguyen T, Chang G (2006) Structure of the multidrug transporter EmrD from Escherichia coli. Science 312(5774):741–744
Zhang W, Guan L, Kaback HR (2002) Helices VII and X in the lactose permease of Escherichia coli: proximity and ligand-induced distance changes. J Mol Biol 315(1):53–62
Zhao M et al (1999a) Nitroxide scanning electron paramagnetic resonance of helices IV, V and the intervening loop in the lactose permease of Escherichia coli. Biochemistry 38:15970–15977
Zhao M, Zen KC, Hubbell WL, Kaback HR (1999b) Proximity between Glu126 and Arg144 in the lactose permease of Escherichia coli. Biochemistry 38(23):7407–7412
Zheng H, Wisedchaisri G, Gonen T (2013) Crystal structure of a nitrate/nitrite exchanger. Nature 497(7451):647–651
Zhou Y, Guan L, Freites JA, Kaback HR (2008) Opening and closing of the periplasmic gate in lactose permease. Proc Natl Acad Sci USA 105(10):3774–3778
Zhou Y, Jiang X, Kaback HR (2012) Role of the irreplaceable residues in the LacY alternating access mechanism. Proc Natl Acad Sci USA 109(31):12438–12442

Chapter 7
Alternating Access Within the POT Family of Oligopeptide Proton Symporters

Simon Newstead

Abstract The POT family of proton-dependent oligopeptide transporters belongs to the Major Facilitator Superfamily and is widely distributed within prokaryotic and eukaryotic kingdoms. They function as proton:peptide symporters, using the inwardly directed proton electrochemical gradient to drive the uptake of di- and tripeptides into the cell. Mammals contain two members of the POT family, PepT1 and PepT2, which in addition to their physiological role in peptide import also recognise and transport several important drug families, including the β-lactam antibiotics and a growing library of peptide modified pro-drugs. A detailed molecular understanding of peptide recognition and transport within the POT family therefore has an acute medical significance that crystallographic methods are now uniquely placed to address. In the last few years the crystal structures of two prokaryotic POT family transporters have been determined, one with bound ligand. These structures represent key intermediates in the transport cycle and reveal remarkably well-conserved sequence and structural motifs that impact peptide specificity and transport rates. However, key questions remain to be answered, including the site(s) and role of protonation and peptide binding in orchestrating the structural changes that result in transport. In this chapter we discuss the insights recent crystal structures of bacterial POT family transporters have provided and examine how these questions are being addressed.

Keywords Best • Christine Ziegler • Drug Transport • POT Family • PepT1 • Peptide Transport • Salt Bridge Networks • SLC15 • Specificity

S. Newstead (✉)
Department of Biochemistry, University of Oxford, Oxford, UK
e-mail: simon.newstead@bioch.ox.ac.uk

R. Krämer and C. Ziegler (eds.), *Membrane Transport Mechanism*, Springer Series in Biophysics 17, DOI 10.1007/978-3-642-53839-1_7,

7.1 Introduction

Peptide transport is the main route through which the body absorbs and retains dietary protein and as such plays an important role in human physiology (Daniel et al. 2006; Matthews 1991; Steinhardt and Adibi 1986). Ingested protein is broken down into peptide fragments and free amino acids through acid hydrolysis in the stomach and the action of non-specific peptidases in the small intestine. The resultant peptides are then actively transported across the intestinal brush border membrane through the action of the integral membrane peptide transporter, PepT1 (Covitz et al. 1998; Fei et al. 1994; Liang et al. 1995; Knütter et al. 2004; Leibach and Ganapathy 1996). Peptide transport also occurs at the renal epithelium, where PepT2, a functionally related paralogue of PepT1, reabsorbs peptides from the glomerular filtrate (Biegel et al. 2006; Daniel and Rubio-Aliaga 2003). PepT1 and PepT2 recognise a truly diverse library of small peptides, which include most zwiterionic, cationic and anionic di- and tripeptides (Döring et al. 1996; Matthews 1975; Terada et al. 2000), consistent with their role as the only peptide-specific plasma membrane transporters in the human body. In addition to dietary peptide absorption, PepT1 and PepT2 also recognise several important families of drug compounds that exhibit a steric resemblance to peptides. These include the commonly prescribed β-lactam antibiotics (Chen et al. 2000; Luckner and Brandsch 2005; Pieri et al. 2009). In addition, PepT1 and PepT2 also recognise a number of peptide pro-drugs. These are molecules that have been modified through the attachment of either a single amino acid or dipeptide to the active drug molecule resulting in the transport of the modified compound via PepT1 and PepT2 (Meredith and Boyd 2000; Rautio et al. 2008). In vivo studies in mice have shown convincing evidence that PepT1 and PepT2 substantially improve the bioavailability of these modified drug molecules (Uchiyama et al. 2003; Yang and Smith2012). These properties make PepT1 and PepT2 attractive pharmaceutical targets for improving drug pharmacokinetics (Pao et al. 1998; Anderson and Thwaites 2010). However a detailed understanding of the mechanism by which peptide transporters both recognise and transport natural peptides and their related drug molecules is currently absent (Abramson et al. 2003; Brandsch 2009; Huang et al. 2003; Hirai et al. 2002; Dang et al. 2010; Yin et al. 2006).

PepT1 and PepT2 belong to the solute carrier (SLC) 15 gene family and phylogenetically form part of the much larger proton-dependent oligopeptide transporter, or POT family (TC 2.A.17), which is widely distributed within prokaryotes and eukaryotes [reviewed in Daniel et al. (2006) and Yan (2013)]. There exists a remarkably high degree of sequence conservation within the POT family, also referred to as the PTR family (Jardetzky 1966; Steiner et al. 1995). Bioinformatic analysis has identified three conserved sequence motifs, one of which is referred to as the PTR2-1 motif, which can be used to identify POT family transporters (Steiner et al. 1995; Law et al. 2008; Paulsen & Skurray 1994). These motifs are all located within the first 180–250 amino acids of the various PepT homologs and point mutations introduced into these regions typically result in inactive transporters (Pieri et al. 2009; Forrest et al. 2011; Fei et al. 1997; Hauser et al. 2005; Solcan et al. 2012; Jensen et al. 2012).

All members of the POT family are proton (H^+)-driven symporters using the inwardly direct proton electrochemical gradient ($\Delta\mu_H{}^+$) to drive the uptake of peptides across the cell membrane (Newstead et al. 2011; Daniel and Kottra 2004). Available functional data mostly concerns the mechanism of transport within the eukaryotic proteins however. With respect to the stoichiometry of proton-driven peptide transport, evidence from TEVC recordings on rabbit PepT1 expressed in *Xenopus laevis* oocytes suggest a 2:1 proton to anionic dipeptide stoichiometry (Kottra et al. 2002). These studies also propose that anionic dipeptides are transported in their neutral and negatively charged forms, with high and low affinity respectively. Cationic dipeptides by contrast are transported in neutral and positively charged forms. Interestingly, similar studies on PepT2 suggest a 2:1 and 3:1 proton:peptide stoichiometry for neutral and anionic dipeptides (Chen et al. 1999). These results indicate a mechanistically important link between the ionisation state of the peptide and the interaction with the transporter.

The first PepT1 gene was cloned from a rabbit cDNA library in 1994 (Fei et al. 1994). The primary structure of PepT1 and PepT2 predict 12 transmembrane (TM) spanning alpha helices (Covitz et al. 1998). A large extramembrane domain is found inserted between helices 9 and 10, creating an asymmetric structure with the amino and carboxy termini located within the cytosol (Fig. 7.1). Intriguingly PepT1 and PepT2 exhibit different kinetic properties; PepT1 is a high capacity, broad specificity transporter with K_M values in the low mM range for most di-and tripeptides (Knütter et al. 2004), whereas PepT2 exhibits higher affinity for peptides with K_M values in low μM range and also appears to be more selective with respect to side-chain accommodation (Biegel et al. 2006). The source of these kinetic differences has been one of the driving questions in the field of peptide transporter research for a number of years. Partial insights have been gained from chimaeras of PepT1 and PepT2. These studies predicted that the region of the transporter containing helices H7, H8 and the large extracellular domain between helices H9 and H10 play a crucial role in the determination of substrate affinity (Döring et al. 1996; Terada et al. 2000). Indeed, swapping the H7–H10 region results in the transfer of PepT1-like transport kinetics to PepT2 and vice versa. Although clearly difficult to interpret fully without a crystal structure, these results provided the first evidence that POT family transporters contained distinct functional domains that dictated the observed transport properties.

Several conserved residues have been shown to affect peptide transport in site-directed mutagenesis studies with PepT1, and these have been illustrated on the topology diagram in Fig. 7.1. Interestingly these cluster within the first six TM helices and appear to be dominated by aromatic amino acids. A cluster of tyrosine residues (Tyr12, 64, 91 and 167) plays an important role in affecting either the V_{max} or K_M of transport (Chen et al. 2000; Pieri et al. 2009), whereas a conserved tryptophan (Trp294), in helix H7, when mutated to an alanine affects both simultaneously (Meredith and Boyd 2000). Of the charged residues only one has been identified that is essential for transport, Glu595 in helix H10. Within the

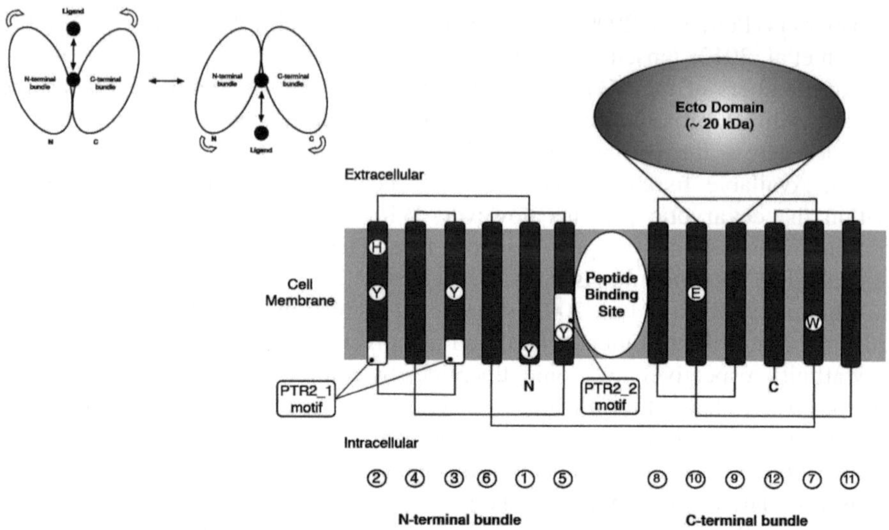

Fig. 7.1 *Structure of PepT1, a mammalian POT family transporter.* PepT1 has 12 transmembrane (TM)-spanning alpha helices with both amino and carboxy termini facing the cytoplasm. The protein is very likely to fold in the membrane as two bundles of 6 TM helices that form a central peptide-binding site, situated approximately in the middle of the membrane. A substantial additional domain is inserted between TM helices H9 and H10, of which little is currently known. Functional studies on several mammalian PepT1 proteins have identified a number of functionally important residues that dictate the kinetics of peptide transport; these are illustrated on the topology diagram. *Inset*, the early two-state 'rocker switch' model for alternating access transport proposed for members of the MFS that involves each of the 6-TM bundles rocking around an axis parallel to the membrane and running through the central ligand binding site

mammalian POT family there exist four histidine residues, His57, 111, 121 and 261. Of these only His57 is essential for PepT1 function and is predicted to be the site of initial protonation during transport (Uchiyama et al. 2003).

The POT family sits within the much larger and functionally diverse major facilitator superfamily (MFS) of secondary active transporters (Pao et al. 1998). Crystal structures from several MFS transporters have now revealed a common fold consisting of two 6-TM bundles that assemble together in the membrane to form a 'V'-shaped transporter with a central substrate-binding site formed between the two bundles (Abramson et al. 2003; Huang et al. 2003; Hirai et al. 2002; Dang et al. 2010; Yin et al. 2006) [recently reviewed in Yan (2013)]. Within the MFS a general model of transport has been proposed that is consistent with the principles of alternating access transport originally proposed by Jardetzky (1966). Now this is often referred to as the rocker switch model (Law et al. 2008). The rocker switch model broadly describes transport occurring through the movement of each 6-TM bundle around a central binding site in a two state process that switches the binding site between each face of the membrane (Fig. 7.1 inset).

Substantial structural insight has recently been gained into the process of alternating access mechanisms operating within other integral membrane transporters (Forrest et al. 2011). These suggest that all transporters will contain regions of their structure that gate access to the respective binding sites. There should minimally be an both 'outward-facing' and an 'inward-facing' gate or, as we will refer to them in this chapter, as the extracellular or intracellular gates, that alternately control access to a central ligand-binding site from either side of the membrane. The alternating access model posits that one gate can only be open at any time; if both gates were free to open simultaneously a non-specific channel would form through which the membrane ion gradients would rapidly dissipate and lead to cell death. This implies a strict coupling mechanism exists to ensure only one gate is open at any time in the transport cycle. Understanding how the MFS fold operates with respect to peptide transport will therefore require identifying these gates and then understanding how their movement is linked to both peptide and proton binding. X-ray crystallography as a biophysical method is uniquely placed to identify the peptide-binding site and the gates that control access to this site during transport. However, membrane protein crystallisation is inherently very challenging as the protein molecules themselves are extremely dynamic, often adopting multiple conformations in solution and thereby failing to form well-ordered, well-diffracting protein crystals. Additionally, caution should also be used when interpreting membrane protein structures using X-ray crystallography, and any hypotheses based on such structures should always, where practically possible, be supported with in vitro and in vivo biochemical data. Artefacts due to packing and detergent solubilisation are always possible and at the resolution of many membrane protein structures (3–4 Å) identifying these in a novel fold or family can be tricky. Although beyond the scope of this chapter, many excellent reviews have been written on the challenges and solutions to membrane protein structure determination (Moraes et al. 2013; Kunji et al. 2011).

The first crystal structure of a POT transporter from the bacterium *Shewanella oneidensis*, $PepT_{So}$, was determined in 2011, revealing a novel occluded conformation for an MFS transporter (Newstead et al. 2011). $PepT_{So}$ was captured with a ligand bound within a central cavity that was itself closed to either side of the membrane. The structure enabled identification of the extracellular gate and uniquely for the MFS, a partially closed intracellular gate. This structure was shortly followed by a second, from the bacterium *Streptococcus thermophilus*, $PepT_{St}$ (Solcan et al. 2012). Both $PepT_{So}$ and $PepT_{St}$ were identified as suitable candidates for protein crystallisation studies due to their high expression levels in *E. coli* (>5 mg L^{-1}) and their high thermal stability in the detergent LDAO, which was used in combination with a cysteine reactive fluorescent dye to screen membrane protein stability in different detergent solutions (Sonoda et al. 2011). In contrast to the occluded structure of $PepT_{So}$, the structure of $PepT_{St}$ revealed an inward-facing conformation that provided significant new insight into the release mechanism of the intracellular gate. Structural comparison between the two proteins enabled the identification of conserved salt bridge pairs that appeared to act as locks, stabilising the gates in their closed conformations, particularly with respect to the extracellular gate that was closed in both structures. The role of these conserved salt bridge interactions in orchestrating the structural changes during transport will be discussed and placed into the context of a wider alternating access mechanism operating within PepT1 and PepT2, their mammalian counterparts.

7.1.1 Architecture of the POT Family

Both $PepT_{So}$ and $PepT_{St}$ contain 14 TM helices (Fig. 7.2), of which helices H1–H12 adopt the overall fold observed previously for the MFS transporters LacY, GlpT, EmrD, FucP and XylE (Matthews 1991; Steinhardt and Adibi 1986; Leibach and Ganapathy 1996; Abramson et al. 2003; Dang et al. 2010; Yin et al. 2006) (see also Chap. 6). The arrangement of the helices is also consistent with the EM projection structure of a POT transporter, DtpD from *E. coli* (Newman et al. 1981; Casagrande et al. 2009). The two structures represent different states of the transport cycle. $PepT_{So}$ adopts an inward-occluded state, whereas $PepT_{St}$ is inward open. Like previous MFS transporter structures, the N- and C-terminal six-helix bundles, formed by helices H1–H6 and H7–H12, come together to form a 'V-'shaped transporter, related by a pseudo twofold symmetry axis running perpendicular to the membrane plane. The current POT family structures have two additional transmembrane helices, HA and HB, which are inserted into the cytoplasmic loop connecting the N- and C-terminal bundles. These form a hairpin in the membrane that packs against the periphery of the protein, although the arrangement is slightly different between the two structures (Fig. 7.2). Their role is currently unclear, as a detailed study of POT family peptide transporters in their native organisms has not yet been carried out. The apparent absence of these helices in the fungal, plant and mammalian protein sequences, however, suggests they do not contribute to a conserved transport mechanism and are specific for the prokaryotic members of the POT family.

$PepT_{So}$ revealed two hydrophilic cavities, a larger central cavity and a smaller extracellular cavity, both of which are hydrophilic (Solcan et al. 2012; Newstead et al. 2011). The larger cavity, which is also observed in the inward open $PepT_{St}$ structure, is situated within the centre of the membrane and is very likely the main peptide-binding site (Solcan et al. 2012; Brandsch et al. 2008). Indeed, the residues extending into the central cavity in both $PepT_{So/St}$ are all known to affect peptide binding and/or transport in the mammalian proteins (Newstead 2011). As further conformation electron density for a bound ligand was also observed sitting within this cavity in the crystal structure (Newstead et al. 2011). The extracellular cavity, which is only observed in the occluded $PepT_{So}$ structure, is also located at the interface between the N- and C-terminal domains. It is roughly cone shaped, with the apex at the bottom near the central cavity, opening outward. The overall dimensions of the cavity are approximately $16 \times 8 \times 8$ Å, too small to accept a peptide though hydrophilic enough to accommodate water molecules. The role of this cavity is currently unclear, but as discussed below it forms as a result of structural flexibility within the C-terminal bundle as a result of closing the intracellular gate.

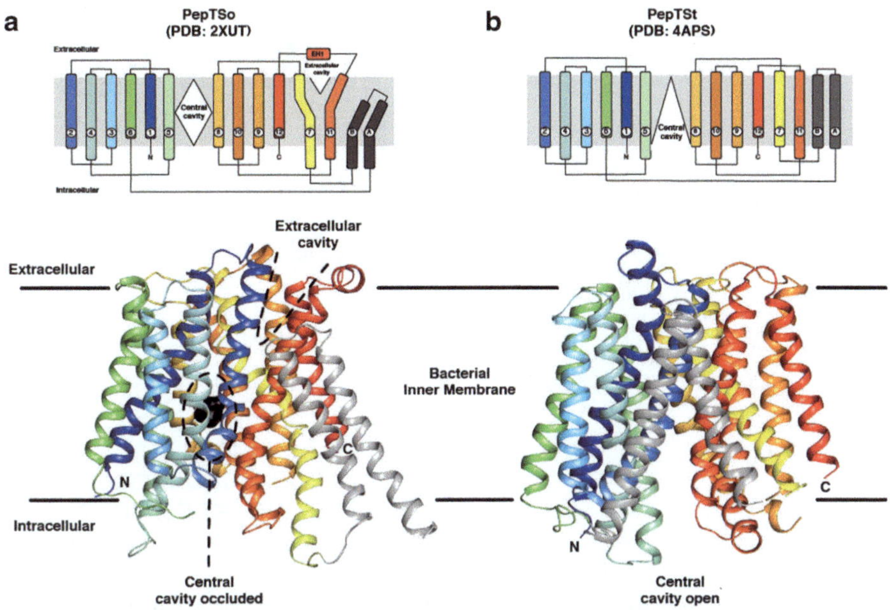

Fig. 7.2 *Architecture of a POT family transporter*. (**a**) Crystal structure of $PepT_{So}$, the POT family transporter from the bacterium *Shewanella oneidensis*, determined to 3.6 Å resolution (PDB: 2XUT). The figure shows the topology of $PepT_{So}$ and a *cartoon* representation in the plane of the membrane. The central peptide-binding site is illustrated along with a second, extracellular cavity opening up in the C-terminal half of the transporter. (**b**) Crystal structure of $PepT_{St}$, a second POT family member from the bacterium *Streptococcus thermophilus*, determined to 3.3 Å resolution (PDB: 4APS). Topology diagrams of both POT family members are shown coloured as the structures. Unlike $PepT_{So}$, $PepT_{St}$ adopts an inward-facing state with the central cavity open and fully accessible to the interior of the cell

7.1.2 Extracellular and Intracellular Gates Control Access to the Central Cavity

The two available structures for the POT family represent different states of the transport cycle, a currently unique situation within the MFS. What do the comparisons between these two structures tell us about possible mechanisms for transport within the POT family? Comparing $PepT_{So}$ with $PepT_{St}$ the identity of the extracellular and intracellular gates has been determined. The extracellular gate is constructed from helices H1 and H2 in the N-terminal bundle and helices H7 and H8 in the C-terminal bundle, which in both $PepT_{So}$ and $PepT_{St}$ pack together forming a constriction at the extracellular side of the peptide-binding site (Fig. 7.3). Two salt bridge interactions can be identified that facilitate the close packing of the gate helices. In $PepT_{St}$ these are Arg53 (H1) with Glu312 (H7), which forms distal to the central peptide-binding site and Arg33 (H1) with Glu300 (H7), which forms directly above the binding cavity and that we have called the

a

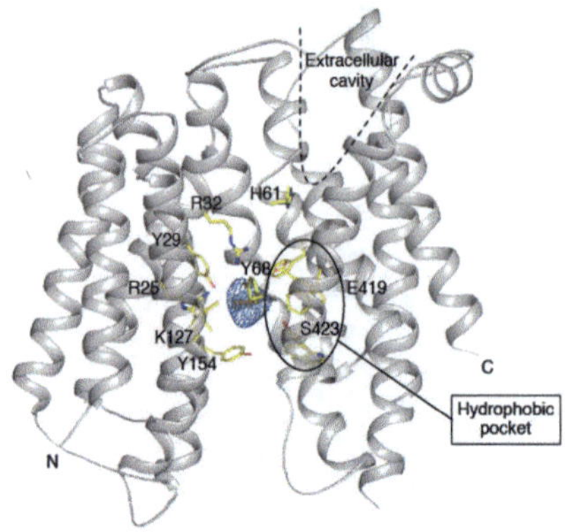

b

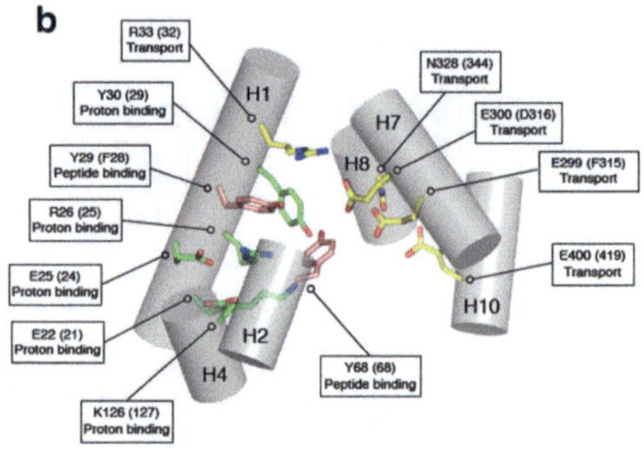

c

IC_{50} (mM)	WT	Y29F	Y68F
Glu-Glu	0.56 +/- 0.07	0.48 +/- 0.05	1.63 +/- 0.09
Phe-Phe	0.005 +/- 0.002	0.006 +/- 0.004	0.02 +/- 0.002
Ala-Ala	0.36 +/- 0.02	0.50 +/- 0.06	0.51 +/- 0.05
Ala-Ala-Ala	0.40 +/- 0.06	1.40 +/- 0.02	0.35 +/- 0.003

Fig. 7.3 *Structural comparison between $PepT_{So}$ and $PepT_{St}$ revealed the nature of the extracellular and intracellular gates.* Centre of the figure shows a slice through the volume of $PepT_{So}$ and $PepT_{St}$ positioned as in Fig. 7.2. The extracellular, central cavity, extracellular cavity and intracellular gate regions are identified. *Top left*, zoomed in view of the extracellular gate region in $PepT_{So}$, illustrating the salt bridge interaction between Arg32 on H1 and D316 on H7 and their position with respect to the extracellular cavity. *Top right*, zoomed in view of the equivalent region in $PepT_{St}$, showing the proximal and distal salt bridges stabilising the interaction between helices H1, H2 from the N-terminal and H7, H8 from the C-terminal bundles, respectively

proximal salt bridge. In comparison, $PepT_{So}$ only has the proximal salt bridge formed between Arg31 (H1) and Asp316 (H7), the distal salt bridge being broken, presumably as a result of adopting a more occluded state. The function of these salt bridge interactions in peptide binding and transport was duly studied and discussed below.

However before we discuss the role of these salt bridge interactions, what can we further deduce from the structural comparison about the intracellular gate and how this may function during transport? Comparing $PepT_{St}$ and $PepT_{So}$, helices H1–H6, which makes up the N-terminal bundle, superimpose with an r.m.s.d of 1.57 Å for 167 Cα atoms clearly adopting a similar arrangement between the two structures. In adopting the occluded state therefore, the main difference between $PepT_{So}$ and $PepT_{St}$ occurs within the C-terminal bundle, where helices H7–H12 superimpose with an r.m.s.d of 2.30 Å for 127 Cα atoms. In the occluded structure, an intracellular gate constructed from conserved side-chain interactions between Ser131 and Phe150 on helices H4, H5 with Leu427 and Met443 on H10, H11 prevents the bound ligand from exiting the binding site (Figs. 7.3 and 7.4a) (Newstead et al. 2011). Structural comparisons with the inward open state of LacY had previously identified helices H7, H10 and H12 as forming a distinct sub-bundle

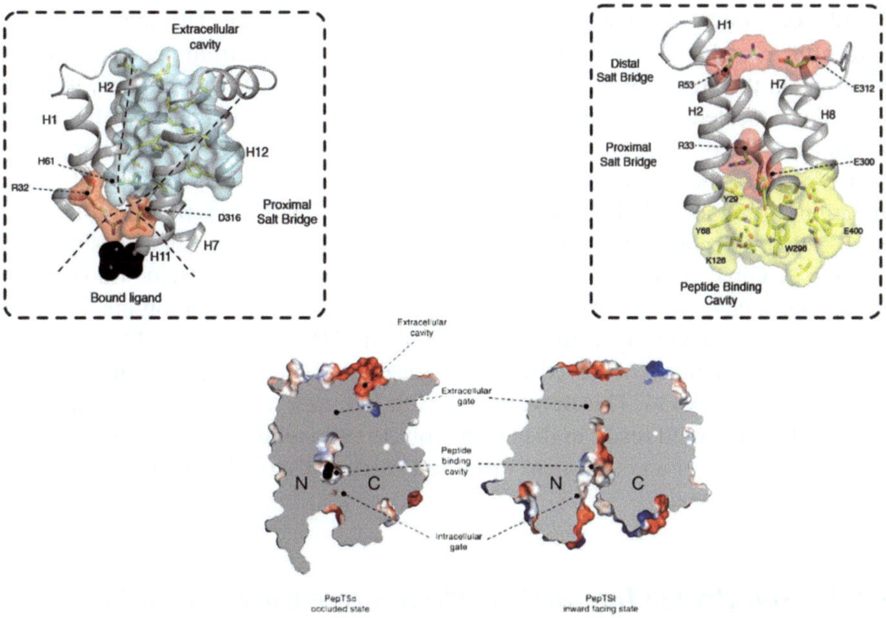

Fig. 7.4 *Intracellular gate mechanism.* (**a**) Intracellular gate in PepT_{So} viewed in the plane of the membrane. Residues forming the obstruction are shown as *stick models* with transparent CPK surfaces. LacY helices (*grey*) are superposed on their equivalents in PepT_{So}. Bound ligand is shown as a black CPK model as in (**a**). (**b**) Rotation of sub-bundle C1 between the occluded conformation of PepT_{So} and inward open conformation of LacY. Superposed transmembrane helices of PepT_{So} and LacY as viewed in the plane of the membrane. The N-terminal six helices of LacY were superposed onto the equivalent helices in PepT_{So}. Helices comprising the C1 sub-bundle are coloured *red*, those from LacY cyan. The central and extracellular cavities are highlighted and the modelled di-alanine peptide is shown as a CPK model. (**c**) Comparison between the occluded structure of PepT_{So} (*coloured*) and inward open structure of PepT_{St} (*grey*), with *arrows* showing the hinge-like movement that releases the obstruction observed in PepT_{So}. Residue numbers are for PepT_{So} and central cavity is indicated. (**d**) View of the intracellular gate helices H7, H10 and H11 showing the location of the ball and socket hinge formed by Gly407 packing against Trp427 in PepT_{St}

within the C-terminal domain (Newstead et al. 2011). We conjectured that helices H7, H10 and H12 formed a functionally important unit that operated independently to control the intracellular gate and which we termed sub-bundle C1. Rotation of this sub-bundle by approximately 18° would release the intracellular gate and simultaneously close the extracellular cavity (Fig. 7.4b).

Following the comparison with PepT_{St} we gained substantial new insight into the release mechanism of the intracellular gate and were able to expand our earlier model (Solcan et al. 2012). We discovered that opening of the intracellular gate does involve movement of the entire C1 sub-bundle, as previously conjectured, but that movement is instead localised towards the cytoplasmic halves of helices H7, H10 and H11 (Fig. 7.4c), consistent with a hinge-like displacement at the apex of the H10–H11 hairpin. Specifically, the gate opens due to bending at Gly407 and

Trp427 on helices H10 and H11, respectively (Fig. 7.4d). This results in the intracellular ends of these helices moving ~13 Å away from their respective positions in the structure of $PepT_{So}$ and an approximate 10° rotation. Gly407 and Trp427 are located at the same point within the H10–H11 helix hairpin, effectively forming a hinge or pivot point, which would control whether the intracellular gate is open or closed between the occluded and inward-facing conformation. Indeed, mutation of either Gly407 or Trp427 to alanine abolished transport (Solcan et al. 2012). Of note is that glycine residues are found at equivalent positions to Gly407 in GlpT (Lemieux et al. 2004), FucP (Dang et al. 2010), EmrD (Yin et al. 2006) and LacY (Abramson et al. 2003) (Chap. 6), implying a wider significance for flexibility within helices H10 and H11 within the MFS. A further localised movement is observed at the extracellular end of H11, at another conserved glycine, Gly434. This movement results in closer packing of helix H11 with H2 and H7 and would result in the closure of the extracellular cavity and we would predict formation of an equivalent distal salt bridge observed in $PepT_{St}$.

7.1.3 Salt Bridge Interactions Orchestrate Communication Between the N- and C-Terminal Bundles During Transport

There exists a remarkably high degree of sequence conservation that spans between the prokaryotic and eukaryotic members of the POT family, suggesting the mechanism of proton-coupled peptide transport evolved early on in cellular life and has remained relatively unchanged (Daniel et al. 2006). Indeed, between the human PepT1 protein and $PepT_{So}$, there exists an ~30 % identity overall across the primary structure, with all of the functionally important residues previously identified in the mammalian proteins present in many of their bacterial counterparts (Newstead 2011). However, to establish the bacterial proteins as models for understanding transport within the wider POT family in our laboratory, purified protein was reconstituted into an in vitro proteoliposome system for functional analysis (Solcan et al. 2012). We wished to address specific questions in regard to peptide transport. In particular, did the conserved salt bridge interactions play roles in proton coupling or peptide binding? What roles, if any, did the conserved tyrosine residues identified in the central cavity of $PepT_{So}$ have in peptide binding? Two in vitro proteoliposome assays were developed to address these questions. The first assay was used to test the role of a particular residue in proton-driven peptide uptake and uses an artificially imposed electrochemical potential to drive the uptake of tritiated L-Alanine, or glycyl-sarcosine, a non-hydrolyzable peptide analogue into the lumen of the liposomes. Uptake is linear and saturatable, enabling accurate kinetic parameters to be determined (Solcan et al. 2012). If a point mutant is inactive in this assay it could be the result of either disruption in the proton-coupling mechanism or an inability to bind peptide and/or undergo the conformational changes that result in transport. To delineate these two possibilities the same mutant protein is reconstituted into proteoliposomes again, but this time assayed for its ability to transport under counterflow conditions. Counterflow-driven transport was shown conclusively for reconstituted LacY in 1981 and is an elegant method for discriminating against point mutants that can no longer couple transport to the proton gradient, but can still bind and transport ligand (Newman et al. 1981) (see also Chap. 6). In this assay transport is driven by an outwardly directed ligand gradient, rather than through an inwardly directed proton gradient. For the POT family this would be a high concentration (10 mM) of cold L-di-Alanine peptide. If the mutation only affects the proton-coupling mechanism, then we would expect the protein to be active under counterflow conditions, which is independent of proton transport. However, if our mutant is still inactive under counterflow, it suggests the residue in question plays a role in peptide binding and/or transport. Using these two assays the residues identified as coordinating the interaction between helices H1, H2 and H7, H8 in the extracellular gate were duly investigated in the $PepT_{St}$ POT family transporter.

Alanine mutations at the proximal salt bridge, between Arg33 and Glu300, had a substantial effect on transport. Mutation of Glu300 to alanine abolished uptake in both the proton-driven and proton-independent counterflow assays, whereas transport in the Arg33 mutant was only abolished in the proton-driven assay. This result suggests that Arg33 has an important role in proton coupling but does not affect peptide recognition. In contrast Glu300 is more likely to be required for peptide recognition. This is an intriguing result, as it demonstrates a clear subdivision of labour between the two residues at this structurally important region of the transporter. For any mechanism to be coupled, one component must rely on the other for function. The result obtained from these assays suggests these two residues both have important complementary roles in proton binding and peptide transport. Clearly the identified salt bridge interactions between Arg33–Glu300 ($PepT_{St}$) and Arg53–Glu312 ($PepT_{So}$) participate in both proton binding and peptide transport and facilitate closure of the extracellular gate region. As discussed below, we propose these interactions play an important role in orchestrating alternating access within the POT family. Perhaps unsurprisingly the distal salt bridge between Arg53 and Glu312 was shown to play a supportive, rather than essential role during transport (Solcan et al. 2012).

7.1.4 The Peptide-Binding Site

The peptide-binding site in the crystal structures is located at the apex of the elongated central cavity, which in $PepT_{St}$ opens outwards from the interior of the protein towards the intracellular side of the membrane (Fig. 7.2). In $PepT_{St}$ the cavity has overall dimensions of approximately 14 Å × 13 Å × 20 Å, whereas in the occluded $PepT_{So}$ structure, the equivalent dimensions are 13 × 12 × 11 Å. These dimensions are sufficient to accommodate both di- and tripeptides, although would be sterically restrictive for larger tetrapeptide ligands and would likely be too

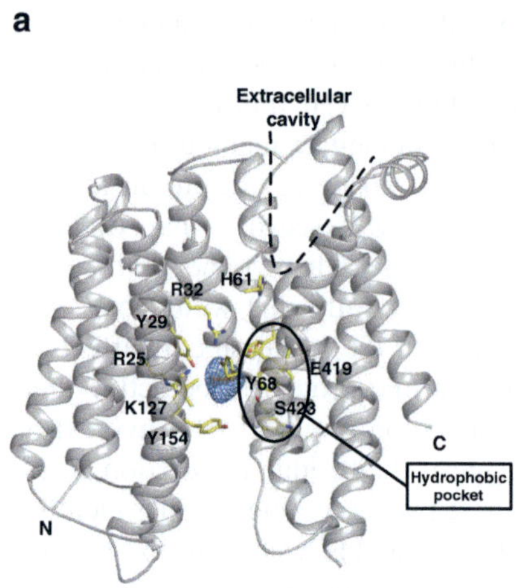

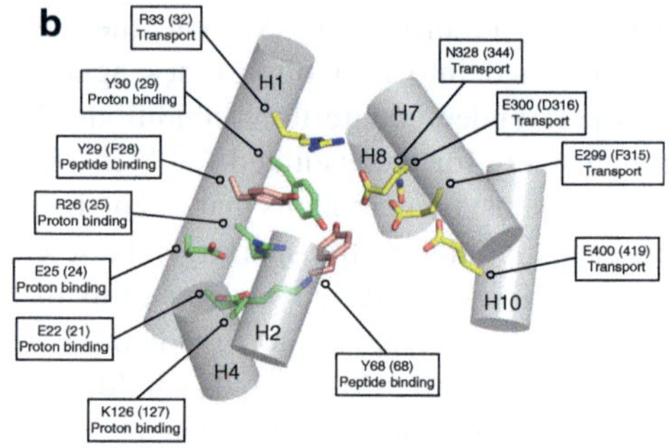

c

IC_{50} (mM)	WT	Y29F	Y68F
Glu-Glu	0.56 +/- 0.07	0.48 +/- 0.05	1.63 +/- 0.09
Phe-Phe	0.005 +/- 0.002	0.006 +/- 0.004	0.02 +/- 0.002
Ala-Ala	0.036 +/- 0.02	0.050 +/- 0.06	0.051 +/- 0.05
Ala-Ala-Ala	0.40 +/- 0.06	1.40 +/- 0.02	0.35 +/- 0.003

Fig. 7.5 *Peptide-binding site*. (**a**) The peptide-binding site in $PepT_{So}$ is shown in the plane of the membrane. Conserved residues between $PepT_{So}$ and PepT1 are labelled and shown as *sticks*. A dipeptide-sized C_α baton (*orange*) has been placed into the mFo-DFc electron density observed in the central cavity (*blue mesh*), contoured at 4σ. (**b**) Zoomed in view of the central cavity in $PepT_{St}$, with the helices represented as *cylinders*. Side chains observed within the cavity are labelled, with the equivalent residue numbers for $PepT_{So}$ shown in *brackets*. The function of these residues determined from the in vitro assays is indicated. (**c**) Table showing the calculated IC_{50} values for different peptides in the WT and Y29F and Y68F variants of $PepT_{St}$

large to selectively recognise a single amino acid. However, the dimensions do not suggest a possible orientation of the peptide within the binding site, as one could model a peptide into the cavity either vertically or horizontally with respect to the membrane.

Sitting within the centre of the cavity in $PepT_{So}$ was observed strong (>4σ) electron density for an unidentified ligand, quite possibly carried through from the purification or crystallisation conditions (Fig. 7.5a). The density has approximately the same dimensions as a dipeptide, being approximately 5 Å in length. A C_α model of a di-alanine peptide was modelled into the density as a useful marker to indicate the size of the cavity with respect to a natural ligand. In this position the peptide is modelled in a horizontal position, but the reader should be aware that we currently do not have any structural evidence supporting this and the vertical orientation is equally likely. The binding site is formed by residues from helices H1, H2, H4 and H5 from the N-terminal six-helix bundle and from helices H7, H8, H10 and H11 from the C-terminal bundle.

In $PepT_{So}$ on the N-terminal side of the binding site sit three conserved positively charged residues, Arg25, Arg32 and Lys127. Two conserved tyrosine residues, Tyr29 and Tyr68 are positioned close to this positively charged cluster. On the C-terminal side of the binding site, at a distance of approximately 13 Å from Lys127, are two further strictly conserved residues, Glu419 and Ser423, located in close proximity to Tyr154. Various mutants of Glu419 in PepT1 have been reported to drastically reduce transport activity, except where mutation was to an aspartatic acid (Xu et al. 2009), indicating the importance of a negatively charged residue at this position. The arrangement of opposite charges within the binding site may play an important role in the recognition and orientation of peptides through the creation of a dipole moment. The presence of several possible hydrogen bond donors and acceptors could be advantageous in adapting to peptides of various lengths, sequences and charges. Most of the other residues in the binding site are conserved hydrophobic residues, including Ile157, Trp312, Phe315 and Trp446. These residues are likely to provide a suitable environment for peptide side chains that in general are more hydrophobic than the peptide backbone. This idea is supported by the fact that mutation of Trp312 of PepT1 to alanine reduces substrate uptake in HEK293 cells (Bolger et al. 1998). The presence of hydrophobic pockets formed by these residues in the peptide-binding site of PepT1 has also been predicted through biochemical studies (Bailey et al. 2000).

To test the function of these residues in proton-coupled peptide symport the equivalent residues seen in the binding pocket of PepTSt were subjected to study using the in vitro proteoliposome assays described above (Solcan et al. 2012). The results of these studies are shown in Fig. 7.5b. The conclusion from this study was that the conserved ExxERFxYY motif on helix H1 has an important role in the proton-coupling mechanism within the POT family. Mutation of Glu22, Glu25, Arg26 or Tyr30 to alanine resulted in a transporter that could no longer transport under proton-driven conditions but could still recognise and transport peptide during counterflow. Within the N-terminal bundle only one residue was found to abolish transport in both assays, Arg33, consistent with the observation that this residue plays a role in stabilising the proximal salt bridge in the extracellular gate (Fig. 7.3). On the C-terminal side of the transporter all the conserved residues assayed produced transport-deficient proteins. Of note is that the conserved asparagine, Asn328, appears to play an important role and interacts closely with Glu300 and Arg33, perhaps playing a role in regulating the extracellular gate. These observations are consistent with our earlier hypothesis that the C-terminal domain within the POT family is more dynamic, perhaps containing the more mobile elements of the extra and intracellular gates (Steiner et al. 1995; Newstead et al. 2011).

7.1.5 *Role of Conserved Tyrosine Residues in Regulating Peptide Specificity*

As noted the peptide-binding site contains three prominent tyrosine side chains that are conserved within the POT family (Fig. 7.1). In $PepT_{St}$ these are Tyr29 (30) and Tyr30 (31), contributed from helix H1, and Tyr68 (64) from helix H2, numbers in brackets refer to the human PepT1 protein. The hydroxyl groups of both Tyr29 and

Tyr68 do not contribute to proton binding, as their mutation to phenylalanine had little impact on proton-driven uptake compared to WT. The Tyr29Ala mutant did show reduced uptake in counterflow, however, demonstrating the conserved ExxERFxYY motif on helix H1 has additional roles to proton binding during transport. To investigate the role of Tyr29 and Tyr68 further, a series of competition experiments were performed with a library of di- and tripeptides. The phenylalanine mutants showed distinct changes in their affinity for di-Glu and tri-Ala peptides compared to the WT protein, with the alanine mutants losing their ability to transport these peptides altogether whilst still retaining affinity for di-Phe and di-Ala peptides. This difference in the observed affinity between peptides, in particular for the alanine mutations, suggests Tyr29 and Tyr68 are important in determining peptide affinity. To quantify the contribution made by Tyr29 and Tyr68 to peptide affinity, IC50 values for these peptides were calculated for each of the phenylalanine mutants (Fig. 7.5c). The Tyr29Phe mutant had a decreased affinity for tri-alanine, IC50 of 1.4 mM compared with 0.4 mM for the WT, whilst still retaining WT affinity levels for di-Glu. The Tyr68Phe mutant displays a decreased affinity for di-Glu, IC50 of 1.63 mM compared with 0.56 mM for the WT protein whilst retaining the same affinity for tri-alanine. Taken together, these results identify helices H1 and H2 within the $PepT_{St}$ as critical sites of interaction with the di-peptides during transport.

7.1.6 Modelling the Outward Open State

In the last decade substantial progress has been made in the structural determination of secondary active transporters (Forrest et al. 2011; Daniel and Kottra 2004). These structures have revealed the presence, within the transmembrane helices, of pseudo-symmetrically related structural repeats. Within the MFS these symmetry-related repeats are contained within each of the 6-TM bundles, such that helices H1, H2 and H3 are symmetrically related to helices H4, H5 and H6 via a pseudo-twofold symmetry axis running orthogonal to the membrane plane (Nicholls et al. 2002; Radestock and Forrest 2011). This observation has led to the development of a new way to model alternate states in the transport cycle of several secondary active transporters and provides evidence that alternating access is the result of symmetrically equivalent structures: outward open being the symmetry equivalent state of inward open. The implication is that the gates, which control access to the binding sites, form part of these repeat units and that during the structural transition between outward open and inward open, these gates move relative to these pseudosymmetry axes. This strategy has been used to model the outward open state of lactose permease, LacY (Chap. 6) and reveals a structure that agrees favourably with biophysical studies that predicted such a state should exist (Hauser et al. 2005; Solcan et al. 2012; Radestock and Forrest 2011; Ernst et al. 2009; Weitz et al. 2007; Fang et al. 2000; Chiang et al. 2004). A similar strategy was used to model the outward-facing state of the POT family, using the structure of $PepT_{St}$ as the starting model (Ms. in preparation) (Fig. 7.6a). The resulting model supports our earlier prediction that helices H1, H2 and H7, H8 move apart and the proximal salt bridge connecting H2 and H7 is broken (Fig. 7.6b). The model also predicts how the intracellular gate may close fully in the outward open state. Close association between helices H4, H5 from the N-terminal bundle and H10, H11 from the C-terminal half of the transporter is observed. This is in agreement with the outward open crystal structure observed for the fucose symporter, FucP (Dang et al. 2010; Smith et al. 2013). Intriguingly, our model of the outward open state also predicts another salt bridge interaction will form during the transport cycle, between a conserved lysine on helix H4, Lys126 and the glutamate on helix H10, Glu418. Mutation of Lys126 results in loss of transport under both proton-driven and counterflow conditions, whereas mutation of Glu400 to alanine produces a protein that can still transport under counterflow, suggesting the latter is involved in di-peptide recognition and the former involved in proton coupling (Solcan et al. 2012; Kottra et al. 2002).

7.1.7 A Model for Alternating Access Within the POT Family

How do the available crystal structures and their associated functional data extend our perspective on alternating access within the POT family? Functional studies on the intracellular salt bridge between Lys126 (H4) and Glu 400 (H10) show a striking symmetry with the results obtained for Arg33 (H1) and Glu300 (H7) in the extracellular gate. A picture emerges from these observations that begins to delineate the structural mechanics of alternating access within the POT family (Fig. 7.6c). In this model the outward-facing and inward-facing states are symmetrical, and quite possibly energetically equivalent. Our current understanding of the structures and identification of the gates within the POT family now reveal that movement between outward and inward open states is very likely composed of smaller, more nuanced helical rearrangements within functional sub-bundles of helices with each of the N- and C-terminal domains. In the outward open state (Fig. 7.6a), the intracellular gate is constructed from helices H4, H5 packing against helices H10, H11 and stabilised through the salt bridge between Lys126 (H4) and Glu400 (H10). The extracellular gate, constructed from helices H1, H2 and H7, H8 remain apart and open. Upon binding proton and peptide within the binding site, helices H1, H2 and H7, H8 begin to close (Fig. 7.6b). This movement is very likely reinforced through the establishment of an electrostatic interaction between Arg31 (H2) and Glu300/Asp316 (H7) observed in the crystal structures. This movement must then couple with the weakening and eventual breaking of the intracellular salt bridge, which must result in the peeling away of helices H4, H5 from H10, H11. This is most likely achieved through an intermediate state similar to the crystal structure of $PepT_{So}$ and resulting in the release of bound proton and peptide into the interior of the cell. Particularly interesting with respect to the proposed model is the role of the inverted symmetry repeats within each of the 6-TM bundles. In this

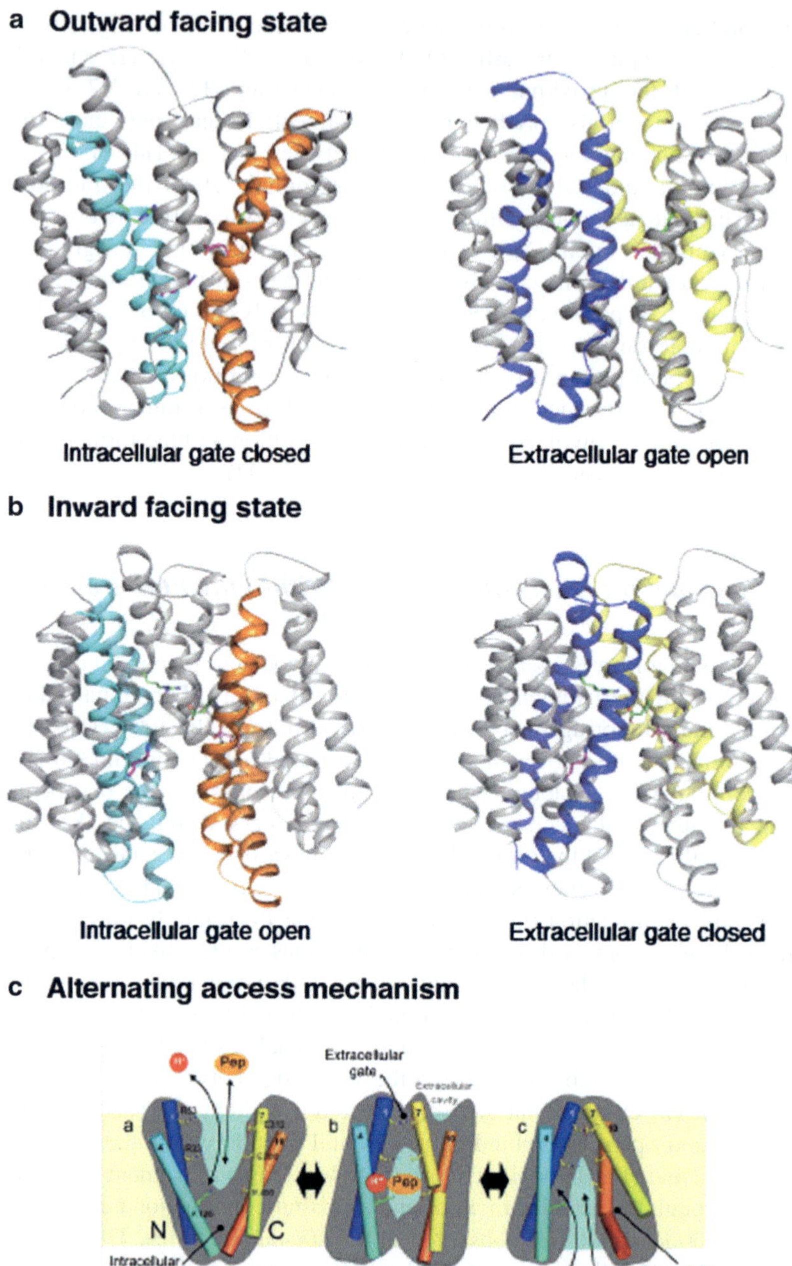

Fig. 7.6 *Alternating access mechanism within the POT family*. (**a**) Cartoon representation of the outward open model of $PepT_{St}$ built using the repeat swapped methodology developed by the Forrest laboratory for the MFS (Fei et al. 1994; Radestock and Forrest 2011). *Left*, the intracellular

model, the first repeat from each bundle contributes to the extracellular gate, whereas the second repeat contributes to the intracellular gate, with two helices acting as support units, helices H3, H6 in the N-terminal bundle and H9, H12 in the C-terminal bundle. This produces a blueprint for alternating access that places helices H1, H2, H4, H5 and H7, H8, H10, H11 in dynamic equilibrium between presumably energetically equivalent states.

7.2 Future Perspectives

The current structures of bacterial POT family members are a huge leap forward in our understanding of transport within this family, particularly with respect to PepT1 and PepT2 (Chen et al. 1999; Terada and Inui 2012). However, there exist considerable differences that are currently not reconciled. Principle among these is the structure and role of the extracellular domain inserted between helices H9 and H10 in the mammalian proteins, but absent in their prokaryotic and fungal counterparts. The role of the conserved histidine, H57, on helix H2 is also unclear. This residue is essential in both PepT1 and $PepT_{So}$, yet absent from many of the bacterial proteins, including $PepT_{St}$. The nature of the binding site in PepT1 and PepT2 is now however much clearer following the structural determination of the bacterial proteins. The peptide-binding site explains the observed preference for dipeptides being ~5 Å in length; longer peptides would be unable to coordinate between the basic and acidic charges present and identifies the location of the predicted hydrophobic pocket. The role of the dipole across the binding site will still require further

Fig. 7.6 (continued) gate constructed from helices H4, 5 and H10, H11 is seen tightly sealed and stabilised by a salt bridge between Lys126 from the N-terminal bundle and Glu400 from the opposing C-terminal bundle. *Right*, the symmetry equivalent extracellular gate, made from helices H1, H2 from the N-terminal bundle and helices H7, H8 from the C-terminal bundle are open, allowing access to the central cavity from the extracellular side of the membrane for a peptide and proton to bind. (**b**) Cartoon representation of the inward open state, taken from the crystal structure of $PepT_{St}$ (PDB: 4APS). *Left*, the intracellular gate is now open, with the Lys126 and Glu400 salt bridge broken. *Right*, the extracellular gate is now tightly sealed, stabilised by the proximal and distal salt bridge described in the text. (**c**) A model for proton-driven peptide transport by $PepT_{St}$ built using a combination of structural comparison and mutagenesis data. Here alternating access is achieved through the binding of proton and peptide to the outward-facing state A that initiates closing of the extracellular gate to occlude the peptide and proton within the central cavity B. In the occluded state(s), of which there may be more than one, rearrangement of internal salt bridge and hydrogen bond networks are very likely responsible for weakening the intracellular salt bridge while simultaneously strengthening the proximal salt bridge in the extracellular gate. Helices H4, H5 and H10, H11 must then move apart allowing exit of the proton and release of the peptide into the interior of the cell. (**c**) This is likely to occur following a mechanism detailed in Fig. 7.5 and involving small structural rearrangements within the transmembrane helices. Return to the outward-facing state must occur via an empty carrier, through an as yet undetermined mechanism

investigation, as it is unclear whether this may help to orientate peptides within the binding site or serve another function in relation to proton coupling. Structural determination of a peptide-bound POT family transporter will yield substantial new insight and is eagerly awaited by the peptide transport community. Being able to identify whether peptides are bound vertically or horizontally will answer many questions regarding the observed affinity differences for modified side chains and have important implications for drug design efforts focused on PepT1 and PepT2.

X-ray crystallography is a powerful technique for probing and understanding protein structure. However, apart from the atomic B-factors, which often cannot be refined at low resolution, it provides much less insight into the dynamics of the protein; which is often essential for fully understanding their function. This is especially true for membrane transporters, which are expected to undergo substantial conformational changes during their reaction cycles. The future however is looking bright for studies on membrane protein dynamics. The advance of spectroscopic methods like double electron-electron resonance to study conformational dynamics (Mchaourab et al. 2011) and single molecule FRET methods in both detergent and energised lipid bilayers should prove very complementary to the growing structural data from crystallography (Zhao et al. 2010, 2011). It should hopefully not be too long before we can study the individual events in our respective transporters; such as the rate of ligand binding and couple these measurements to the conformational changes they induce to reveal the key mechanistic switches that make these fascinating proteins work.

References

Abramson J, Smirnova I, Kasho V, Verner G, Kaback HR, Iwata S (2003) Structure and mechanism of the lactose permease of Escherichia coli. Science 301(5633):610–615

Anderson CMH, Thwaites DT (2010) Hijacking solute carriers for proton-coupled drug transport. Physiology 25(6):364–377

Bailey P, Boyd C, Bronk J, Collier I, Meredith D, Morgan K, Temple C (2000) How to make drugs orally active: a substrate template for peptide transporter PepT1. Angew Chem Int Ed Engl 39:505–508

Biegel A, Knutter I, Hartrodt B, Gebauer S, Theis S, Luckner P et al (2006) The renal type H+/peptide symporter PEPT2: structure-affinity relationships. Amino Acids 31(2):137–156

Bolger MB, Haworth IS, Yeung AK, Ann D, von Grafenstein H, Hamm-Alvarez S, Okamoto CT, Kim KJ, Basu SK, Wu S, Lee VH (1998) Structure, function, and molecular modeling approaches to the study of the intestinal dipeptide transporter PepT1. J Pharm Sci 87: 1286–1291

Brandsch M (2009) Transport of drugs by proton-coupled peptide transporters: pearls and pitfalls. Expert Opin Drug Metab Toxicol 5(8):887–905

Brandsch M, Knütter I, Bosse-Doenecke E (2008) Pharmaceutical and pharmacological importance of peptide transporters. J Pharm Pharmacol 60(5):543–585

Casagrande F, Harder D, Schenk A, Meury M, Ucurum Z, Engel A et al (2009) Projection structure of DtpD (YbgH), a prokaryotic member of the peptide transporter family. J Mol Biol 394 (4):708–717

Chen X-Z, Zhu T, Smith DE, Hediger MA (1999) Stoichiometry and kinetics of the high-affinity H +-coupled peptide transporter PepT2. J Biol Chem 274(5):2773–2779

Chen XZ, Steel A, Hediger MA (2000) Functional roles of histidine and tyrosine residues in the H (+)-peptide transporter PepT1. Biochem Biophys Res Commun 272(3):726–730

Chiang C-S, Stacey G, Tsay Y-F (2004) Mechanisms and functional properties of two peptide transporters, AtPTR2 and fPTR2. J Biol Chem 279(29):30150–30157

Covitz KM, Amidon GL, Sadée W (1998) Membrane topology of the human dipeptide transporter, hPEPT1, determined by epitope insertions. Biochemistry 37(43):15214–15221

Dang S, Sun L, Huang Y, Lu F, Liu Y, Gong H et al (2010) Structure of a fucose transporter in an outward-open conformation. Nature 467(7316):734–738

Daniel H, Kottra G (2004) The proton oligopeptide cotransporter family SLC15 in physiology and pharmacology. Pflugers Arch 447(5):610–618

Daniel H, Rubio-Aliaga I (2003) An update on renal peptide transporters. Am J Physiol Renal Physiol 284(5):F885–F892

Daniel H, Spanier B, Kottra G, Weitz D (2006) From bacteria to man: archaic proton-dependent peptide transporters at work. Physiology 21:93–102

Döring F, Dorn D, Bachfischer U, Amasheh S, Herget M, Daniel H (1996) Functional analysis of a chimeric mammalian peptide transporter derived from the intestinal and renal isoforms. J Physiol (Lond) 497(Pt 3):773–779

Ernst HA, Pham A, Hald H, Kastrup JS, Rahman M, Mirza O (2009) Ligand binding analyses of the putative peptide transporter YjdL from E. coli display a significant selectivity towards dipeptides. Biochem Biophys Res Commun 389(1):112–116

Fang G, Konings WN, Poolman B (2000) Kinetics and substrate specificity of membrane-reconstituted peptide transporter DtpT of Lactococcus lactis. J Bacteriol 182(9):2530–2535

Fei YJ, Kanai Y, Nussberger S, Ganapathy V, Leibach FH, Romero MF et al (1994) Expression cloning of a mammalian proton-coupled oligopeptide transporter. Nature 368(6471):563–566

Fei YJ, Liu W, Prasad PD, Kekuda R, Oblak TG, Ganapathy V et al (1997) Identification of the histidyl residue obligatory for the catalytic activity of the human H+/peptide cotransporters PEPT1 and PEPT2. Biochemistry 36(2):452–460

Forrest LR, Krämer R, Ziegler C (2011) The structural basis of secondary active transport mechanisms. Biochim Biophys Acta 1807(2):167–188

Hauser M, Kauffman S, Naider F, Becker JM (2005) Substrate preference is altered by mutations in the fifth transmembrane domain of Ptr2p, the di/tri-peptide transporter of Saccharomyces cerevisiae. Mol Membr Biol 22(3):215–227

Hirai T, Heymann JAW, Shi D, Sarker R, Maloney PC, Subramaniam S (2002) Three-dimensional structure of a bacterial oxalate transporter. Nat Struct Biol 9(8):597–600

Huang Y, Lemieux MJ, Song J, Auer M, Wang D-N (2003) Structure and mechanism of the glycerol-3-phosphate transporter from Escherichia coli. Science 301(5633):616–620

Jardetzky O (1966) Simple allosteric model for membrane pumps. Nature 211(5052):969–970

Jensen JM, Ismat F, Szakonyi G, Rahman M, Mirza O (2012) Probing the putative active site of YjdL: an unusual proton-coupled oligopeptide Transporter from E. coli. PLoS ONE 7(10): e47780

Knütter I, Hartrodt B, Theis S, Foltz M, Rastetter M, Daniel H et al (2004) Analysis of the transport properties of side chain modified dipeptides at the mammalian peptide transporter PEPT1. Eur J Pharm Sci 21(1):61–67

Kottra G, Stamfort A, Daniel H (2002) PEPT1 as a paradigm for membrane carriers that mediate electrogenic bidirectional transport of anionic, cationic, and neutral substrates. J Biol Chem 277(36):32683–32691

Kunji E, Michel H, Neutze R, Newstead S (2011) Overcoming barriers to membrane protein structure determination. Nat Biotechnol 29(4):335–340

Law CJ, Maloney PC, Wang D-N (2008) Ins and outs of major facilitator superfamily antiporters. Annu Rev Microbiol 62:289–305

Leibach FH, Ganapathy V (1996) Peptide transporters in the intestine and the kidney. Annu Rev Nutr 16:99–119

Lemieux MJ, Huang Y, Wang D-N (2004) Glycerol-3-phosphate transporter of Escherichia coli: structure, function and regulation. Res Microbiol 155(8):623–629

Liang R, Fei YJ, Prasad PD, Ramamoorthy S, Han H, Yang-Feng TL et al (1995) Human intestinal H+/peptide cotransporter. Cloning, functional expression, and chromosomal localization. J Biol Chem 270(12):6456–6463

Luckner P, Brandsch M (2005) Interaction of 31 beta-lactam antibiotics with the H+/peptide symporter PEPT2: analysis of affinity constants and comparison with PEPT1. Eur J Pharm Biopharm 59(1):17–24

Matthews DM (1975) Intestinal absorption of peptides. Physiol Rev 55(4):537–608

Matthews DM (1991) Protein absorption: development and present state of the subject. Wiley-Liss, New York

Mchaourab HS, Steed PR, Kazmier K (2011) Toward the fourth dimension of membrane protein structure: insight into dynamics from spin-labeling EPR spectroscopy. Structure 19(11):1549–1561

Meredith D, Boyd CA (2000) Structure and function of eukaryotic peptide transporters. Cell Mol Life Sci 57(5):754–778

Moraes I, Evans G, Sanchez-Weatherby J, Newstead S, Stewart PDS (2013) Membrane protein structure determination – The next generation. Biochim Biophys Acta 1838(1):78–87

Newman MJ, Foster DL, Wilson TH, Kaback HR (1981) Purification and reconstitution of functional lactose carrier from Escherichia coli. J Biol Chem 256(22):11804–11808

Newstead S (2011) Towards a structural understanding of drug and peptide transport within the proton-dependent oligopeptide transporter (POT) family. Biochem Soc Trans 39(5):1353–1358

Newstead S, Drew D, Cameron AD, Postis VLG, Xia X, Fowler PW et al (2011) Crystal structure of a prokaryotic homologue of the mammalian oligopeptide-proton symporters, PepT1 and PepT2. EMBO J 30(2):417–426

Nicholls DG, Ferguson SJ, Ferguson S (2002) Bioenergetics. Academic, London

Pao SS, Paulsen IT, Saier MH (1998) Major facilitator superfamily. Microbiol Mol Biol Rev 62 (1):1–34

Paulsen IT, Skurray RA (1994) The POT family of transport proteins. Trends Biochem Sci 19 (10):404

Pieri M, Gan C, Bailey P, Meredith D (2009) The transmembrane tyrosines Y56, Y91 and Y167 play important roles in determining the affinity and transport rate of the rabbit proton-coupled peptide transporter PepT1. Int J Biochem Cell Biol 41(11):2204–2213

Radestock S, Forrest LR (2011) The alternating-access mechanism of MFS transporters arises from inverted-topology repeats. J Mol Biol 407(5):698–715

Rautio J, Kumpulainen H, Heimbach T, Oliyai R, Oh D, Järvinen T et al (2008) Prodrugs: design and clinical applications. Nat Rev Drug Discov 7(3):255–270

Smith DE, Clémençon B, Hediger MA (2013) Proton-coupled oligopeptide transporter family SLC15: physiological, pharmacological and pathological implications. Mol Aspects Med 34 (2–3):323–336

Solcan N, Kwok J, Fowler PW, Cameron AD, Drew D, Iwata S et al (2012) Alternating access mechanism in the POT family of oligopeptide transporters. EMBO J 31(16):3411–3421

Sonoda Y, Newstead S, Hu N-J, Alguel Y, Nji E, Beis K et al (2011) Benchmarking membrane protein detergent stability for improving throughput of high-resolution X-ray structures. Structure 19(1):17–25

Steiner H-Y, Naider F, Becker JM (1995) The PTR family: a new group of peptide transporters. Mol Microbiol 16(5):825–834

Steinhardt HJ, Adibi SA (1986) Kinetics and characteristics of absorption from an equimolar mixture of 12 glycyl-dipeptides in human jejunum. Gastroenterology 90(3):577–582

Terada T, Inui K-I (2012) Chapter Eight – recent advances in structural biology of peptide transporters. In: Bevensee MO (ed) Current topics in membranes. Elsevier, Amsterdam, pp 257–274

Terada T, Saito H, Sawada K, Hashimoto Y, Inui K (2000) N-terminal halves of rat H+/peptide transporters are responsible for their substrate recognition. Pharm Res 17(1):15–20

Uchiyama T, Kulkarni AA, Davies DL, Lee VHL (2003) Biophysical evidence for His57 as a proton-binding site in the mammalian intestinal transporter hPepT1. Pharm Res 20(12):1911–1916

Weitz D, Harder D, Casagrande F, Fotiadis D, Obrdlik P, Kelety B et al (2007) Functional and structural characterization of a prokaryotic peptide transporter with features similar to mammalian PEPT1. J Biol Chem 282(5):2832–2839

Xu L, Haworth IS, Kulkarni AA, Bolger MB, Davies DL (2009) Mutagenesis and cysteine scanning of transmembrane domain 10 of the human dipeptide transporter. Pharm Res 26:2358–2366

Yan N (2013) Structural advances for the major facilitator superfamily (MFS) transporters. Trends Biochem Sci 38(3):151–159

Yang B, Smith D (2012) Significance of Peptide Transporter 1 in the intestinal permeability of Valacyclovir in wild-type and PepT1 knockout mice. Drug Metab Dispos 41(3):608–614

Yin Y, He X, Szewczyk P, Nguyen T, Chang G (2006) Structure of the multidrug transporter EmrD from Escherichia coli. Science 312(5774):741–744

Zhao Y, Terry D, Shi L, Weinstein H, Blanchard SC (2010) Single-molecule dynamics of gating in a neurotransmitter transporter homologue. Nature 465(7295):188–193

Zhao Y, Terry DS, Shi L, Quick M, Weinstein H, Blanchard SC et al (2011) Substrate-modulated gating dynamics in a Na + -coupled neurotransmitter transporter homologue. Nature 474 (7349):109–113

Chapter 8
The Maltose ABC Transporter: Where Structure Meets Function

Cédric Orelle, Michael L. Oldham, and Amy L. Davidson

Abstract ATP-binding cassette (ABC) transporters mediate active transport of a variety of substrates across biological membranes. The long studied maltose importer from *Escherichia coli* has provided a wealth of biochemical and genetic information, making it an attractive target for structure determination. The maltose transporter is a complex composed of a heterodimer of transmembrane subunits and a nucleotide-binding subunit homodimer that together function in concert with a periplasmic-binding protein, responsible for delivery of maltose to the transporter. The complex has now been crystallized in multiple states along the translocation pathway. Functional studies were key for stabilizing the transporter in these conformational states, validating the structures and understanding the molecular mechanism of transport. ATP hydrolysis and substrate transport are coupled via concerted conformational changes spanning the bilayer that result in alternating access of the substrate-binding site to either side of the membrane. Although the detailed dynamics of the complex remain to be revealed, we illustrate here how the interplay between biochemical, biophysical, and structural studies established the maltose transporter as one of the best characterized proteins of the ABC family. While many of these molecular insights are relevant to the entire ABC transporter family, some mechanistic features may differ between subclasses, highlighting the extraordinary diversity of ABC proteins and the need for further structure/function analyses.

C. Orelle
Department of Chemistry, Purdue University, West Lafayette, IN, USA

Center for Pharmaceutical Biotechnology, University of Illinois, Chicago, IL, USA
e-mail: corelle@uic.edu

M.L. Oldham (✉)
Department of Biological Sciences, Purdue University, Howard Hughes Medical Institute, West Lafayette, IN, USA
e-mail: mloldham@purdue.edu

A.L. Davidson
Department of Chemistry, Purdue University, West Lafayette, IN, USA

R. Krämer and C. Ziegler (eds.), *Membrane Transport Mechanism*, Springer Series in Biophysics 17, DOI 10.1007/978-3-642-53839-1_8,

Keywords ATP hydrolysis • Electron Paramagnetic Resonance (EPR) • Transition state • Inducer exclusion • Maltodextrin • Allostery

8.1 The Maltose Transporter MalFGK$_2$: A Member of the ABC Protein Superfamily

8.1.1 Three Classes of ABC Proteins

ATP-binding cassette (ABC) proteins comprise one of the largest protein superfamilies with representatives from all kingdoms of life (Higgins 2001). These proteins use the action of several conserved sequence motifs in deriving energy from ATP binding and hydrolysis in order to perform diverse mechanical functions. Based on phylogeny, function, and general architecture, ABC proteins can be segregated into three classes: exporters, importers, and non-transporters (Dassa 2011).

Most ABC proteins are membrane protein transporters that translocate a diverse array of substrates across cellular membranes, including ions, metals, peptides, proteins, polysaccharides, lipids, fatty acids, and cholesterol (Rees et al. 2009). ABC transporters minimally contain two transmembrane domains (TMD) that form the substrate translocation pathway and two cytoplasmic nucleotide-binding domains (NBD) that energize the system. These four core domains can occur in multiple arrangements on a single polypeptide or as separate subunits (Biemans-Oldehinkel et al. 2006). ABC transporters function either as exporters or importers and are usually specific for a particular substrate class. There is poor sequence conservation among the TMDs, which are adapted to the binding and translocation of their specific substrates, whereas there is much greater sequence conservation among the NBDs that share the common ATP hydrolytic function (Isenbarger et al. 2008; Higgins 2001). While exporters are found in both eukaryotes and prokaryotes, importers are found exclusively in prokaryotic organisms (Davidson et al. 2008).

Exporters are not only critical in the transport of diverse physiological substrates but are also involved in protective functions through the export of toxins and xenobiotics. Mutations in many of the 48 human ABC transporters have been linked to various diseases or disorders such as cystic fibrosis or adrenoleukodystrophy [see for reviews Linton and Holland (2011) and Dean (2005)]. In addition, other exporters participate in chemotherapeutic resistance in cancer cells by exporting cytotoxic drugs (Sharom 2008). For instance, P-glycoprotein is a multidrug transporter that translocates a myriad of structurally unrelated drugs.

Other soluble ABC proteins couple ATP hydrolysis to processes unrelated to transport, including chromosome organization (SMC proteins), DNA repair (e.g., UvrA, Rad50), gene regulation (e.g., eEF3, Gcn20p, Uup), translation (e.g.,

ABCE1), and drug resistance (e.g., UvrA, VgA proteins) [see for review Dassa (2011)]. A well-known example is the DNA double-strand break repair Rad50 protein. Rad50 contains a single NBD, in which two 40 nm long coiled-coil domains are separated by a zinc hook. Two Rad50 molecules assemble to form the functional unit with a head-to-tail arrangement of the NBDs. Importantly, the head-to-tail arrangement of the ATPase domains, which is conserved in all ABC systems, was historically discovered with the crystal structure of Rad50 (Hopfner et al. 2000).

8.1.2 The Importer Class

Importers mediate the uptake of a variety of nutrients in bacteria (Cui and Davidson 2011), including ions, amino acids, peptides, saccharides, vitamins, nucleosides, hemes, and siderophores. There are two main classes of importers. The larger class contains importers that utilize an extracytoplasmic substrate-binding protein (SBP) for high-affinity capture of substrates. In these binding protein-dependent (BP-dependent) transport systems, the accessory proteins not only deliver substrate to the transporter but also promote conformational changes necessary for transport. In Gram-negative bacteria, SBPs are soluble proteins located in the periplasmic space. In Gram-positive bacteria, SBPs are either tethered to the membrane as lipoproteins or fused to the TMDs (van der Heide and Poolman 2002). The BP-dependent transporters can be further divided into type I and type II importers (Locher 2009). First established by the structure of the molybdate/tungstate transporter $ModB_2C_2A$ from *Archaeoglobus fulgidus* and later seen for the *E. coli* maltose ($MalFGK_2$) and methionine transporters (MetNI), the TMDs of type I importers, which contain a core of five TM helices, dimerize to form the translocation pathway. Type II importers contain larger TMDs with a distinct architecture and generally transport larger substrates (metal chelates, heme, and vitamin B_{12}). Type II importers were first established by the structure of the *E. coli* vitamin B_{12} transporter BtuCD.

Members of the second class of ABC importers do not utilize SBPs and have been recently termed as energy-coupling factor (ECF) transporters (Rodionov et al. 2009). These transporters uptake micronutrients, including vitamins (biotin, thiamine, riboflavin) and transition-metal ions (nickel, cobalt). They contain a pair of ABC nucleotide-binding domains (A_2), a high-affinity substrate-binding transmembrane subunit (S), and a transmembrane-coupling subunit (T) (Eitinger et al. 2011). Given their picomolar to nanomolar dissociation constants to micronutrients, S components are thought to function as scavengers (Erkens et al. 2012). Two ECF transporters have been recently crystallized (Wang et al. 2013; Xu et al. 2013).

8.2 Early Genetic and Biochemical Studies Established the Maltose Transporter as an Attractive Candidate for Crystallization

A wealth of knowledge on the maltose transporter from both genetic and biochemical studies long preceded successful structural studies of ABCs (Boos and Shuman 1998). The *E. coli* maltose system was part of the earliest work on bacterial genetics from Jacob and Monod. Interest in the system continued with its use as a model system of a positively regulated regulon and the enigmatic connection between *E. coli* maltose metabolism and susceptibility to λ phage infection. The λ phage receptor was later identified as a homotrimeric outer membrane porin (LamB or maltoporin) responsible for entrance of maltose and other maltodextrins into the periplasm (Luckey and Nikaido 1980; Schirmer et al. 1995; Szmelcman and Hofnung 1975).

Bacterial transport systems mediating the uptake of nutrients were extensively studied during the 1960s and 1970s (Higgins 2001). Three types of systems were revealed: phosphotransferase systems, secondary transporters energized by an electrochemical gradient, and primary transporters energized directly by the hydrolysis of ATP. The primary transporters were sensitive to osmotic shock due to the loss of a SBP from the periplasm. In the 1980s, gene sequences for each maltose transporter component were determined (Duplay et al. 1984; Froshauer and Beckwith 1984; Dassa and Hofnung 1985; Gilson et al. 1982). The bacterial maltose/maltodextrin-import system is referred to as $MalEFGK_2$, where MalE is the periplasmic maltose-binding protein (MBP), MalF and MalG are the TMDs, and two MalKs form a homodimer of NBDs.

MBP was shown to be essential for maltose uptake in vivo (Shuman 1982), and it was the first transporter component to be extensively characterized and crystallized (Spurlino et al. 1991). MBP binds maltose and other maltodextrins with high affinity (Kd ~1 μM) (Kellermann and Szmelcman 1974) and its expression can be induced by maltose to very high concentrations in the periplasm (~1 mM) (Dietzel et al. 1978). MBP contains two globular lobes (N and C lobes) that sandwich the bound substrate (Fig. 8.1). Upon sugar binding, the lobes rotate from an open, apo form to a closed, holo form. MBP closure has been robustly characterized by biochemical, biophysical, and structural studies [see for review Davidson et al. (2008)]. In addition to its role in substrate delivery and activation of the maltose transporter, MBP is also involved in maltose chemotaxis by interacting with the Tar chemoreceptor localized at the cytoplasmic membrane (Hazelbauer 1975; Falke and Hazelbauer 2001).

Studies on histidine and maltose transporters demonstrated that ATP hydrolysis was the energy source for ABC transport systems (Dean et al. 1989; Bishop et al. 1989). Overexpression, purification, and reconstitution of maltose transporter into liposomes indicated that ATPase activity and maltose transport were strongly dependent on the presence of MBP (Davidson and Nikaido 1990, 1991). These observations suggested that periplasmic MBP initiates a critical transmembrane

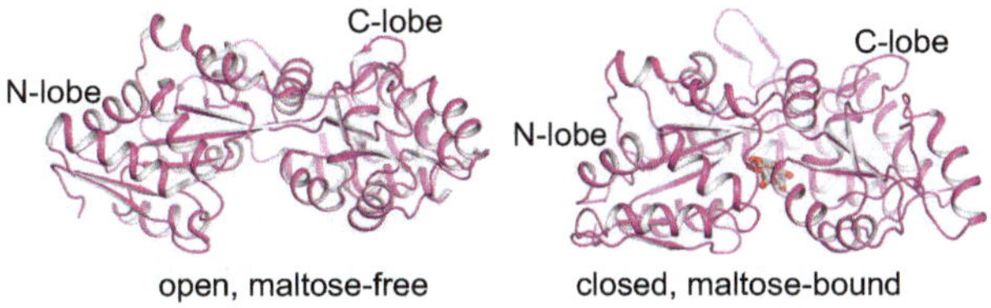

Fig. 8.1 *Binding of maltose results in closure of the N and C lobes of MBP*. Structures of open (maltose free; PDB 1OMP) and closed (maltose bound; PDB 1ANF) MBP

signaling event to MalF and MalG resulting in an ATP hydrolysis competent form of the cytosolic MalKs. The subunit stoichiometry of the transporter, $MalFGK_2$, was determined after immunoprecipitation of the complex from detergent-solubilized *E. coli* membranes (Davidson and Nikaido 1991). Moreover, positive cooperativity for transporter ATPase activity (Hill number ~1.4) indicated coordination in the binding of two ATP molecules by the two MalK subunits (Davidson et al. 1996). In contrast to the SBP dependence of the wild-type transporter, mutants that do not require MBP for maltose transport were discovered from in vivo selection (Shuman 1982). *E. coli* revertants of a strain carrying a *malE* deletion were selected on maltose as the sole carbon source. Binding protein-independent (BPI) revertant strains carried a pair of mutations that mapped to either *malF* or *malG* (Treptow and Shuman 1985). These strains transported maltose specifically but with a much higher Km (~2 mM) than wild type (~1 μM) (Manson et al. 1985; Szmelcman et al. 1976), suggesting that the TMDs might contain a low affinity substrate-binding site. In addition, mutant $MalFGK_2$ complexes reconstituted in liposomes maintained ATP hydrolysis activity in the absence of MBP (Davidson et al. 1992).

Although cooperativity in ATP hydrolysis was only observed in a subset of other intact ABC transporters (e.g., histidine permease, BmrA, ABCC3), several mutational and biochemical studies suggested that the NBDs do not function independently. Mutation of the Walker A lysine in a single MalK subunit drastically reduced transporter activity (Davidson and Sharma 1997), and similar observations were reported with P-glycoprotein using mutagenesis and chemical modification (Loo and Clarke 1995; Urbatsch et al. 1998). Early structural work on the ABCs provided unclear assignment of the physiological ABC dimer; crystal structures of ABC NBDs showed five different dimeric arrangements (Yuan et al. 2001). It was suggested that ABCs might form dimers in a head-to-tail arrangement (Jones and George 1999), similar to that later observed for the ABC homolog, Rad50 (Hopfner et al. 2000). Dimerization of Rad50 is dependent on the presence of ATP. In the dimer, two ATP molecules are bound between the Walker A motif of one subunit and the LSGGQ motif of the other. Proper assignment of a Rad50-like ABC dimer was later achieved for MJ0796 ABC in which an E171Q mutation was introduced (Smith et al. 2002). This glutamate, which is immediately adjacent to the Walker B motif, was proposed as the catalytic base for ATP hydrolysis in ABC proteins (Geourjon et al. 2001). Its mutation was shown to stabilize the NBD dimer in the presence of ATP, suggesting that binding of ATP promotes the dimerization of the NBDs, while ATP hydrolysis dissociates the dimer (Moody et al. 2002).

The utilization of ATP hydrolysis transition state analogs by Senior and collaborators to study P-glycoprotein (Urbatsch et al. 1995) provided further insights into the catalytic cycle of ABC transporters. Orthovanadate (Vi or VO_4^{3-}) is an analog of inorganic phosphate that mimics the trigonal pyramidal geometry of the γ-phosphate of the ATP hydrolysis transition state (Smith and Rayment 1996). It inhibits ATPase activity by trapping the ADP product at the catalytic site. Interestingly, the intact maltose transporter was shown to be sensitive to orthovanadate (Hunke et al. 1995; Sharma and Davidson 2000), while the isolated MalK subunit was not (Morbach et al. 1993). Upon vanadate trapping, MBP became tightly associated to the $MalFGK_2$ transporter and simultaneously lost its high affinity for maltose (Chen et al. 2001). Tightly bound MBP was proposed to be in an open conformation, which could explain its reduced affinity for maltose. These observations suggested that MBP stimulated the ATPase activity of the transporter by stabilizing the ATP hydrolysis transition state conformation of $MalFGK_2$. In addition, fluorescent studies showed that the ATP-binding sites of MalK were less accessible to solvent in the transition state, consistent with an activation of ATPase activity associated with conformational changes that bring the two MalK subunits closer together (Mannering et al. 2001). Finally, the head-to-tail arrangement of the ABC domains (Hopfner et al. 2000) was experimentally validated on the maltose transporter from determination of photoactive cleavage of the polypeptide backbone by orthovanadate at both the Walker A and LSGGQ motifs, thereby supporting a head-to-tail arrangement in which the ATP is sandwiched between these two motifs from neighboring MalK subunits (Fetsch and Davidson 2002). These results formed the basis of a concerted model of maltose transport (see inward-facing and outward-facing states, left and right cartoons Fig. 8.2).

To ascertain the site of interaction between the NBDs and TMDs of BP-dependent importers, Dassa and coworkers identified a conserved sequence (EAA—G——————I-LP) in the TMDs (Dassa and Hofnung 1985) predicted to be localized to an intracellular region formed by two amphipathic α-helices connected by a loop containing the invariant glycine (Saurin et al. 1994). Paired mutations of this region in both MalF and MalG resulted in mutants that were unable to transport maltose in vivo with MalK free in the cytoplasm (Mourez et al. 1997). Suppressor mutations in MalK that restored maltose transport were mapped to the helical subdomain, which is the most variable region among the NBDs (Bouige et al. 2002; Dassa and Bouige 2001). Cross-linking studies suggested that direct contact between the EAA loop and the MalK helical subdomain forms the TMD–NBD interface (Hunke et al. 2000). Moreover, because ATP modulates this interaction, it was proposed that the helical subdomain is involved in the coupling between ATP-binding/hydrolysis and transport (Mourez et al. 1998; Hunke et al. 2000).

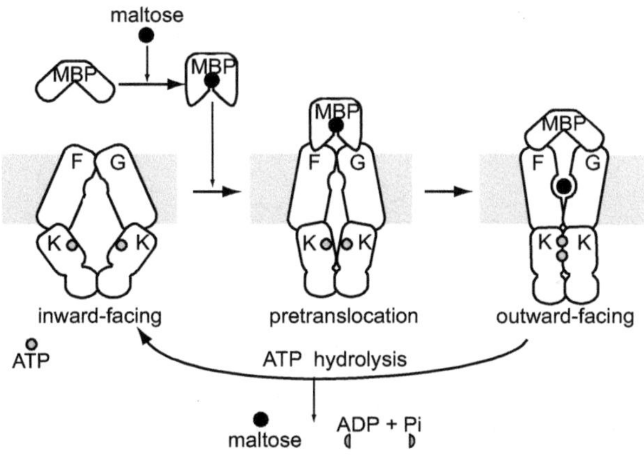

Fig. 8.2 *Model for maltose transport*. In the resting state, the NBDs are held open by the inward-facing TMDs. Upon maltose binding, MBP undergoes a transition from an open conformation to a closed conformation, generating a high-affinity substrate-binding site. Docking of substrate-bound MBP results in partial closure of the TMDs and NBDs and lowers the activation energy for reaching the outward-facing state. In the outward-facing state, MBP is tightly bound to $MalFGK_2$ reducing its affinity for maltose and facilitating transfer of the substrate to a TMD-binding site. Following ATP hydrolysis in the closed MalK dimer, the TMD binding site is re-exposed to the cytoplasm allowing maltose to be transferred into the cytosol while MBP is released from its docking atop $MalFGK_2$

8.3 Crystal Structures of $MalK_2$: First Molecular Details on the Catalytic Cycle

The MalK subunit is comprised of a nucleotide-binding domain (NBD) and a regulatory domain (RD). The NBD can be further divided into the RecA-like subdomain that is common to many ATPases and the helical subdomain that is unique to ABC proteins. The RecA domain contains several conserved sequence motifs required for ATP hydrolysis: the Walker A motif with a consensus sequence of GxxGxGKST, where x represents any amino acid; the Walker B motif, which has four hydrophobic residues followed by a negatively charged residue; and the switch region (or H-loop), Q-loop, and D-loop motifs named after their conserved residues. The helical subdomain contains the LSGGQ loop diagnostic of ABC ATPases.

Crystal structures of MalK were solved in different conformations along the proposed transport cycle (Fig. 8.3) (Chen et al. 2003; Lu et al. 2005). The fortuitous association of the RDs throughout the transport cycle maintains the MalK dimer regardless of its nucleotide-bound state. In agreement, cross-linking the two RDs had no effect on the ATPase activity of the intact transporter, demonstrating that the RDs do remain associated throughout the transport cycle (Samanta et al. 2003). Thus, proper association of the MalK NBDs could be assessed in the different nucleotide-bound states. In the absence of nucleotide, the MalK dimer was

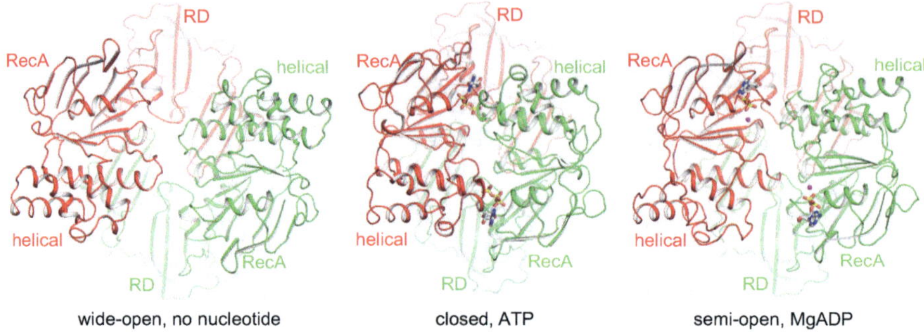

Fig. 8.3 *Conformation of the MalK dimer is nucleotide dependent.* Structures of isolated $MalK_2$ in three nucleotide-dependent conformations (nucleotide free; PDB 1Q1E, ATP bound; PDB 1Q12; MgADP bound; PDB 2AWO)

crystallized in two conformations: one with well-separated NBDs (open state), and one with an intermediate degree of opening (semi-open state). In the presence of ATP, $MalK_2$ forms a closed dimer in which two ATP molecules are sandwiched between the Walker A and LSGGQ motifs of the neighboring subunits in a head-to-tail arrangement equivalent to that Rad50 and MJ0796 ABC. In the presence of ADP, the MalK dimer was semi-open (Lu et al. 2005). Thus, the MalK dimer resembles a pair of tweezers in which the RDs act as the fixed end and the NBDs act as the pincers.

The three different conformations are mainly achieved by a rotation of the entire NBD relative to the RD, but this interdomain movement is also accompanied by a second rigid-body rotation between the RecA-like and helical subdomains. During closure, this inward rotation results in the proper orientation of the LSGGQ motif at the closed dimer interface. Closing of the MalK dimer in the presence of ATP and its reopening following ATP hydrolysis suggested a mechanism by which the conformational state of the NBD dimer might be coupled to substrate translocation through its association with the TMDs. These observations supported the model based on the studies of Rad50 and MJ0796, in which ATP-driven dimerization of the NBDs is directly coupled to transport or other mechanical motions.

8.4 Crystal Structures and Biophysical Studies of the Intact Maltose Transporter

8.4.1 Inward-Facing Resting State

The inward-facing resting state of the intact transporter was crystallized in the absence of nucleotide and MBP (Fig. 8.4a) (Khare et al. 2009). Similar to the most open structure of isolated $MalK_2$, the NBDs of the intact transporter are well separated, while the MalK subunits remain connected by the dimeric C-terminal RDs. In this state, the TMDs form an inward-facing cavity that exposes a MalF transmembrane maltose-binding site to the cytoplasm. MalF and MalG form a gate that closes this inward-facing cavity to the periplasm. In addition, the maltose transporter has a unique extended periplasmic domain connecting the third and fourth TM helices of MalF, named the P2 loop; in subsequent transport states, the P2 loop helps mediate interactions between MBP and the TMDs. In the absence of MBP, the bulk of the flexible P2 loop is absent from the resting state crystal structure. The affinity of the P2 loop to MBP has been determined as 10–20 μM (Jacso et al. 2009); given a maltose-induced 1 mM concentration of MBP in the periplasm, the P2 loop is likely to be bound to MBP. In agreement, Electron Paramagnetic Resonance (EPR) spectroscopy experiments have determined that MBP is bound to the P2 loop throughout the transport cycle (Grote et al. 2009; Jacso et al. 2009; Daus et al. 2007a, 2009). The NBD/TMD interface agrees with the proposal from Dassa and collaborators (Mourez et al. 1997). TMD-coupling helices that contain the EAA motifs dock into largely hydrophobic clefts formed between the RecA and helical subdomains of each MalK subunit (Fig. 8.4b). This transmission interface is accompanied by a salt bridge interaction in which the conserved glutamate from the EAA motif forms a salt bridge to a conserved arginine from MalK.

8.4.2 *Outward-Facing State*

The outward-facing state was initially obtained from co-crystallizing an ATP hydrolysis incompetent mutant, $MalFGK_2$(E159Q), with ATP, maltose, and MBP (Oldham et al. 2007). The mutation of the catalytic base was introduced to prevent ATP hydrolysis and stabilize a tight ATP-bound NBD dimer (Moody et al. 2002; Orelle et al. 2003; Smith et al. 2002), while addition of maltose-loaded MBP was predicted to stabilize the outward-facing conformation (Austermuhle et al. 2004; Chen et al. 2001). The same state was later obtained in the presence of AMP-PNP-Mg^{2+}, maltose, and MBP (Fig. 8.4a). The structures of the outward-facing conformation were in good agreement with the proposed model of transport (Chen et al. 2001). As observed in the isolated, closed $MalK_2$ structure, two ATP molecules are sandwiched between the Walker A and LSGGQ motifs. In the outward-facing state, a large cavity formed by MBP, MalF, and MalG extends halfway across the predicted membrane bilayer. The cavity can sufficiently accommodate maltoheptaose, the longest maltodextrin that can be transported.

At the bottom of the transmembrane cavity, maltose forms hydrogen bond and van der Waals interactions with MalF T295, Y325, S329, N376, L379, G380, S433, N437, and N440, while the reducing end participates in stacking interactions with MalF residues Y383 and F436 (Fig. 8.4c). Six of these residues had been identified through genetic studies to severely decrease or eliminate maltose transport. No residue from MalG forms direct interactions to maltose. The identification of a

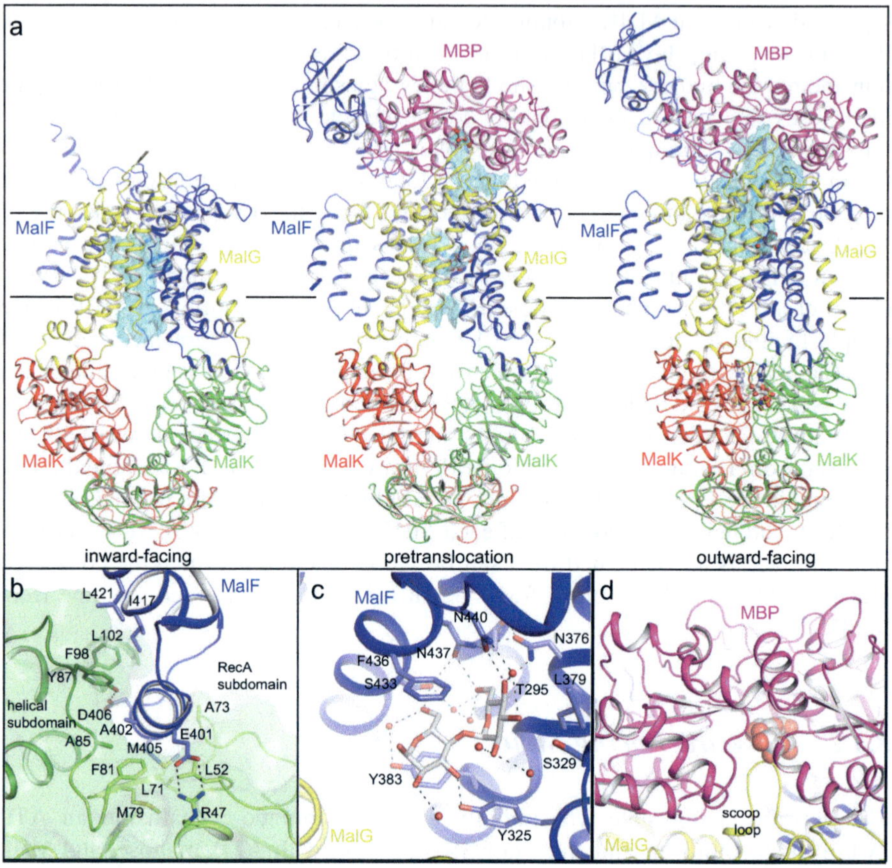

Fig. 8.4 *Structures of the intact transporter*. (**a**) Overall structures of the intact transporter in the inward-facing (PDB 3FH6), pretranslocation (PDB 3PV0), and outward-facing (PDB 3RLF) conformations. Accompanying each structure is a transparent, cyan surface that indicates the location of the TMD cavity. Solvent accessible surfaces were calculated using a 1.4 Å probe radius. (**b**) Docking of the MalF coupling helix containing the EAA loop into the hydrophobic cleft formed between the RecA and helical subdomains of MalK. *Dashed lines* indicate hydrogen bonds and salt bridges. (**c**) MalF maltose-binding site in the outward-facing state. MalF residues within van der Waals distance to maltose are shown. (**d**) Insertion of the MalG scoop loop between the open lobes of MBP in the outward-facing state. Maltose in sphere model is taken from a structure of maltose-MBP in an open form (PDB 1JW5)

single transmembrane maltose-binding site suggests that only one maltose is transported per cycle. The large outward-facing cavity is capped by the opened MBP thus shielding release of the substrate back into the periplasm. The MalG P3 loop, the "scoop" loop, that connects the fifth and sixth TM helices is inserted into the substrate-binding site at the cleft between the opened lobes of MBP and ensures release of maltose from the substrate-binding protein upon formation of the outward-facing state (Fig. 8.4d). The unanticipated function of this loop was later verified by EPR spectroscopy and mutagenesis, in which its deletion abolished maltose transport but did not strongly affect ATPase activity (Cui et al. 2010). A similar uncoupled phenotype was observed for mutation of a MalF residue at the transmembrane maltose-binding site.

8.4.3 An Alternating Access Mechanism

The maltose transporter invokes a classical alternating access mechanism, in which it alternates between inward- and outward-facing conformations, concomitantly exposing the substrate-binding site to opposite sides of the membrane (Jardetzky 1966). Closure of the MalK dimer results in rotation of the coupled TMDs and conversion to an outward-facing conformation, in which a MalF maltose-binding site is closed off from the cytoplasm and opened toward the periplasmic space. In turn, rotation of the TMDs results in the opening of MBP and release of the substrate-binding protein cargo to the transmembrane cavity. ATP hydrolysis opens the MalK subunits and thus converts the transporter back to an inward-facing conformation.

Specifically, global conformational changes for maltose transporter are a function of coordinated motions of several rigid-body groups (see also Chaps. 1, 3, and 4). MBP alternates between an open (apo) and closed (holo) conformation in which the N and C lobes of MBP form two rigid-body groups. Given a maintained interaction between the Ig-like subdomain of MalF P2 loop and the N lobe of MBP, they can be grouped as one rigid-body group. The cores of MalF and MalG, which form the translocation cavity, are composed of five TM helices. MalF and MalG are intertwined such that MalF TM1-3/MalG TM2-6 and MalF TM4-8/MalG TM1 form two independent rigid-body groups. The MalG scoop loop that inserts between the lobes of MBP in the outward-facing state is part of the MalF TM1-3/ MalG TM2-6 rigid-body group. Thus, release of maltose from the opened MBP and concurrent insertion of the scoop loop in its place is definitional to formation of the outward-facing state. The RecA and helical subdomains of MalK also act as independent rigid body groups. Rotations of the two TM rigid-body groups are transmitted through the coupling helices to the MalK subunits and reciprocally. In progression from the inward-facing state to the outward-facing state, the only interdomain interfaces that are maintained are that of the RD dimer and that of the Ig-like subdomain of the MalF P2 to the N lobe of MBP.

8.4.4 Progression from the Inward-Facing State to the Outward-Facing State

While crystal structures revealed molecular details about the maltose transporter mechanism, biochemical and biophysical studies attempted to further decipher the dynamics of transport and ATP hydrolysis. Taking advantage of the crystal structures, site-directed spin labeling and EPR spectroscopy were employed to study conformational changes at strategic positions for MalEFGK$_2$. Distance measurements were used to track the opening and closure events of both MBP (Austermuhle et al. 2004) and MalK (Orelle et al. 2008) during the catalytic cycle. In the intact transporter, MalK was observed in three different conformations: open, closed, and semi-open. As seen in the MalFGK$_2$ resting state structure in the absence of ligand (Khare et al. 2009), MalK was in an open conformation. In contrast to the isolated MalK dimer, binding of ATP alone did not induce MalK closure in the intact transporter; however, the addition of both ATP and MBP stabilized a closed MalK dimer. These results indicated that MBP stimulated ATPase activity by promoting NBD closure (Orelle et al. 2008). Conversely, ATP binding was shown to open MBP (Austermuhle et al. 2004). These data were consistent with the concerted mechanism (Chen et al. 2001) recapitulated in the inward- and outward-facing crystal structures (Fig. 8.2) (Khare et al. 2009; Oldham et al. 2007), in which closing of the MalK dimer is concomitant with opening MBP and vice versa. Interestingly, MalK$_2$ conformational states also correlated with distinct conformations of the P2 loop (Grote et al. 2009), consistent with concerted motions of the periplasmic and intracellular components of the transporter (Bohm et al. 2013).

To further analyze the progression from the inward-facing state to the outward-facing state, the conformation of MalK$_2$ was analyzed in the presence of a mutant MBP (G69C/S337C) that is locked in a closed conformation due to a disulfide cross-link between its two lobes (Zhang et al. 1996). The transporter, in detergent or functionally reconstituted in phospholipid nanodiscs (Alvarez et al. 2010), was spin labeled to measure either the closure of the MalK dimer or the relative rotation between the helical and RecA subdomains (Orelle et al. 2010). The addition of AMP-PNP-Mg^{2+} to the mutant complex induced a semi-open conformation of MalK$_2$. In contrast, the addition of AMP-PNP-Mg^{2+} to the wild-type complex triggered MalK$_2$ closure. These observations indicated that closed MBP promotes the transition from open to semi-open MalK$_2$, whereas open MBP stabilizes the closed configuration of MalK$_2$. Thus, both MBP conformations participate in the stimulation of MalFGK$_2$ ATPase activity. In addition, this study showed that complete closure of MalK$_2$ (from semi-open to closed) is accompanied by a relative rotation of the helical subdomain toward the RecA subdomain. These conformational changes are coupled to the reorientation of the TMDs and thus the opening of MBP events that promote transfer of maltose to the transporter.

The intermediate between the inward- and outward-facing states, called the pretranslocation state, was successfully crystallized in the presence of maltose-bound MBP and the absence of nucleotide (Fig. 8.4a) (Oldham and Chen 2011a). Upon docking of substrate-loaded MBP atop the periplasmic surface of MalF and MalG, the inward conformation of the TMDs narrows, resulting in segmenting of the translocation pathway into three parts. The lowest, inward-facing cavity is accessible to the cytoplasm but inaccessible to the now-shielded TM substrate-binding site. The middle semi-occluded cavity is shielded from both the cytoplasm and the periplasm and is of sufficient size to accommodate maltoheptaose. Substrate occupies the TM-binding site in the structure of the pretranslocation state given its accessibility to either side of the protein during detergent-based crystallization; however, during normal maltose transport this site would likely be free of substrate. The upper, occluded cavity is newly formed between the substrate-bound, closed MBP and the periplasmic surface of MalFG. A similar structure was previously obtained for the molybdate/tungstate transporter (Hollenstein et al. 2007). As in the outward-facing state, the P2 loop is extended from the membrane and wraps around the N-lobe of MBP.

Partial closure of the TMDs results in a coupled partial closure of the MalK NBDs. In the structure of the nucleotide-free, pretranslocation state containing a semi-open MalK dimer, several catalytic residues that are required for ATP hydrolysis are brought in contact with the neighboring NBD. Specifically, the conserved D loop of one MalK subunit hydrogen bonds with both the Walker motif S38 and the switch motif H192 from the other NBD. Molecular dynamic studies also supported the importance of these interactions in initiating MalK closure (Oloo et al. 2006). Given that both S38 and H192 also interact with the γ-phosphate of ATP in the closed MalK dimer poised for ATP hydrolysis (Fig. 8.5a), binding of two ATP molecules at the MalK dimer interface would disrupt the pretranslocation intermediate (Fig. 8.5b). In the absence of nucleotides, this conformation was not seen by EPR studies (Orelle et al. 2010; Bohm et al. 2013), perhaps due to the low affinity of maltose-bound MBP for the transporter. Given that the affinity of ATP to the open MalK dimer of the inward-facing state is unknown, it is unclear whether productive maltose transport proceeds through a nucleotide-free, pretranslocation intermediate or rather this conformation is along the trajectory between nucleotide-bound, inward-, and outward-facing states.

Given the high concentration of maltose-bound MBP and ATP with respect to transporter in vivo, it is likely that productive maltose transport proceeds through the pretranslocation conformation. The pretranslocation state was crystallized both with wild-type MBP and with the G69C/S337C mutant form where the N and Clobes of MBP are disulfide cross-linked. Soaking crystals of the wild-type and mutant complexes with AMP-PNP resulted in two different conformational states (Oldham and Chen 2011a). AMP-PNP soaking of the wild-type complex resulted in conversion within the crystal to an outward-facing conformation with clear electron density for the entire nucleotide. In contrast, AMP-PNP soaking of crystals of the mutant complex maintained a pretranslocation state conformation, as seen with EPR (Orelle et al. 2010), with nucleotides bound to the Walker A motifs of each MalK subunit. Electron density was obtained for the adenosine and the α- and β-phosphates but absent for the γ-phosphate suggesting that the γ-phosphate is less

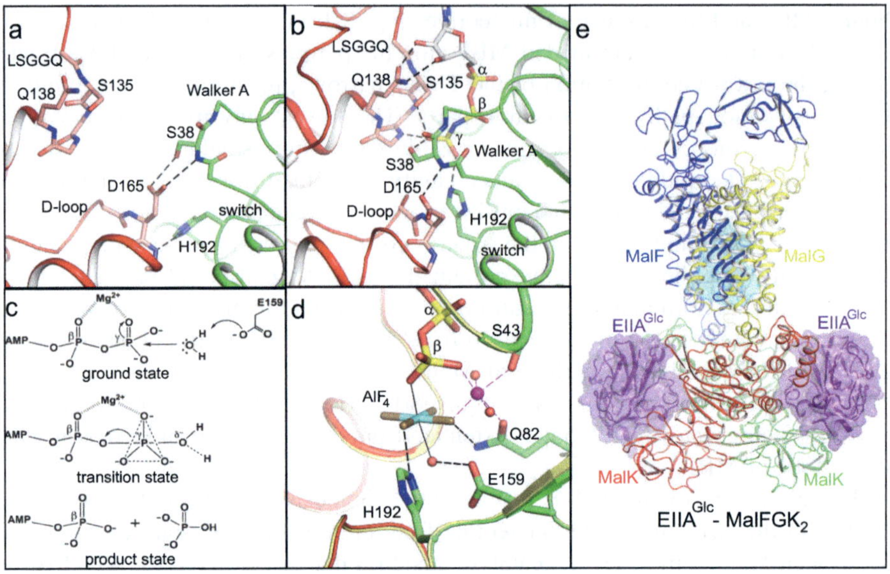

Fig. 8.5 *Mechanism of ATP hydrolysis and its regulation*. Progression from the pretranslocation state (**a**) to the outward-facing state (**b**) is accompanied by closure of the MalK dimer bringing the LSGGQ motif in contact with ATP. (**c**) General base mechanism for ATP hydrolysis. (**d**) Superposition of the ADP-AlF_4 (PDB 3PUW) and ADP-BeF_3 (PDB 3PUX) structures of the outward-facing $MalEFGK_2$ complex. The catalytic base E159, attacking water, AlF_4 ligand, and a subset of residues that interact with Mg^{2+} or the γ-phosphate analog are shown in stick format. (**e**) Structure of the $EIIA^{Glc}$–$MalFGK_2$ complex. Each $EIIA^{Glc}$ is shown as both a ribbon and a transparent surface docked between the NBD and RD of MalK

ordered in the pretranslocation state. That is, formation of a closed NBD dimer may be necessary to orient the γ-phosphate for ATP hydrolysis. In agreement, high-resolution structures of monomeric ABCs differ in the position of the γ-phosphate. Consistent with the concerted model of transport (Fig. 8.2) (Chen et al. 2001), the disulfide linking the N and C lobes of the closed MBP disallows rotation of the TMD and thus prevents nucleotide-dependent closure of the coupled NBDs.

8.4.5 Mechanism of ATP Hydrolysis

Three different ATP hydrolysis mechanisms had been proposed for the ABC transporter family. The most popular was general base catalysis invoking a conserved glutamate at the end of the Walker B motif to activate a water molecule for nucleophilic attack of the ATP γ-phosphate. In the general base mechanism, ATP hydrolysis is achieved by in-line nucleophilic attack of the ATP γ-phosphate by a water molecule in the formation of a pentacovalent transition state intermediate (Fig. 8.5c). The second mechanism was substrate-assisted catalysis in which the polarization of the attacking water is performed by the ATP γ-phosphate itself and the conserved switch region histidine stabilizes the resulting transition state. The third mechanism was akin to that suggested for myosin and kinesin in which, in the absence of a candidate catalytic base, a conserved serine, proposed to be from the LSSGQ motif for ABC transporters, bonds to the attacking water molecule.

The use of metallofluorides and the aforementioned orthovanadate (Vi) have been an invaluable tool in biochemical and structural studies of ATPases. During progression from the ground state to the transition state, the geometry of the γ-phosphate changes from tetrahedral to trigonal bipyramidal, in which the three γ-phosphate oxygens lie at the equatorial plane and the β/γ phosphate bridging oxygen and the attacking water are at axial positions. These analogs form tight complexes with ADP and can be used as mimics of the ATP hydrolysis ground and transition states. Beryllium fluoride (BeF_3) forms a complex with ADP with tetrahedral geometry that mimics the ground state. Vi and Aluminum tetrafluoride (AlF_4) form trigonal bipyramidal and octahedral complexes, respectively, with ADP. The longer apical bond from ADP-O to V or Al mimics the transition state. Structures of the maltose transporter were obtained with all three analogs in addition to AMPPNP (Oldham and Chen 2011b). Consistent with a general base mechanism, in the ADP-AlF_4 and ADP-Vi structures, the general base candidate, E159, was respectively bonded to a properly positioned attacking water molecule or the equivalent Vi apical oxygen, respectively (Fig. 8.5d). While both the switch region histidine and the LSGGQ serine interact with the nucleotide, neither was bonded to the attacking water molecule, precluding a substrate-assisted or serine-assisted mechanism. In the closed NBD dimer, the LSGGQ motif, which is required for ATP hydrolysis, is directly across the ATP-binding site. The position of the LSGGQ motif is the same as the "arginine finger" found in other RecA-like ATPases (Ye et al. 2004). For example, in F1-ATPase, the arginine finger (R373) from the α subunit inserts into the catalytic β subunit forming interactions with the ATP γ-phosphate. In these structures, the Q loop interacts with both Mg^{2+} and the γ-phosphate. The orientation of the helical subdomain with respect to the RecA subdomain is related to the coupling of the TMDs. Thus, the low ATPase activities measured for isolated NBDs are likely a function of improper positioning of the helical subdomain harboring the LSSGQ motif and the Q loop motif that links the two NBD subdomains.

The overall structures of the maltose transporter using mimics of both the ground (ADP-BeF_3, AMP-PNP) and transition states (ADP-VO_4, ADP-ALF_4) were equivalent (Fig. 8.4d), indicating that no structural rearrangements accompany formation of the transition state. Accordingly, the transition state is only a chemical intermediate and has no special properties regarding substrate translocation. Thus, reversion of the transporter back to an inward-facing state is likely a function of release of inorganic phosphate and/or ADP.

8.4.6 Post-hydrolysis

It is unclear whether one or two ATP molecules are hydrolyzed per transport cycle. If hydrolysis of two ATP is required, are they hydrolyzed concomitantly or sequentially? A molecular dynamics study performed on $MalK_2$ suggested that ATP hydrolysis at either site is sufficient to trigger dimer opening, although the hydrolysis of two ATP may occur stochastically before opening (Wen and Tajkhorshid 2008). In $MalEFGK_2$, opening of both sites of the $MalK_2$ dimer was determined to be necessary for return to the inward-facing state (Wen and Tajkhorshid 2011); however, the latter study does not exclude the possibility that on longer timescales hydrolysis of a single ATP could drive the opening of both sites. Finally, it is also unclear at which catalytic step maltose is translocated into the cell, i.e., after hydrolysis or release of products (Pi and/or ADP). EPR studies suggested that the ADP-bound post-hydrolysis conformation is inward facing with semi-open $MalK_2$ (Bohm et al. 2013; Orelle et al. 2008). Thus, the post-hydrolysis conformation might resemble the pretranslocation state, with a semi-open MalK dimer coupled to inward-facing TMDs and a transiently reclosed MBP; the resulting substrate-free, closed MBP docked to an inward-facing transporter would be consistent with a higher energy state that would favor rapid conversion back to the resting state after disengaging. Once MalK has started to open, ADP and inorganic phosphate would be free to exchange with ATP.

8.5 $MalFGK_2$ Interaction with Regulatory Proteins: EIIA and MalT

The maltose transporter interacts through the RD domains of MalK with regulatory proteins: MalT, the transcription activator of maltose regulon (Richet et al. 2012), and $EIIA^{Glc}$, the carbon catabolite repressor enzyme in *E. coli* (Bluschke et al. 2006; Dean et al. 1990; Kuhnau et al. 1991; Stein et al. 2002).

A recent crystal structure of $MalFGK_2$ complexed with $EIIA^{Glc}$ reveals how the maltose transporter is inhibited by the phosphotransferase transport system (PTS) (Fig. 8.5e) (Chen et al. 2013). $EIIA^{Glc}$, a member of the glucose phosphotransferase transport system (PTS), inhibits the maltose transporter by a mechanism termed inducer exclusion. When glucose or other PTS substrates are available, the PTS inhibits transporters and metabolic enzymes required for the utilization of less efficient carbon sources, including maltodextrins. Specifically the action of the PTS leads to dephosphorylation of $EIIA^{Glc}$, a form of the regulator that is able to bind to MalK (Magasanik 1970). In the crystal structure of the inhibited transporter complex, two copies of $EIIA^{Glc}$ are bound at symmetrically equivalent positions of the MalK dimer. In the inhibited complex, the maltose transporter is in a conformation similar to that of the resting state determined in the absence of MBP. Each $EIIA^{Glc}$ is wedged between the NBD of one MalK and the RD of the opposing

MalK subunit. Progression from the inward-facing, resting state to the pretranslocation and outward-facing states requires rotation of each NBD with respect to the RD and toward the NBD of its neighboring MalK subunit. Binding of $EIIA^{Glc}$ at the NBD/RD dimer interface locks the transporter in the resting state conformation by preventing rotations of the NBD with respect to the RD and acts as a physical barrier from closure of the MalK dimer. Consistent with the crystal structure, mutations that prevent inducer exclusion occur in both the NBD and RD at the MalK interface with $EIIA^{Glc}$ (Dean et al. 1990; Kuhnau et al. 1991).

8.6 How MBP Stimulates the ATPase Activity of $MalFGK_2$

Based on the findings previously described, MBP stimulation of $MalFGK_2$ ATPase activity would be the product of two steps (Fig. 8.2). In the absence of MBP, the TMDs impose structural constraints that prevent $MalK_2$ closure (Khare et al. 2009), lowering the likelihood of reaching the hydrolytically competent closed state (Fetsch and Davidson 2002) and resulting in a low-basal ATPase activity. Interaction with maltose-bound MBP decreases the energy barrier between the inward- and outward-facing states, promoting closure of the ATP-bound MalK dimer (Mannering et al. 2001; Grote et al. 2009; Orelle et al. 2008; Oldham et al. 2007). The crystal structure of the pretranslocation state shows that docking of maltose-bound MBP facilitates partial rotations of both the TMDs and the MalK dimer, bringing certain catalytic residues to the NBD dimer interface (Oldham and Chen 2011a). Given that the NBDs of the pretranslocation state are closer together (Oldham and Chen 2011a; Orelle et al. 2010), ATP can complete the progression to the outward-facing state in which MBP is opened to deliver the substrate. Maltose-free, open MBP also participates in the stimulation of ATPase activity by stabilizing the ATP hydrolysis transition state (Austermuhle et al. 2004; Bohm et al. 2013; Chen et al. 2001; Gould and Shilton 2010). In agreement, several studies have suggested that maltose transporter binding protein-independent (BPI) mutants may shift the equilibrium from the inward-facing state to the outward-facing state (Daus et al. 2006, 2007b; Mannering et al. 2001; Hall et al. 1997), thereby explaining how these mutants hydrolyze ATP in the absence of MBP. Accordingly, most of the primary BPI mutations lie at the inward-facing interface between MalF and MalG and are predicted to destabilize the inward-facing conformation (Khare et al. 2009).

Therefore, the stimulation of ATPase activity found in productive maltose transport is governed by direct interactions of maltose-bound MBP with the transporter; however, maltose-free MBP is capable of stimulating low levels of ATPase activity suggesting that open MBP might interact with a small population of transporters in the outward-facing conformation (Gould et al. 2009). In contrast, maltose-bound MBP is considerably more efficient in stimulating the ATPase activity of $MalFGK_2$ because the substrate directly affects the conformation of MBP and thus its capacity to interact productively with the transporter (Oldham and Chen 2011a).

8.7 How the Maltose Transporter Mechanism Relates to Other ABC Members

8.7.1 Type I Versus Type II Importers

As mentioned earlier, Locher proposed that BP-dependent importers could be divided into two classes. While maltose transporter is a type I representative member, type II importers typically include a larger number of TM helices (up to 20) and mostly transport larger substrates, such as metal chelates, heme, or vitamin B_{12}. A structural difference in the NBD/TMD transmission interface suggests a possible variation in the coupling mechanism of type II importers (Wen and Tajkhorshid 2011). The best-studied type II member is BtuCD, which transports vitamin B_{12}. Several biochemical, biophysical, and structural studies suggested that BtuCD functions differently than the maltose transporter (Joseph et al. 2011; Lewinson et al. 2010; Locher et al. 2002). In contrast to $MalFGK_2$, BtuCD displays high-basal ATPase activity (Borths et al. 2005) and predominantly rests in an outward-facing conformation. Based on several crystal structures of BtuCD(F) in different states, a model of transport was proposed (Korkhov et al. 2012). Interaction of vitamin B_{12}-bound BtuF with the outward-facing conformation of BtuCD initiates an ATP-dependent closure of BtuD. This conformational change promotes the release of vitamin B_{12} into an internal transmembrane cavity. Similar to the maltose transporter, periplasmic loops from BtuC occupy the vitamin B_{12}-binding site of BtuF; however, vitamin B_{12} does not seem to have a fixed transmembrane-binding site. Instead, the vitamin is likely transiently trapped in the translocation cavity in which both cytoplasmic and periplasmic gates are closed. After ATP hydrolysis, the NBD dimer dissociates and pulls the coupling helices apart to open the lower gate, thereby permitting substrate release into the cytoplasm.

A recent study comparing two importers with identical substrate specificity (molybdate/tungsten), one from type I and the other from type II, further supported mechanistic differences between the two types (Vigonsky et al. 2013). The first system has low affinity for its binding protein, but the addition of substrate and nucleotides increases the interaction between them. The second system formed a high affinity complex with the binding protein, and the complex is destabilized in the presence of ligands.

8.7.2 Structural and Functional Features in Exporters

The architecture of importers and exporters differ significantly. While there is structural conservation among the NBDs of importers and exporters, the TMDs are quite different. The TMDs of exporters extend well beyond the lipid bilayer with intracellular loops that contact the NBDs. Each TMD carries two coupling helices (CH1 and CH2) (Dawson and Locher 2007). Similar to the coupling helix in importers, the CH2 lies in a groove between the RecA-like and helical subdomains; however this helix comes from the opposite TMD. The other coupling helix, CH1, primarily interacts with regions near the adenine ring of ATP bound in the "cis" subunit. Progression toward the intertwined conformation of outward-facing state involves distinct conformational changes compared to importers (Oldham et al. 2008; Rees et al. 2009). Transport substrates are thought to bind to sites localized at the inner membrane (Gutmann et al. 2010; Hinz and Tampe 2012) stimulating the dimerization of the NBDs (Loo et al. 2003; Tombline and Senior 2005). Interestingly, some exporters contain asymmetric ATPase sites with degenerate ABC motifs in one of the ATP-binding sites (Hohl et al. 2012; Procko et al. 2009). In these transporters, one NBD is hydrolytically active, whereas the other one is poorly active, suggesting a possible different mechanism of transport.

8.7.3 Two General Transport Models Proposed for ABC Transporters

Two transport models have been proposed for ABC transporters [see for review George and Jones (2012)]. In the switch model, binding of two ATP molecules drives the closure of the NBDs in generation of the outward-facing conformation. Sequential hydrolysis of both ATP molecules results in full dissociation of the NBDs and a return to the inward-facing conformation. The constant contact model invokes alternating cycles of hydrolysis in each NBD. During the catalytic cycle, constant contact is maintained between the NBDs through the ATP-bound site. Hydrolysis of ATP in the other site promotes an opening at that site denoted by an outward rotation of the RecA-like subdomain relative to the helical subdomain.

Although studies on the maltose transporter are often reported as supporting the switch model, several uncertainties remain. While the crystal structures show symmetrical states, the existence of asymmetrical intermediate states, especially during ATP hydrolysis, is not excluded. The presence of several conformational populations in EPR studies was described as compatible with the constant contact model (Jones and George 2012), but spin labels themselves can adopt several conformations or orientations (called rotamers) depending on their immediate environment. The existence of asymmetric transporters, in which one NBD is hydrolytically defective, has also been suggested to support the constant contact model in ABC transporters; however, these transporters do not rule out either model, since they are unlikely to proceed via strict alternative site catalysis. In addition, ATP hydrolysis could be stochastic between the NBDs, with one ATP hydrolysis being sufficient to disrupt the dimer (Wen and Tajkhorshid 2008).

Further work will be necessary to decipher a precise mechanistic model of ABC transporters, possibly by using underexploited techniques such as single molecule fluorescence, rapid kinetics, or NMR. Given the structural and functional diversity of ABC members, it may be difficult to reveal an entirely unifying mechanism. However, the structural and dynamic characterization of the maltose transporter deeply contributed to the general understanding of ABC proteins, and the thorough study of other ABC proteins will be as valuable in achieving this goal.

References

Alvarez FJ, Orelle C, Davidson AL (2010) Functional reconstitution of an ABC transporter in nanodiscs for use in electron paramagnetic resonance spectroscopy. J Am Chem Soc 132 (28):9513–9515

Austermuhle MI, Hall JA, Klug CS, Davidson AL (2004) Maltose-binding protein is open in the catalytic transition state for ATP hydrolysis during maltose transport. J Biol Chem 279 (27):28243–28250

Biemans-Oldehinkel E, Doeven MK, Poolman B (2006) ABC transporter architecture and regulatory roles of accessory domains. FEBS Lett 580(4):1023–1035

Bishop L, Agbayani R Jr, Ambudkar SV, Maloney PC, Ames GF (1989) Reconstitution of a bacterial periplasmic permease in proteoliposomes and demonstration of ATP hydrolysis concomitant with transport. Proc Natl Acad Sci U S A 86(18):6953–6957

Bluschke B, Volkmer-Engert R, Schneider E (2006) Topography of the surface of the signal-transducing protein EIIA(Glc) that interacts with the MalK subunits of the maltose ATP-binding cassette transporter (MalFGK2) of Salmonella typhimurium. J Biol Chem 281 (18):12833–12840. doi:10.1074/jbc.M512646200

Bohm S, Licht A, Wuttge S, Schneider E, Bordignon E (2013) Conformational plasticity of the type I maltose ABC importer. Proc Natl Acad Sci U S A 110(14):5492–5497. doi:10.1073/pnas.1217745110

Boos W, Shuman H (1998) Maltose/maltodextrin system of Escherichia coli: transport, metabolism, and regulation. Microbiol Mol Biol Rev 62(1):204–229

Borths EL, Poolman B, Hvorup RN, Locher KP, Rees DC (2005) In vitro functional characterization of BtuCD-F, the Escherichia coli ABC transporter for vitamin B12 uptake. Biochemistry 44(49):16301–16309

Bouige P, Laurent D, Piloyan L, Dassa E (2002) Phylogenetic and functional classification of ATP-binding cassette (ABC) systems. Curr Protein Pept Sci 3(5):541–559

Chen J, Sharma S, Quiocho FA, Davidson AL (2001) Trapping the transition state of an ATP-binding cassette transporter: evidence for a concerted mechanism of maltose transport. Proc Natl Acad Sci U S A 98(4):1525–1530

Chen J, Lu G, Lin J, Davidson AL, Quiocho FA (2003) A tweezers-like motion of the ATP-binding cassette dimer in an ABC transport cycle. Mol Cell 12(3):651–661

Chen S, Oldham ML, Davidson AL, Chen J (2013) Carbon catabolite repression of the maltose transporter revealed by X-ray crystallography. Nature 499(7458):364–368

Cui J, Davidson AL (2011) ABC solute importers in bacteria. Essays Biochem 50(1):85–99

Cui J, Qasim S, Davidson AL (2010) Uncoupling substrate transport from ATP hydrolysis in the Escherichia coli maltose transporter. J Biol Chem 285(51):39986–39993

Dassa E (2011) Natural history of ABC systems: not only transporters. Essays Biochem 50(1):19–42

Dassa E, Bouige P (2001) The ABC of ABCS: a phylogenetic and functional classification of ABC systems in living organisms. Res Microbiol 152(3–4):211–229

Dassa E, Hofnung M (1985) Sequence of gene malG in E. coli K12: homologies between integral membrane components from binding protein-dependent transport systems. EMBO J 4 (9):2287–2293

Daus ML, Landmesser H, Schlosser A, Muller P, Herrmann A, Schneider E (2006) ATP induces conformational changes of periplasmic loop regions of the maltose ATP-binding cassette transporter. J Biol Chem 281(7):3856–3865
Daus ML, Berendt S, Wuttge S, Schneider E (2007a) Maltose binding protein (MalE) interacts with periplasmic loops P2 and P1 respectively of the MalFG subunits of the maltose ATP binding cassette transporter (MalFGK(2)) from Escherichia coli/Salmonella during the transport cycle. Mol Microbiol 66(5):1107–1122
Daus ML, Grote M, Muller P, Doebber M, Herrmann A, Steinhoff HJ, Dassa E, Schneider E (2007b) ATP-driven MalK dimer closure and reopening and conformational changes of the "EAA" motifs are crucial for function of the maltose ATP-binding cassette transporter (MalFGK2). J Biol Chem 282(31):22387–22396
Daus ML, Grote M, Schneider E (2009) The MalF P2 loop of the ATP-binding cassette transporter MalFGK2 from Escherichia coli and Salmonella enterica serovar typhimurium interacts with maltose binding protein (MalE) throughout the catalytic cycle. J Bacteriol 191(3):754–761
Davidson AL, Nikaido H (1990) Overproduction, solubilization, and reconstitution of the maltose transport system from Escherichia coli. J Biol Chem 265(8):4254–4260
Davidson AL, Nikaido H (1991) Purification and characterization of the membrane-associated components of the maltose transport system from Escherichia coli. J Biol Chem 266(14):8946–8951
Davidson AL, Sharma S (1997) Mutation of a single MalK subunit severely impairs maltose transport activity in Escherichia coli. J Bacteriol 179(17):5458–5464
Davidson AL, Shuman HA, Nikaido H (1992) Mechanism of maltose transport in Escherichia coli: transmembrane signaling by periplasmic binding proteins. Proc Natl Acad Sci U S A 89 (6):2360–2364
Davidson AL, Laghaeian SS, Mannering DE (1996) The maltose transport system of Escherichia coli displays positive cooperativity in ATP hydrolysis. J Biol Chem 271(9):4858–4863
Davidson AL, Dassa E, Orelle C, Chen J (2008) Structure, function, and evolution of bacterial ATP-binding cassette systems. Microbiol Mol Biol Rev 72(2):317–364, table of contents
Dawson RJ, Locher KP (2007) Structure of the multidrug ABC transporter Sav 1866 from Staphylococcus aureus in complex with AMP-PNP. FEBS Lett 581(5):935–938
Dean M (2005) The genetics of ATP-binding cassette transporters. Methods Enzymol 400:409–429
Dean DA, Davidson AL, Nikaido H (1989) Maltose transport in membrane vesicles of Escherichia coli is linked to ATP hydrolysis. Proc Natl Acad Sci U S A 86(23):9134–9138
Dean DA, Reizer J, Nikaido H, Saier MH Jr (1990) Regulation of the maltose transport system of Escherichia coli by the glucose-specific enzyme III of the phosphoenolpyruvate-sugar phosphotransferase system. Characterization of inducer exclusion-resistant mutants and reconstitution of inducer exclusion in proteoliposomes. J Biol Chem 265(34):21005–21010
Dietzel I, Kolb V, Boos W (1978) Pole cap formation in Escherichia coli following induction of the maltose-binding protein. Arch Microbiol 118(2):207–218
Duplay P, Bedouelle H, Fowler A, Zabin I, Saurin W, Hofnung M (1984) Sequences of the malE gene and of its product, the maltose-binding protein of Escherichia coli K12. J Biol Chem 259 (16):10606–10613
Eitinger T, Rodionov DA, Grote M, Schneider E (2011) Canonical and ECF-type ATP-binding cassette importers in prokaryotes: diversity in modular organization and cellular functions. FEMS Microbiol Rev 35(1):3–67
Erkens GB, Majsnerowska M, ter Beek J, Slotboom DJ (2012) Energy coupling factor-type ABC transporters for vitamin uptake in prokaryotes. Biochemistry 51(22):4390–4396
Falke JJ, Hazelbauer GL (2001) Transmembrane signaling in bacterial chemoreceptors. Trends Biochem Sci 26(4):257–265
Fetsch EE, Davidson AL (2002) Vanadate-catalyzed photocleavage of the signature motif of an ATP-binding cassette (ABC) transporter. Proc Natl Acad Sci U S A 99(15):9685–9690

Froshauer S, Beckwith J (1984) The nucleotide sequence of the gene for malF protein, an inner membrane component of the maltose transport system of Escherichia coli. Repeated DNA sequences are found in the malE-malF intercistronic region. J Biol Chem 259(17):10896–10903

George AM, Jones PM (2012) Perspectives on the structure-function of ABC transporters: the Switch and Constant Contact models. Prog Biophys Mol Biol 109(3):95–107

Geourjon C, Orelle C, Steinfels E, Blanchet C, Deleage G, Di Pietro A, Jault JM (2001) A common mechanism for ATP hydrolysis in ABC transporter and helicase superfamilies. Trends Biochem Sci 26(9):539–544

Gilson E, Nikaido H, Hofnung M (1982) Sequence of the malK gene in E.coli K12. Nucleic Acids Res 10(22):7449–7458

Gould AD, Shilton BH (2010) Studies of the maltose transport system reveal a mechanism for coupling ATP hydrolysis to substrate translocation without direct recognition of substrate. J Biol Chem 285(15):11290–11296

Gould AD, Telmer PG, Shilton BH (2009) Stimulation of the maltose transporter ATPase by unliganded maltose binding protein. Biochemistry 48(33):8051–8061

Grote M, Polyhach Y, Jeschke G, Steinhoff HJ, Schneider E, Bordignon E (2009) Transmembrane signaling in the maltose ABC transporter MalFGK2-E: periplasmic MalF-P2 loop communicates substrate availability to the ATP-bound MalK dimer. J Biol Chem 284(26):17521–17526

Gutmann DA, Ward A, Urbatsch IL, Chang G, van Veen HW (2010) Understanding polyspecificity of multidrug ABC transporters: closing in on the gaps in ABCB1. Trends Biochem Sci 35(1):36–42

Hall JA, Ganesan AK, Chen J, Nikaido H (1997) Two modes of ligand binding in maltose-binding protein of Escherichia coli. Functional significance in active transport. J Biol Chem 272 (28):17615–17622

Hazelbauer GL (1975) Maltose chemoreceptor of Escherichia coli. J Bacteriol 122(1):206–214

Higgins CF (2001) ABC transporters: physiology, structure and mechanism–an overview. Res Microbiol 152(3–4):205–210

Hinz A, Tampe R (2012) ABC transporters and immunity: mechanism of self-defense. Biochemistry 51(25):4981–4989

Hohl M, Briand C, Grutter MG, Seeger MA (2012) Crystal structure of a heterodimeric ABC transporter in its inward-facing conformation. Nat Struct Mol Biol 19(4):395–402

Hollenstein K, Frei DC, Locher KP (2007) Structure of an ABC transporter in complex with its binding protein. Nature 446(7132):213–216

Hopfner KP, Karcher A, Shin DS, Craig L, Arthur LM, Carney JP, Tainer JA (2000) Structural biology of Rad50 ATPase: ATP-driven conformational control in DNA double-strand break repair and the ABC-ATPase superfamily. Cell 101(7):789–800

Hunke S, Dose S, Schneider E (1995) Vanadate and bafilomycin A1 are potent inhibitors of the ATPase activity of the reconstituted bacterial ATP-binding cassette transporter for maltose (MalFGK2). Biochem Biophys Res Commun 216(2):589–594

Hunke S, Mourez M, Jehanno M, Dassa E, Schneider E (2000) ATP modulates subunit-subunit interactions in an ATP-binding cassette transporter (MalFGK2) determined by site-directed chemical cross-linking. J Biol Chem 275(20):15526–15534

Isenbarger TA, Carr CE, Johnson SS, Finney M, Church GM, Gilbert W, Zuber MT, Ruvkun G (2008) The most conserved genome segments for life detection on Earth and other planets. Orig Life Evol Biosph 38(6):517–533

Jacso T, Grote M, Daus ML, Schmieder P, Keller S, Schneider E, Reif B (2009) Periplasmic loop P2 of the MalF subunit of the maltose ATP binding cassette transporter is sufficient to bind the maltose binding protein MalE. Biochemistry 48(10):2216–2225

Jardetzky O (1966) Simple allosteric model for membrane pumps. Nature 211(5052):969–970

Jones PM, George AM (1999) Subunit interactions in ABC transporters: towards a functional architecture. FEMS Microbiol Lett 179(2):187–202

Jones PM, George AM (2012) Mechanism of the ABC transporter ATPase domains: catalytic models and the biochemical and biophysical record. Crit Rev Biochem Mol Biol 48(1):39–50

Joseph B, Jeschke G, Goetz BA, Locher KP, Bordignon E (2011) Transmembrane gate movements in the type II ATP-binding cassette (ABC) importer BtuCD-F during nucleotide cycle. J Biol Chem 286(47):41008–41017

Kellermann O, Szmelcman S (1974) Active transport of maltose in Escherichia coli K12. Involvement of a "periplasmic" maltose binding protein. Eur J Biochem 47(1):139–149

Khare D, Oldham ML, Orelle C, Davidson AL, Chen J (2009) Alternating access in maltose transporter mediated by rigid-body rotations. Mol Cell 33(4):528–536

Korkhov VM, Mireku SA, Locher KP (2012) Structure of AMP-PNP-bound vitamin B12 transporter BtuCD-F. Nature 490(7420):367–372

Kuhnau S, Reyes M, Sievertsen A, Shuman HA, Boos W (1991) The activities of the Escherichia coli MalK protein in maltose transport, regulation, and inducer exclusion can be separated by mutations. J Bacteriol 173(7):2180–2186

Lewinson O, Lee AT, Locher KP, Rees DC (2010) A distinct mechanism for the ABC transporter BtuCD-BtuF revealed by the dynamics of complex formation. Nat Struct Mol Biol 17(3):332–338

Linton KJ, Holland IB (2011) The ABC transporters of human physiology and disease, 1st edn. World Scientific, Singapore

Locher KP (2009) Review. Structure and mechanism of ATP-binding cassette transporters. Philos Trans R Soc Lond B Biol Sci 364(1514):239–245

Locher KP, Lee AT, Rees DC (2002) The E. coli BtuCD structure: a framework for ABC transporter architecture and mechanism. Science 296(5570):1091–1098

Loo TW, Clarke DM (1995) Covalent modification of human P-glycoprotein mutants containing a single cysteine in either nucleotide-binding fold abolishes drug-stimulated ATPase activity. J Biol Chem 270(39):22957–22961

Loo TW, Bartlett MC, Clarke DM (2003) Drug binding in human P-glycoprotein causes conformational changes in both nucleotide-binding domains. J Biol Chem 278(3):1575–1578

Lu G, Westbrooks JM, Davidson AL, Chen J (2005) ATP hydrolysis is required to reset the ATP-binding cassette dimer into the resting-state conformation. Proc Natl Acad Sci U S A 102 (50):17969–17974

Luckey M, Nikaido H (1980) Specificity of diffusion channels produced by lambda phage receptor protein of Escherichia coli. Proc Natl Acad Sci U S A 77(1):167–171

Magasanik B (1970) Glucose effects: inducer exclusion and repression. In: The Lactose operon, vol 1. Cold Spring Harbor Monograph Archive, pp. 189–219

Mannering DE, Sharma S, Davidson AL (2001) Demonstration of conformational changes associated with activation of the maltose transport complex. J Biol Chem 276(15):12362–12368

Manson MD, Boos W, Bassford PJ Jr, Rasmussen BA (1985) Dependence of maltose transport and chemotaxis on the amount of maltose-binding protein. J Biol Chem 260(17):9727–9733

Moody JE, Millen L, Binns D, Hunt JF, Thomas PJ (2002) Cooperative, ATP-dependent association of the nucleotide binding cassettes during the catalytic cycle of ATP-binding cassette transporters. J Biol Chem 277(24):21111–21114

Morbach S, Tebbe S, Schneider E (1993) The ATP-binding cassette (ABC) transporter for maltose/maltodextrins of Salmonella typhimurium. Characterization of the ATPase activity associated with the purified MalK subunit. J Biol Chem 268(25):18617–18621

Mourez M, Hofnung M, Dassa E (1997) Subunit interactions in ABC transporters: a conserved sequence in hydrophobic membrane proteins of periplasmic permeases defines an important site of interaction with the ATPase subunits. EMBO J 16(11):3066–3077

Mourez M, Jehanno M, Schneider E, Dassa E (1998) In vitro interaction between components of the inner membrane complex of the maltose ABC transporter of Escherichia coli: modulation by ATP. Mol Microbiol 30(2):353–363

Oldham ML, Chen J (2011a) Crystal structure of the maltose transporter in a pretranslocation intermediate state. Science 332(6034):1202–1205

Oldham ML, Chen J (2011b) Snapshots of the maltose transporter during ATP hydrolysis. Proc Natl Acad Sci U S A 108(37):15152–15156

Oldham ML, Khare D, Quiocho FA, Davidson AL, Chen J (2007) Crystal structure of a catalytic intermediate of the maltose transporter. Nature 450(7169):515–521

Oldham ML, Davidson AL, Chen J (2008) Structural insights into ABC transporter mechanism. Curr Opin Struct Biol 18(6):726–733

Oloo EO, Fung EY, Tieleman DP (2006) The dynamics of the MgATP-driven closure of MalK, the energy-transducing subunit of the maltose ABC transporter. J Biol Chem 281(38):28397–28407. doi:10.1074/jbc.M513614200

Orelle C, Dalmas O, Gros P, Di Pietro A, Jault JM (2003) The conserved glutamate residue adjacent to the Walker-B motif is the catalytic base for ATP hydrolysis in the ATP-binding cassette transporter BmrA. J Biol Chem 278(47):47002–47008

Orelle C, Ayvaz T, Everly RM, Klug CS, Davidson AL (2008) Both maltose-binding protein and ATP are required for nucleotide-binding domain closure in the intact maltose ABC transporter. Proc Natl Acad Sci U S A 105(35):12837–12842

Orelle C, Alvarez FJ, Oldham ML, Orelle A, Wiley TE, Chen J, Davidson AL (2010) Dynamics of alpha-helical subdomain rotation in the intact maltose ATP-binding cassette transporter. Proc Natl Acad Sci U S A 107(47):20293–20298

Procko E, O'Mara ML, Bennett WF, Tieleman DP, Gaudet R (2009) The mechanism of ABC transporters: general lessons from structural and functional studies of an antigenic peptide transporter. FASEB J 23(5):1287–1302

Rees DC, Johnson E, Lewinson O (2009) ABC transporters: the power to change. Nat Rev Mol Cell Biol 10(3):218–227

Richet E, Davidson AL, Joly N (2012) The ABC transporter MalFGK(2) sequesters the MalT transcription factor at the membrane in the absence of cognate substrate. Mol Microbiol 85 (4):632–647

Rodionov DA, Hebbeln P, Eudes A, ter Beek J, Rodionova IA, Erkens GB, Slotboom DJ, Gelfand MS, Osterman AL, Hanson AD, Eitinger T (2009) A novel class of modular transporters for vitamins in prokaryotes. J Bacteriol 191(1):42–51

Samanta S, Ayvaz T, Reyes M, Shuman HA, Chen J, Davidson AL (2003) Disulfide cross-linking reveals a site of stable interaction between C-terminal regulatory domains of the two MalK subunits in the maltose transport complex. J Biol Chem 278(37):35265–35271

Saurin W, Koster W, Dassa E (1994) Bacterial binding protein-dependent permeases: characterization of distinctive signatures for functionally related integral cytoplasmic membrane proteins. Mol Microbiol 12(6):993–1004

Schirmer T, Keller TA, Wang YF, Rosenbusch JP (1995) Structural basis for sugar translocation through maltoporin channels at 3.1 A resolution. Science 267(5197):512–514

Sharma S, Davidson AL (2000) Vanadate-induced trapping of nucleotides by purified maltose transport complex requires ATP hydrolysis. J Bacteriol 182(23):6570–6576

Sharom FJ (2008) ABC multidrug transporters: structure, function and role in chemoresistance. Pharmacogenomics 9(1):105–127

Shuman HA (1982) Active transport of maltose in Escherichia coli K12. Role of the periplasmic maltose-binding protein and evidence for a substrate recognition site in the cytoplasmic membrane. J Biol Chem 257(10):5455–5461

Smith CA, Rayment I (1996) X-ray structure of the magnesium(II).ADP.vanadate complex of the Dictyostelium discoideum myosin motor domain to 1.9 A resolution. Biochemistry 35 (17):5404–5417

Smith PC, Karpowich N, Millen L, Moody JE, Rosen J, Thomas PJ, Hunt JF (2002) ATP binding to the motor domain from an ABC transporter drives formation of a nucleotide sandwich dimer. Mol Cell 10(1):139–149

Spurlino JC, Lu GY, Quiocho FA (1991) The 2.3-A resolution structure of the maltose- or maltodextrin-binding protein, a primary receptor of bacterial active transport and chemotaxis. J Biol Chem 266(8):5202–5219

Stein A, Seifert M, Volkmer-Engert R, Siepelmeyer J, Jahreis K, Schneider E (2002) Functional characterization of the maltose ATP-binding-cassette transporter of Salmonella typhimurium by means of monoclonal antibodies directed against the MalK subunit. Eur J Biochem 269 (16):4074–4085

Szmelcman S, Hofnung M (1975) Maltose transport in Escherichia coli K-12: involvement of the bacteriophage lambda receptor. J Bacteriol 124(1):112–118

Szmelcman S, Schwartz M, Silhavy TJ, Boos W (1976) Maltose transport in Escherichia coli K12. Eur J Biochem 65(1):13–19

Tombline G, Senior AE (2005) The occluded nucleotide conformation of p-glycoprotein. J Bioenerg Biomembr 37(6):497–500

Treptow NA, Shuman HA (1985) Genetic evidence for substrate and periplasmic-binding-protein recognition by the MalF and MalG proteins, cytoplasmic membrane components of the Escherichia coli maltose transport system. J Bacteriol 163(2):654–660

Urbatsch IL, Sankaran B, Weber J, Senior AE (1995) P-glycoprotein is stably inhibited by vanadate-induced trapping of nucleotide at a single catalytic site. J Biol Chem 270 (33):19383–19390

Urbatsch IL, Beaudet L, Carrier I, Gros P (1998) Mutations in either nucleotide-binding site of P-glycoprotein (Mdr3) prevent vanadate trapping of nucleotide at both sites. Biochemistry 37 (13):4592–4602

van der Heide T, Poolman B (2002) ABC transporters: one, two or four extracytoplasmic substrate-binding sites? EMBO Rep 3(10):938–943

Vigonsky E, Ovcharenko E, Lewinson O (2013) Two molybdate/tungstate ABC transporters that interact very differently with their substrate binding proteins. Proc Natl Acad Sci U S A 110 (14):5440–5445. doi:10.1073/pnas.1213598110

Wang T, Fu G, Pan X, Wu J, Gong X, Wang J, Shi Y (2013) Structure of a bacterial energy-coupling factor transporter. Nature. doi:10.1038/nature12045

Wen PC, Tajkhorshid E (2008) Dimer opening of the nucleotide binding domains of ABC transporters after ATP hydrolysis. Biophys J 95(11):5100–5110

Wen PC, Tajkhorshid E (2011) Conformational coupling of the nucleotide-binding and the transmembrane domains in ABC transporters. Biophys J 101(3):680–690

Xu K, Zhang M, Zhao Q, Yu F, Guo H, Wang C, He F, Ding J, Zhang P (2013) Crystal structure of a folate energy-coupling factor transporter from Lactobacillus brevis. Nature. doi:10.1038/nature12046

Ye J, Osborne AR, Groll M, Rapoport TA (2004) RecA-like motor ATPases–lessons from structures. Biochim Biophys Acta 1659(1):1–18

Yuan YR, Blecker S, Martsinkevich O, Millen L, Thomas PJ, Hunt JF (2001) The crystal structure of the MJ0796 ATP-binding cassette. Implications for the structural consequences of ATP hydrolysis in the active site of an ABC transporter. J Biol Chem 276(34):32313–32321

Zhang Y, Mannering DE, Davidson AL, Yao N, Manson MD (1996) Maltose-binding protein containing an interdomain disulfide bridge confers a dominant-negative phenotype for transport and chemotaxis. J Biol Chem 271(30):17881–17889

Chapter 9
Cooperative Transport Mechanism and Proton-Coupling in the Multidrug Efflux Transporter Complex ArcAB-TolC

Hi-jea Cha and Klaas Martinus Pos

Abstract Cooperativity and allostery within catalytically active protein complexes are important concepts for control of phenotypic outcome within a living system. In this chapter, the putative molecular determinants for cooperativity and allostery of a multi-subunit antibiotic efflux pump complex are discussed. While resisting multiple antibiotic stresses, Gram-negative bacteria deploy a network of single and multicomponent drug efflux pumps to reduce the concentration of the drugs in the cytoplasm and periplasm. The dual membrane setup imposes a particular challenge for transport of drugs from the cytoplasm to the medium, and current hypothesis describes a multistep transport path of drugs across the inner and outer membrane by the action of different drug efflux pumps. Drug transport from the cytoplasm to the periplasm is catalyzed by single component drug efflux systems belonging to the ABC-transporter superfamily, the Major Facilitator Superfamily (MFS), The Multi Antimicrobial Extrusion (MATE) Family, and the Small Multidrug Resistance (SMR) family. Transport from the periplasm across the outer membrane is catalyzed by a tripartite transport complex consisting of an inner membrane Resistance-Nodulation-cell Division (RND) transporter, an adaptor protein, and an outer membrane channel. Cooperative effects occur at multiple levels within this three component RND system: Anticipated allostery for binding of ligands (drugs and H^+) at the level of the protomer, interdependence of the protomers within the trimer, and cooperative effects between the three components within the entire transport complex. The molecular determinants of multiple substrates binding at different sites within the inner membrane transporter and its coupling to H^+ binding and transport are being described here for the paradigm tripartite transport machinery AcrAB-TolC from *Escherichia coli*. High resolution structures of the three components, the anticipated three component setup, and a multitude of biophysical and biochemical data are combined to address the overall molecular understanding of secondary H^+/drug antiport. With focus on the inner membrane

---

H. Cha • K.M. Pos (✉)
Institute of Biochemistry, Goethe University Frankfurt, Frankfurt, Germany
e-mail: pos@em.uni-frankfurt.de

R. Krämer and C. Ziegler (eds.), *Membrane Transport Mechanism*, Springer Series in Biophysics 17, DOI 10.1007/978-3-642-53839-1_9,

RND component, drug and H^+ binding and transport cooperativity are exemplified by exploiting a mechanism based on binding change and the strict coupling to the influx of H^+.

Keywords Multidrug efflux • Functional rotation • Cooperativity • Tripartite RND systems • Proton translocation • Drug/H^+ antiporter • Multiple drug binding • Conformational changes • Transporter states • Membrane transport

9.1 Introduction

Multidrug resistance by bacteria is an increasing concern as the arsenal of antibiotics used to treat infections has been described as being increasingly ineffective (Bush et al. 2011). Bacteria defend themselves against stress by deployment of resistance mechanisms which are either constitutively present by moonlighting activities of systems with different physiological functions or induced upon contact with stress signals. For antibiotic stress, common defense mechanisms are modification of the antibiotic by hydrolysis, phosphorylation, acetylation, or adenylation of the specific drugs, sequestering of the toxic compound by proteins with nonessential function in the cell, or modification of the antibiotic target by mutation or methylation (Walsh 2000). These resistance mechanisms can be combined to render the bacteria resistant against multiple antibiotics. Since entry of the toxic compound into the cells is the first event to reach the target(s) inside the cells, the first line of defense for many bacteria is to prevent entry of the drug via the membrane. This is accomplished by transport proteins which extrude the antibiotics from within the cell or cell membrane towards the outside. For Gram-negative bacteria, containing an inner and out membrane, the downregulation of the expression of porins, often a preferred route for drugs to enter the periplasmic space between the two membranes, is another effective means to resist the incoming threat (Davin-Regli et al. 2008). Gram-negative cells face an interesting ambiguity in relation to drug permeation, since the outer membrane presents on the one hand a protective barrier for drugs entering the periplasmic space, but on the other hand, drugs expelled out of the cytoplasm across the inner membrane accumulate in the limited space of the periplasm leading to very high concentrations in this compartment and therefore limited effectiveness of inner membrane transporters to clean the cytoplasm due to the reflux of drugs along the inward gradient. Past research has shown that Gram-negative bacteria install transport complexes which span the inner and outer membrane as to avoid drug accumulation in the periplasmic space (Nikaido and Pagès 2012). The best characterized complexes are the tripartite systems build up out of an inner membrane transporter, connected to an outer membrane channel and stabilized by a periplasmic adaptor molecule. The action of these systems prevents drug accumulation in the periplasm, thereby avoiding large gradients across the inner-membrane and/or lethal concentrations of drugs inside the periplasm (e.g., of β-lactam antibiotics).

Table 9.1 Membrane transporters involved in antibiotic resistance

Name	TCDB* number	Energization	Prominent examples involved in antibiotic resistance	Typical number of TM* helices
Major facilitator superfamily (MFS)	2.A.1	Na^+ or H^+ gradient	NorA of *Staphylcoccus aureus* EmrAB complex of *Escherichia coli*	12 or 14
ATP-binding cassette (ABC) family	3.A.1	ATP hydrolysis	LmrA of *Escherichia coli* MacB of *Escherichia coli*	6
Small multidrug resistance (SMR) family	2.A.7.1	H^+ gradient	EmrE of *Escherichia coli* Smr of *Staphylcoccus aureus*	4
Multidrug and toxic compound extrusion (MATE) family	2.A.66.1	Na^+ or H^+ gradient	NorM of *Vibrio parahaemolyticus* YdhE of *E. coli*	4
Resistance-nodulation-cell division (RND) superfamily	2.A.6	Na^+ or H^+ gradient	AcrB of *Escherichia coli, Klebsiella pneumoniae, and Enterobacter spp.* *Ptchl of Homo sapiens* VexF of *Vibrio cholerae* Mmpl7 of *Mycobacterium tuberculosis*	12

**TCDB* Transporter classification database (Saier et al. 2006), *TM* transmembrane

Infections by (multidrug-resistant) Gram-negative bacteria (e.g., *Escherichia coli*, *Acinetobacter baumanii*, and *Pseudomonas aeruginosa*) became in the last decades a major health issue (Vila and Martínez 2008). The additional outer membrane of Gram-negative bacteria is contributing to a reduced permeability for drugs into the bacterial cytosol. The resistance phenotype in Gram-negative bacteria appears to be largely connected with the expression of efflux pump genes. In *E. coli*, for example, 20 different membrane transporters were identified to be involved in mediating antibiotic resistances (Nishino and Yamaguchi 2001; Sulavik et al. 2001). In general, there are five superfamilies of membrane-embedded transporters that are involved in antibiotic resistances and are summarized in Table 9.1.

Many clinical isolates of multidrug resistant pathogens exhibit overexpression of endogenous membrane transporters, particularly, members of the Resistance-Nodulation-cell Division (RND) superfamily (Piddock 2006). For example, overexpression of the RND transporter genes *acrB* (*E. coli*) or *mexB* (*P. aeruginosa*) is sufficient to render bacterial cells resistant against a huge variety of structurally unrelated molecules, including antibiotics (Nikaido 1998). Disruption of these genes from the clinical isolates chromosome renders these cells susceptible for many antibiotics, indicating the particular importance of these efflux pumps (Piddock 2006).

The tripartite system, consisting of AcrA, AcrB, and TolC (AcrAB-TolC), was shown to play a central role in generating a drug resistant phenotype (Nishino and

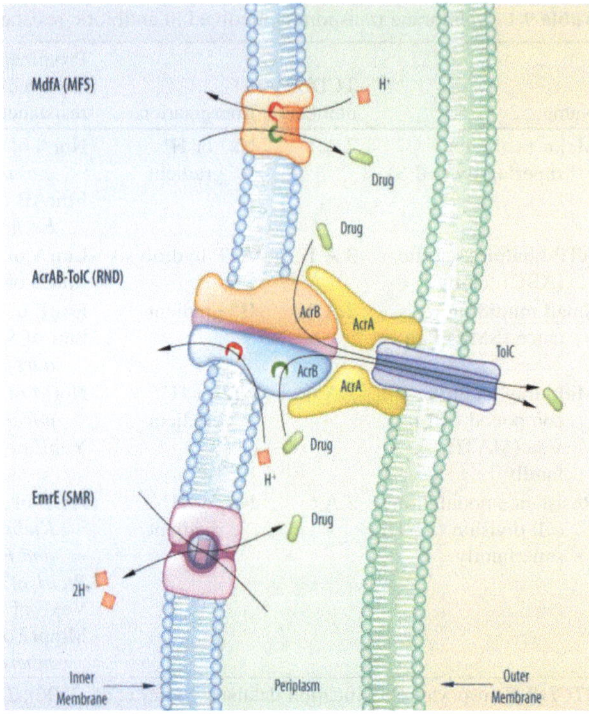

Fig. 9.1 Efflux-mediated resistance is most likely organized in a coordinated network. Transporters of the inner membrane like EmrE and MdfA are transporting substrates from the cytosol into the periplasm. Drugs from the periplasm are then transported by AcrAB-TolC across the outer membrane [figure taken from Tal and Schuldiner (2009)]

Yamaguchi 2001). Consequently, *acrB*-deficient (or *acrA* respectively *tolC*-deficient) *E. coli* mutants were rendered largely susceptible to a wide variety of antibiotics, disinfectants, and dyes (see also Chaps. 10 and 11). The deletion of other transporter genes has, relative to the *acrB*-deficient *E. coli* mutant, a smaller impact on the resistance phenotype. Tal and Schuldiner (Tal and Schuldiner 2009) revealed by combinatorial knock-outs of *emrE*, *mdfA*, and *acrB* in *E. coli* that drug efflux is most likely organized in a coordinated network, where several transporters with overlapping substrate spectra are transporting drugs from the cytosol into the periplasm from where these are transported by AcrAB-TolC across the outer membrane (Fig. 9.1). This cooperative behavior between single component efflux transporters and tripartite efflux systems has been reported also earlier on in the case of resistance against tetracycline and chloramphenicol (de Cristóbal et al. 2006; Lee et al. 2000).

The AcrAB-TolC three-component system has been studied most intensively on a structural and functional level, and considerable research effort has also been focused on other tripartite RND efflux systems, like MexAB-OprM from *P. aeruginosa* as well (Nikaido and Takatsuka 2009). X-ray crystallographic studies have been conducted on several RND systems, which are, apart from the systems mentioned above, CusCBA from *E. coli* (Su et al. 2011a), and the SecDF protein export machinery from *Thermus thermophilus* (Tsukazaki et al. 2011). We will mainly focus on the work on *E. coli* AcrAB-TolC with references to the results obtained from homolog systems. We will shortly introduce AcrA and TolC and the current knowledge on tripartite assembly, while most of this chapter will focus on the functional and structural analysis including allosteric and cooperative activities displayed by the trimeric RND inner membrane antiporter AcrB.

9.1.1 The Membrane Fusion Protein AcrA

AcrA is a member of the membrane fusion protein (MFP) family, and its gene is cotranscribed in *E. coli* with *acrB* (Ma et al. 1995). Like all subunits of the tripartite system, AcrA is indispensable for the AcrAB-TolC-mediated resistance. AcrA is postulated to support interactions between AcrB and TolC in a scaffold-like manner (Tikhonova et al. 2011). However, recent studies indicate alternative functions for AcrA besides its adaptor function and might include transduction of conformational changes between TolC and AcrB, as well as playing a role in substrate specificity, most likely caused by an indirect effect via AcrB (Weeks et al. 2010; Zgurskaya et al. 2009).

AcrA is a periplasmic soluble protein, which is secreted across the inner membrane via a signal peptide-mediated export mechanism (Kawabe et al. 2000). Upon cleavage of the N-terminal signal peptide, the newly N-terminally exposed cysteine residue at position 25 is palmitoylated and anchors AcrA to the inner membrane. Palmitoylation appears not to be essential for the function of AcrA within the tripartite system (Zgurskaya and Nikaido 1999), but surface plasmon resonance (SPR) spectroscopy experiments indicated that the lipidation of AcrA is important for AcrA-dimerization, which resulted in increased affinity for AcrB in comparison to the AcrA monomer (Tikhonova et al. 2011).

The structure of AcrA can be described by three globular domains (membrane proximal domain, β-barrel domain, and lipoyl domain) and a helical domain (α-hairpin domain) (Fig. 9.2); a molecular setup shared by the structurally elucidated close homologues MexA (Higgins et al. 2004; Akama et al. 2004) and CusB (Su et al. 2009).

9.1.2 The Outer Membrane Factor TolC

TolC (tolerance to colicin E1) of *E. coli* is a member of the outer membrane factor (OMF) family of membrane proteins and encoded by a single gene at a location on the chromosome unrelated to the *acrAB* operon. Like other members of the OMF family, TolC is embedded in the outer membrane of Gram-negative bacteria and serves many tripartite systems in *E. coli* including AcrA-AcrB-TolC (Symmons et al. 2009), EmrA-EmrB-TolC (Tanabe et al. 2009), HlyB-HlyD-TolC (Balakrishnan et al. 2001), and MacA-MacB-TolC (Lu and Zgurskaya 2012).

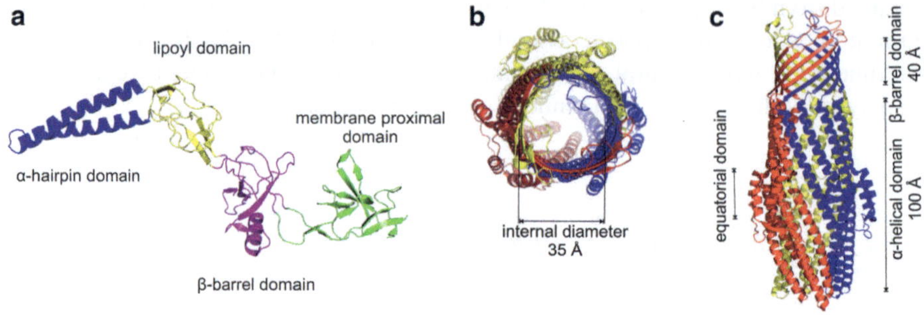

Fig. 9.2 Structures of MexA (AcrA) and TolC. (**a**) Structure of MexA (pdb entry 2V4D, homolog of AcrA) in *cartoon representation*. MexA consists of 4 domains, where the lipoyl domain (*yellow*), the β-barrel domain (*magenta*), and the membrane proximal domain (*green*) are globular. The α-hairpin domain consists of two α-helices (*blue*). The general structures of MexA and AcrA are highly congruent to each other, whereas the α-hairpin domain of MexA is by 7 amino acids shorter than AcrA. (**b, c**) Structure of TolC (pdb entry 2VDE) in *cartoon representation*. The monomers are color-coded in *blue*, *yellow*, and *red*, respectively. (**b**) View on TolC parallel to the membrane plane. (**c**) View from the periplasmic space on TolC in its open state (Koronakis et al. 2000)

Trimeric TolC (Koronakis et al. 2000) contains an outer membrane embedded 12-sheeted β-barrel domain, a cylinder-like α-helical domain protruding 100 Å into the periplasm, and an equatorial domain consisting of laterally positioned short N- and C-terminal helices at the equator of the α-helical cylinder (Fig. 9.2). TolC crystal structures in the closed (Koronakis et al. 2000), open (Bavro et al. 2008) and the transition state (Pei and Hinchliffe 2011) demonstrated that the opening and closing of the TolC α-helical cylinder is accomplished by an iris-like movement, reminiscent of aperture opening/closing of optical devices like cameras or microscopes.

9.1.3 Assembly of RND-MFP-OMF Tripartite Systems

The assembly of the RND-MFP-OMF tripartite system is a heavily researched field and a critically discussed topic in the field of efflux transporters. Due to the distribution of RND and OMF proteins in two distinct membranes, the technical challenges to obtain structural and in vitro data are unparalleled. Nevertheless, several biochemical approaches using appropriate members of RND-MFP-OMF tripartite systems were able to indicate interactions of the single components, whereas the MFP component seems to play a central role in the assembly of tripartite systems. For example, the MFP component MexA was successfully co-purified with OprM (OMF) in the absence of MexB (RND) and also within the complete MexA-MexB-OprM tripartite system in pull down experiments (Mokhonov et al. 2004). MexA, furthermore, co-purified with OprM, whereas

MexB was not able to be co-purified as bipartite complex, neither as MexB-OprM nor as MexA-MexB complex. On the other hand, spatial proximity of AcrB (RND) with AcrA (MFP) (Tikhonova and Zgurskaya 2004; Touzé et al. 2004; Zgurskaya and Nikaido 2000) and AcrB with TolC (OMF) (Tamura et al. 2005) is supported by in vivo cross-linking experiments using dithiosuccinimidylpropionate, which indicates interactions between these components. A recent study using SPR spectroscopy and size exclusion chromatography showed that N-terminal lipidation of AcrA was essential for dimerization. Lipidated and dimeric AcrA, on its part, exhibited a higher binding affinity to AcrB in comparison to unlipidated monomeric AcrA (Tikhonova et al. 2011). The affinity of the AcrA dimer was apparently pH-dependent, where experiments at pH 6 were able to detect interactions of the AcrA dimer with AcrB whereas no interactions could be detected at pH 7.5. Particularly, the stoichiometry of the tripartite system is an intensively debated aspect. So far, ratios of 4:1:1 (Akama et al. 2004), 3:1:1 (Higgins et al. 2004), 2:1:1 (Akama et al. 2004; Su et al. 2011b), and 1:1:1 (Symmons et al. 2009; Xu et al. 2011) for MFP-RND-OMF systems were postulated.

9.2 The Structure of the Resistance-Nodulation-Cell Division Pump AcrB

The gene locus of *acrB* (Acriflavin resistance protein B) was initially identified by (Nakamura et al. (1978). Deletion of *acrB* rendered *E. coli* cells highly susceptible to acriflavine. AcrB is a member of the hydrophobe/amphiphile efflux-1 family (HAE-1, transporter class number (TC#) 2.A.6.2) within the RND superfamily (TC#: 2.A.6) characterized by its typical repeated topology of 1 + 5 transmembrane (TM) helices connected by extensive soluble loops (in Gram-negative bacteria located in the periplasm) between TM helices TM1 and TM2 and TM helices TM7 and TM8, respectively (Fujihira et al. 2002; Murakami et al. 2002). AcrB functions within the tripartite AcrAB-TolC setup and extrudes antibiotics from the periplasmic side over the outer membrane. This process is energized by the proton motive force across the inner membrane. While drug binding/specificity and transport have been shown to be located at the periplasmic loops (Elkins and Nikaido 2002; Mao et al. 2002), proton translocation occurs at the transmembrane domain (TMD) of AcrB (Seeger et al. 2009; Takatsuka and Nikaido 2006). Energy transduction and drug binding/transport are therefore spatially separated, which is atypical for secondary transporters but common for primary transporters like ABC-transporters or F_1F_o ATP-synthases. The substrate specificity spectrum of AcrB (within the AcrAB-TolC complex) includes structurally diverse molecules ranging from organic solvents, dyes, detergents, and antibiotics including hydrophobic β-lactams, but excluding the more hydrophilic aminoglycosides [references in Pos (2009)]. Crystallization and initial X-ray crystal diffraction of AcrB was first reported in 2002 by Pos and Diederichs (2002), and its crystal structure was first

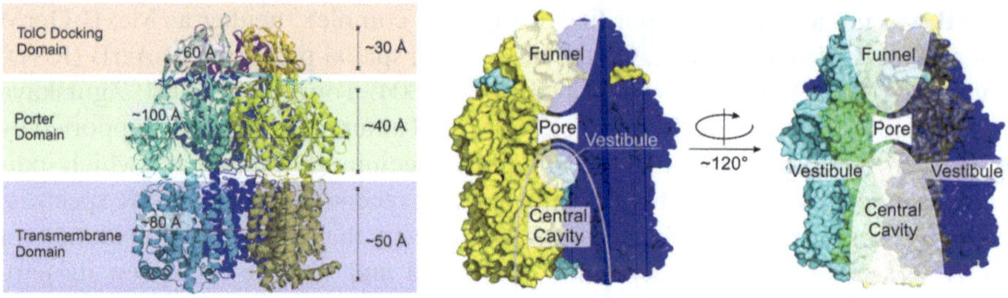

Fig. 9.3 Side view parallel to the membrane plane onto the symmetric AcrB structure [pdb entry 1IWG (Murakami et al. 2002)] in *cartoon representation*. The structure can be subdivided in a TolC docking domain, a porter domain and a transmembrane domain. The sizes of the domains are indicated. (*Center* and *Right*) 120° rotation side views in surface representation. In the *right representation*, the front protomer is displayed in transparent *yellow*. Trimeric AcrB contains two large cavities: the "funnel" in the TolC docking domain and a most likely lipid-filled "central cavity" in the TM domain. The central cavity is connected to the periplasm by "vestibules" at the interfaces of the protomers

solved in 2002 by Murakami et al. (2002) in a R32 crystal lattice at 3.5 Å resolution (pdb entry 1IWG). This trimeric AcrB structure exhibited a threefold symmetry axis perpendicular to the membrane plane and indicated homotrimeric assembly, where all protomers adopted identical conformation. The general structure of AcrB can be subdivided into three major domains: (1) the transmembrane domain, (2) the porter domain, and (3) the TolC-docking domain (Fig. 9.3).

In the TM domain of each protomer, TM helices TM4 and TM10 are positioned in the topological core surrounded by the other 10 TM helices, whereas TM1 and TM7 are positioned in a more peripheral location (Fig. 9.4). Site-directed mutagenesis studies identified a functional H^+-translocation triad of D407 and D408 positioned on TM4 and at equiplanar level K940 on TM10, as well as the additional essential residue R971 on TM11 (Murakami et al. 2002; Seeger et al. 2009; Su et al. 2006; Takatsuka and Nikaido 2006).

The periplasmic porter domain is subdivided in four subdomains: PN1, PN2, PC1, and PC2 (Fig. 9.4). The coupling of the subdomains can be described by their primary sequence proximity (PN1 with PN2 and PC1 with PC2) and by structural proximity of PN1 with PC2 and PN2 with PC1. The PN1/PC2 and PN2/PC1 units are in addition coupled by a shared β-sheet. The PN2, PC1, and PC2 subdomains encircle the PN1 subdomain and are laterally exposed to the periplasm, with a large solvent accessible opening between PC1 and PC2, denoted as "cleft." The PN1 subdomain is located at the interior of the AcrB trimer and exposes a short helix into the center of the trimer, denoted as Nα2 or pore helix. The pore helices of all three protomers are forming a closed pore-like structure (Fig. 9.4). These helices form strong non-covalent contacts and are considered a prime interface for the conformational communication between the protomers. The porter domain is connected at its outer membrane-proximal side to the TolC docking domain. There is experimental evidence that this domain provides the interaction site of AcrB with TolC (Tamura et al. 2005). A long loop is protruding from the TolC-docking domain of

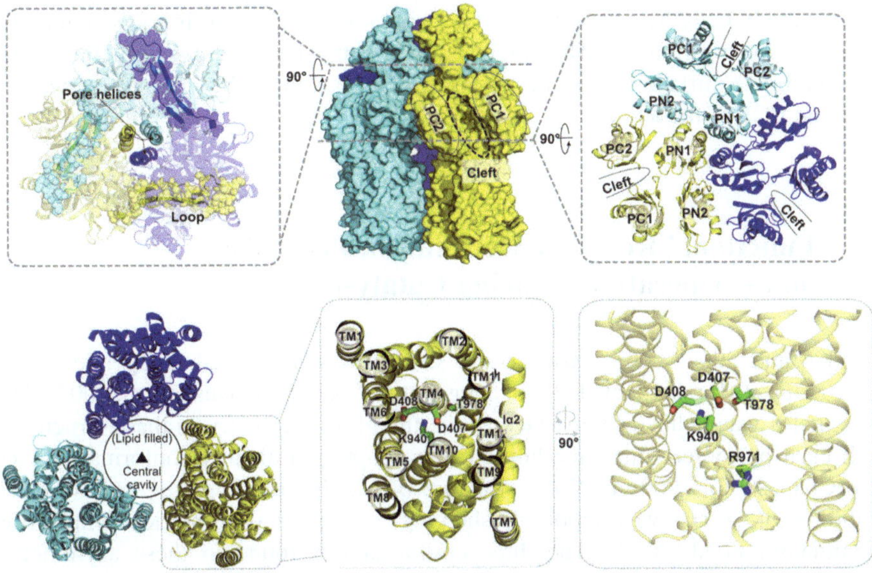

Fig. 9.4 Overview of the AcrB structure. *Top* panel (*Center*): Side view parallel to the membrane plane of the symmetric AcrB structure (pdb entry 1IWG) in surface representation. A large and deep cavity, denoted as the "cleft," (*dotted ellipse*) is visible between subdomains PC1 and PC2. (*Right*) View onto the porter domain from the periplasm in *cartoon representation*. The porter domain of AcrB is organized into PN1, PN2, PC1, and PC2 subdomains. The clefts between the PC1 and PC2 subdomains are indicated by *open ellipses*. The TolC docking domain is not shown in this view. (*Left*) View onto the TolC docking domain is shown in *transparent cartoon representation*. The protruding loops connecting the protomers are highlighted by *transparent spheres*. The Nα2 helices (pore helices) constituting the pore-like structure in the center of the AcrB trimer are displayed as *opaque cartoons*. *Lower* panel (*Left*): Topological view from the periplasm onto the transmembrane domain of the symmetric AcrB structure in *cartoon representation*. The lipid-filled central cavity is indicated by the *closed circle*, as well as the symmetry operator (*triangle*). (*Center*) Detailed view of one protomeric TM domain. TM helices are numbered TM1-TM12. TM4 and TM10 are encircled by the other helices, whereas TM1 and TM7 are located in a more peripheral position. Residues essential for AcrB function and most likely involved in proton translocation are shown in *sticks* (C: *green*, O: *red*, N: blue). (*Right*) Side view (90° rotation view of center figure) of one AcrB protomer with residues essential for proton translocation

one protomer to the adjacent protomer. Studies using laser induced liquid beam ion desorption mass spectrometry (LILBID MS) and phenotypic analysis showed the essential role for this loop in the assembly of AcrB protomers to form a stable trimer (Brandstatter et al. 2011; Yu et al. 2011).

The AcrB structure reveals two large cavities (Fig. 9.4): a large central cavity, open to the cytoplasm, and a funnel-like cavity, open to the periplasm. The most likely lipid-filled "central cavity" is situated in the center of the TM domains of all three protomers and extends into the periplasmic domain. The "funnel"-like opening in the TolC-docking domain of AcrB on its part widens towards TolC and is most likely part of the AcrB exit from where substrates are further transported into the TolC channel. The "central cavity" and "funnel" are separated from each other by the pore helices. The "central cavity," with a 30 Å diameter, is connected to the periplasm by "vestibules." This might provide pathways for drugs tunneling along the membrane surface below the interface of PN1 and PC2 of one protomer and PN2 and PC1 of the adjacent protomer.

9.3 Functional Rotation Mechanism, Peristalsis, and Cooperativity During Catalysis

In 2006 and 2007, asymmetric crystal structures were solved independently by three groups (Murakami et al. 2006; Seeger et al. 2006; Sennhauser et al. 2007). The asymmetry was defined by the lack of the threefold symmetry in these structures, with each protomer within the homotrimer adopting a distinct conformation (see also Chap. 4). In 2009, an X-ray structure of MexB from *P. aeruginosa* (a close homolog of AcrB) was reported, showing a similar asymmetric composition (Sennhauser et al. 2009). The three conformations found in these asymmetric trimers were designated Loose (L), Tight (T), and Open (O) (Seeger et al. 2006), or, Access, Binding, and Extrusion (Murakami et al. 2006), respectively. It has been proposed that the different conformations represent consecutive states in a catalytic cycle for drug export defining a "functional rotation" mechanism in analogy to the binding change mechanism of the F_1F_o ATP-synthase (Boyer 1997). In this model, protomers cycle consecutively through all conformations, i.e., from L to T to O back to L. In general this would suggest that, since all protomers in the trimer adopt the next conformation in this cycle, the overall trimeric composition remains "LTO" at any time of the cycle. The prerequisite conformational flexibility and interdependent conformational changes of the single protomers within the trimer were further biochemically analyzed by engineered disulfide bonds (Seeger et al. 2008; Takatsuka and Nikaido 2007). The cross-linking data showed reporting of the conformational states as predicted by the X-ray structure. Interestingly, the amount of cross-linking expected for the L and T conformational states was far beyond the theoretical one-third maximum of cross-linking expected within one trimer (data show approximately 41 % crosslinking for L or T conformations), suggesting more than one L or T protomer within the trimer. This might indicate structural flexibility of the protomers within the trimer and presence of more than one L or T state at any time of the conformational cycle, e.g., LLT or LTT (Pos 2009). Cross-linked AcrB variants with anticipated conformational restriction (i.e., cross-links in regions with large conformational changes deduced from the X-ray structure) showed strongly reduced activities. Upon reduction of the Cys-disulfide bonds, activity was recovered to a large extent, showing proper folding of the cross-linked AcrB trimers and that loss of activity was due to the conformational restraints caused by the disulfide bridges (Seeger et al. 2008).

Drug transport is proposed to start in the L-protomer, where substrates can enter the protein through the large periplasmic cleft. Indeed, using computational tools

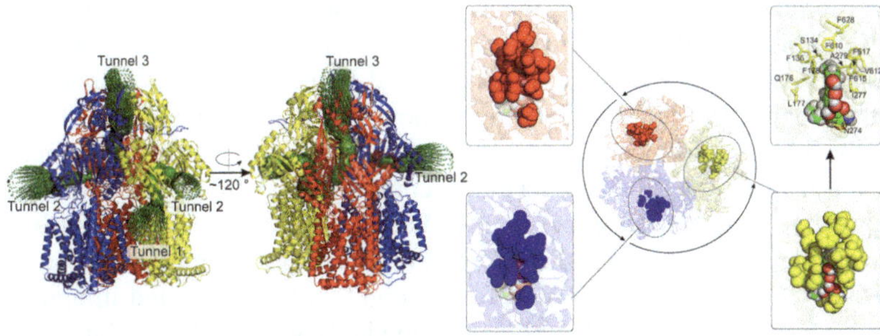

Fig. 9.5 (*Left*) Side view parallel to the membrane plane of asymmetric AcrB [pdb entry 4DX5 (Eicher et al. 2012)]. The protomers are shown in *blue* (L), *yellow* (T), and *red* (O). Drug transport is anticipated to start in the L conformation where Tunnel 2 connects the internal part of AcrB with the periplasm. The entrance of Tunnel 2 is located at the periplasmic "cleft," approximately 15 Å above the membrane plane. After conformational transition from L to T, an additional second tunnel (Tunnel 1) is present. This tunnel connects the deep binding pocket with the periplasmic end of a groove present between TM8 and TM9. In the O protomer, Tunnels 1 and 2 disappear due to closure of the cleft by movements of the PC2 subdomain. In this conformation, Tunnel 3 is generated, connecting the now closed deep binding pocket with the "funnel" leading to the TolC channel. (*Right*) *Top* view onto the membrane plane of asymmetric AcrB (pdb entry 4DX5) in *cartoon representation*. The protomers are shown in *blue* (L), *yellow* (T), and *red* (O). Residues involved in binding of minocycline and doxorubicin in the deep binding pocket, as well as the phenylalanine-rich cluster, are represented as *spheres*. Detailed pictures of the deep binding pocket reveal that the deep binding pocket is closed in the L (*bottom left inset*) and the O (*top left inset*) protomer. Binding of minocycline (*transparent spheres*; C: *green*, O: *red*, H: *grey*) is prevented by sterical clashes as shown. In contrast, the deep binding pocket is widened in the T protomer (*bottom right* inset), which enables binding of minocycline (indicated *sphere*; C: *green*, O: *red*, H: *grey*). Residues involved in binding of minocycline (*sphere representation*) and doxorubicin (not shown) in the deep binding pocket, as well as the phenylalanine cluster, are shown in *stick representation* in the *top right inset*

(Petrek et al. 2007) a tunnel (Tunnel 2) was detected with its lateral entrance at the cleft in the L protomer about 15 Å above the membrane plane, possibly representing an entrance site for substrates from the periplasm (Fig. 9.5). The conformation of the protomers of the symmetric structure, where the large cleft initially was identified, resembles the L protomer in the asymmetric structure. This symmetric "all-Loose" (LLL) state is suggested to be a "resting state," where all protomers are in a state to recruit substrates (Pos 2009; Su et al. 2006).

During the transition from L to T, substrate is transported into a hydrophobic pocket, designated "deep binding pocket" (Fig. 9.5, right) (Eicher et al. 2012). In several co-crystal structures, electron densities in the T protomer deep binding pocket were assigned to minocycline or doxorubicin (Eicher et al. 2012; Murakami et al. 2006; Nakashima et al. 2011). Several residues within the deep binding pocket are involved in interaction with the substrates in the deep binding pocket, including a most prominent phenylalanine cluster (F136 + F178 + F610 + F615 + F617 + F628) (Bohnert et al. 2008). Different residues are involved in the binding of minocycline (N274 + G179 + I277 + A279 + V612 + N176) or doxorubicin (Q176 + F615 + F617 + S134 + F178 + G179 + I277 + L177). The opening of the deep binding pocket during the L to T transition entails motions within the PN2/PC1 unit (RMSD ~2.3), whereas the PN1/PC2 unit remains largely rigid (RMSD < 0.5) (Schulz et al. 2011). A second tunnel (Fig. 9.5, "Tunnel 1") was identified connecting the deep binding pocket with an entrance at the periplasmic end of the TM8/9 interface at the height of the outer leaflet of the inner membrane (Sennhauser et al. 2007). Very recently, computational free energy analysis and corresponding mutagenesis experiments (Yao et al. 2013) indicated that based on the physiochemical properties the different drug substrates preferred either Tunnel 2 or Tunnel 1. Relatively, low molecular weight hydrophobic and lipophilic substrates are proposed to be preferably taken up via Tunnel 1, whereas larger and less hydrophobic substrates are entering AcrB via Tunnel 2. Moreover, an additional pathway was identified, which could not be identified using the X-ray structures.

In analogy to the binding change mechanism of the F_1F_o ATP-synthase (Boyer 1997), binding of substrates in the T conformation is proposed to be prerequisite for the energy-dependent T to O transition step (Pos 2009). Large movements within the porter domain are apparent during the T to O transition on the periplasmic side of AcrB, where the cleft closes from a width of over ~27 Å in L and T to 18 Å in the O protomer (distances from K708 to A654) and thereby close the entrance of tunnel 2. The flexibility of the cleft was confirmed to be essential for functionality of AcrB by disulfide crosslinking experiments (Takatsuka and Nikaido 2007). The O protomer is proposed to represent the conformation where substrates exit through the AcrB funnel region to TolC. Husain and Nikaido (2010) recently reported that cysteines introduced in the vicinity of the cleft (N274C or E273C) were able to covalently link the maleimide-labeled AcrB substrate Bodipy-FL, indicating that this area of the cleft is part of the drug pathway. Additionally, computational calculations indicated a movement of doxorubicin by 8 Å from the deep binding pocket through tunnel 3 during the T to O transition. This study showed that the porter domain of AcrB exhibits a bottom-to-top zipper-like closing motion (Schulz et al. 2010; Vargiu et al. 2011), strongly supporting the peristaltic efflux pump model proposed by Seeger et al. (2006). After the release of substrates to TolC, conformational transition from O back to L is the anticipated final step of one catalytic cycle.

9.4 Proton Translocation and Energy Coupling in the AcrB Protomer

The TM domain of the AcrB protomer is organized in two repeats consisting of Repeat 1 (R1: TM1 + TM3–TM6) and Repeat 2 (R2: TM7 + TM9–TM12). In addition, elongated TM helices, TM2 and TM8, connect the PN2/PC1 unit and the PN1/PC2 unit with R1 and R2, respectively. These helices are proposed to be coupling units for the energy transduction from the TM domain to the substrate transporting porter domain. In fact, TM8 undergoes large structural rearrangements, where the periplasmic N-terminal end of the TM8 is in a coiled state in L, a helix-coil-helix state in T, and exhibits a continuous helix state in the O protomer (Murakami et al. 2006; Seeger et al. 2006; Sennhauser et al. 2007).

The 1.9 Å resolution AcrB/minocycline (pdb entry 4DX5) co-crystal structure reveals the importance of interactions along the interface of R1 and R2 for proton translocation (Eicher et al. 2012). Here two distinct states can be described: an "engaged" state in the L and T protomers in contrast to a "disengaged" state in the O protomer. In the L and T states, residues of the proton translocation triad D407, D408 (TM4), and K940 (TM10) interact directly with each other, i.e., K940 from R2 is sandwiched between D407 and D408 on TM 4 from R1 by ionic interactions (Fig. 9.6, "L"). Additional interactions across R1 and R2 are described, amongst others, by the interaction of D407 of R1 with T978 (TM11) of R2. Further interactions within one repeat are described for example between D408 and S481 (TM6). All interactions from the engaged state in the T conformation are detached in the O conformation, the "disengaged" state. D407, for example, rotates into its own repeat R1 and interacts with the backbone oxygen of G403, while D408 reorients to interact with the backbone of L442 (TM5). Also, K940 of R2 turns away from D407 and D408 and interacts in the disengaged state with N941 (TM10) and T978 (TM11) of R1.

Several observations point towards the central role of the T conformation in the proton translocation. In the T conformation, all triad residues D407/D408/K940 are postulated to be charged. In the high resolution AcrB/minocycline structure, a network of water molecules, exclusively present in the T state, almost continuously connects the periplasm with the titratable residues within the proton translocation triad (Fig. 9.6, "T") (Eicher et al. 2012). Computational simulations in the group of Kandt independently suggested a major role of the water molecule network in the TMD for proton translocation (Fischer and Kandt 2011). We hypothesize that D407 accepts, in a stepwise manner, two periplasmic protons during the T to O transition, where one proton is further transported to D408. Therefore, D407 and D408 are both protonated in the O conformation (Fig. 9.6, "O"). This postulate is supported by the observation that K940H and K940R variants of AcrB are largely active in transport (Seeger et al. 2009). In contrast, K940A and K940M variants of AcrB were inactive. R971 on TM11 is also predicted to be charged in all three states and interacts with the triad in the T monomer by the coordination of a water cluster residing between D407 and R971. Protonation of D407 and D408 during the T to O transition converts the net charge of the triad from -1 to $+1$. This most likely weakens the interaction between R971 and the triad. Reorientation of the R971 side chain in the O conformation to neighboring side chains (e.g., F948 in TM10) and surrounding cytoplasmic bulk water might be a consequence of the electrostatic change in the triad. As a consequence the interactions, which were observable in the T conformation, are disengaged in the O conformation. The actual release of the bound protons on D407 and D408 in the O conformation is proposed to occur during the O to L transition (Fig. 9.6, "O $\rightarrow$ L"), coinciding with the release of drugs in the porter domain of AcrB towards TolC. The release of protons to the

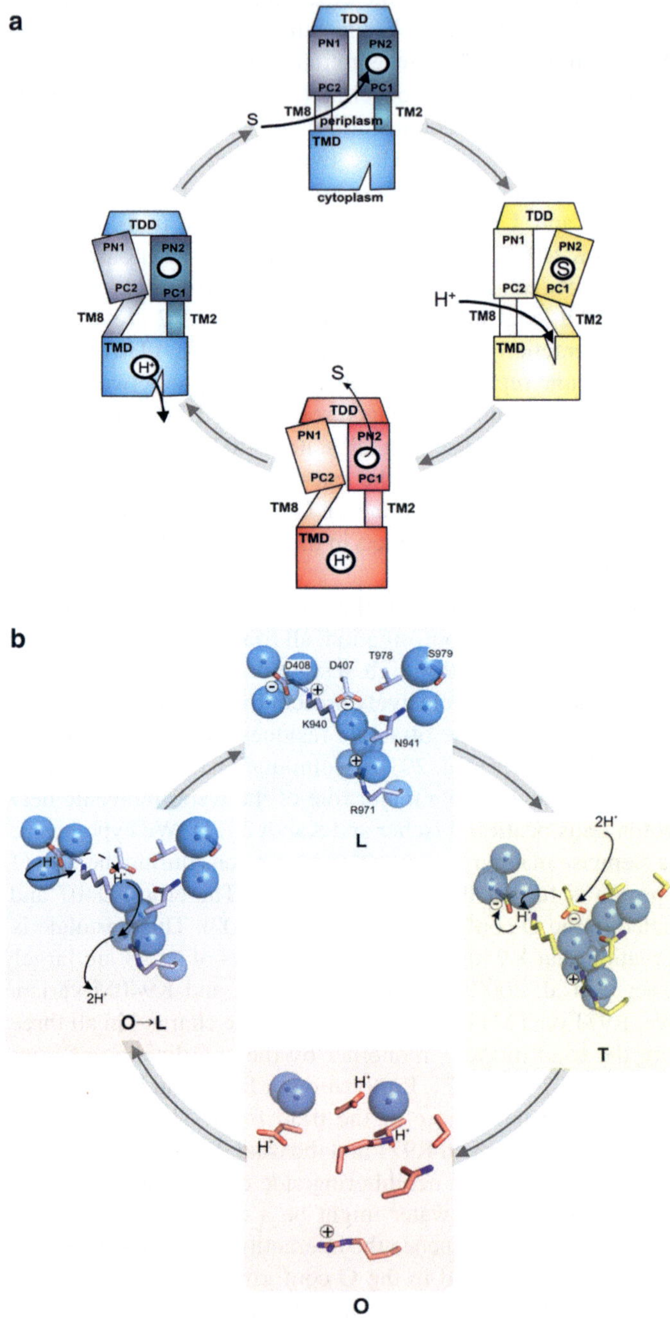

Fig. 9.6 (**a**) Schematic representation of the catalytic cycle of drug/H^+ antiport by the AcrB protomer. A substrate molecule (S) accesses the AcrB L conformer (*blue*). Substrate binding induces a conformational change from L to T involving by TM2. In the T conformation (*yellow*),

cytoplasm during the O to L transition causes the TM domain to revert to its engaged state including reorientation of D407, D408, K940, and R971 (Fig. 9.6).

More support of a directional cycling in the drug/H^+ antiport catalysis and interdependence of the protomers within the RND trimer comes from mutational studies with (1) a covalent trimer of AcrB (Takatsuka and Nikaido 2009), (2) monomeric AcrB (Brandstatter et al. 2011; Yu et al. 2011), and (3) the heterotrimeric MdtBC RND homolog from *E. coli* (Kim and Nikaido 2012; Kim et al. 2010).

Ad (1) The *acrB* gene was triplicated in-frame as to express a single giant covalently linked trimeric AcrB by the Nikaido lab. This setup has the advantage that substitutions at sites essential for function (i.e., in the proton binding/translocation sites D407 and/or D408) can be exclusively introduced in one protomer of the trimer, whereas the other two protomers remain wild type. The experiments gave rise to some background activity due to formation of truncation products (AcrB monomers) able to form non-covalent and active AcrB trimers, but in principle very clearly indicated that inactivation of one of the protomers in the trimer results in a nonfunctional transporter. This observation disfavors a transport mechanism with independent stochastic cycling of the protomers.

Ad (2) If drug efflux activity is dependent on strictly coupled conformational cycling within the AcrB trimer, monomeric AcrB would be inactive. Two cases of monomeric AcrB production have been reported. In one case, the intermonomeric loop (Brandstatter et al. 2011) (Fig. 9.4) was genetically deleted and gave rise to monomeric AcrB which could be purified from membrane preparations in dodecyl-β-maltoside (DDM) and consisted of mainly not only monomers but also a species eluting within a retention volume alike to wild-type trimeric AcrB as shown by a size-exclusion chromatography. Interestingly, some residual activity could be measured for *E. coli ΔacrB* cells expressing the deleted-loop *acrB* gene mutant for some of the substrates tested. This might indicate that deleted-loop monomeric variants retain some ability to form functional trimeric AcrB or that residual activity is caused by monomeric species. Another interesting monomeric

Fig. 9.6 (continued) the transmembrane domain is accessible for protons from the periplasm. Protonation of the D407 and D408 residues of the proton translocation triad results in a conformational change from T to O mediated by TM8. In analogy to catalytic cycle of the F_1F_o ATP-synthase, the energy consuming step is proposed to be the release of substrate molecules from the porter domain towards the TolC channel. Release of substrate triggers the reverse of the protomer to the L conformation releasing the bound protons into the cytoplasm. (**b**) Detailed view on proton translocation residues within the TMD of the AcrB protomer. In the L (*blue*) and T (*yellow*) conformation, proton translocation triad residues exhibit an "engaged" state, where K940 is sandwiched between D407 and D408. All three residues are charged in this "engaged" state. Substrate binding in the porter domain of the L protomer triggers the L to T transition based on structural rearrangements in the TMD. In the T conformation, residues D407, D408, and K940 of the proton translocation triad are connected to the periplasm by a continuous cluster of water molecules leading to a stepwise protonation. This triggers a substantial reorientation of the TMD helices towards the protonated "disengaged" state of the O conformer (*red*). Bound protons on D407 and D408 are released towards the cytoplasm during the O to L transition recovering the engaged state of the proton translocation triad (*TDD* TolC docking domain, *TM2* transmembrane helix 2, *TM8* transmembrane helix 8, *TMD* transmembrane domain)

variant was created in the Wei Lab containing a P223G substitution [within the (tip) intermonomeric loop] (Yu et al. 2011). Here again, the activity was severely reduced, but some residual activity was maintained for some of the substrates. Via BN-PAGE it was shown that the P223G variant was monomeric; however, using Cys-crosslinking of the loop with the PC1 domain [shown previously to have no effect on the activity of trimeric AcrB (Seeger et al. 2008)], much of the activity could be recovered, showing the structural integrity of the monomeric species and the capability to assemble into a trimeric setup for function. The situation for the AcrB trimer appears to be different from another trimeric transporter, BetP. In the case of this betaine/Na^+ symporter, the monomeric species still transports its substrate efficiently; however, the observed regulation by hyperosmotic shock in case of BetP trimer, is absent (Perez et al. 2011). It appears therefore that trimerization for AcrB and BetP might serve different regulatory roles, the former showing catalytic cooperativity and the latter regulatory cooperativity.

Ad (3) The *E. coli* heterotrimeric RND transporter homolog MdtBC offers a unique experimental flexibility which has been used by the Nikaido lab to address the interdependence of the protomers within the $MdtB_2C$ trimer (Kim and Nikaido 2012; Kim et al. 2010). Purified native MdtBC showed a clear $MdtB_2C$ complex formation. As with the in-gene triple fusions of *acrB*, combinations of *mdtB* and *mdtC* were fused to obtain trimers with different MdtB and/or MdtC combinations, showing that an $N_{terminus}$-BCB-$C_{terminus}$ construct was most active. Within this heterotrimeric complex, MdtB and MdtC appeared to have different and complementing roles concerning the energy transduction (H^+ binding/transport) and substrate binding/transport. MdtB with one of the D410A, D411A, K933A, R964A, or T971A substitutions (within the TMD proton translocation site, homologous to the D407, D408, K940, R971, and T978 residues in AcrB) yielded an inactive $MdtB_2C$ transporter trimer, but these substitutions in MdtC (D401A, D402A, K919A, R950A, or T957A) did not severely affect activity of the trimer. Interestingly, in $MdtB_2C$ mutants with an D410A substitution in MdtB appeared to influence the conformation of MdtC and its ability to accommodate substrates, since labeling with the MdtBC substrate Fluorescein-5-maleimide in MdtB_D410A_MdtC_F610C (or F598C, both located in the deep binding pocket region of MdtC) was enhanced compared to labeling of MdtB_wt_MdtC_F610A (or F598C). On the other hand, Cys-substitutions in the periplasmic deep binding pocket of MdtC affected the cloxacillin efflux activity substantially, whereas the Cys-subtitutions in homologous positions in MdtB did in general not affect the efflux activity. Moreover, the deep binding pocket Cys-substituted residues were labeled effectively in MdtC, whereas in MdtB at homolog positions only 10–20 % (two exceptions of 50 %) of the MdtC labeling efficiency could be achieved. These data indicate a clear separation of the functional roles of MdtB (energy transduction) and MdtC (drug binding/transport) within the $MdtB_2C$ heterotrimer and also points towards a clear interdependence of these subunits. It also shows that energy is not only transduced within the protomers as depicted in Fig. 9.6 for AcrB but also transduced interprotomerically, explaining the activity of the heterotrimer containing H^+-binding-deficient MdtC. Conformational cycling and interprotomeric energy transduction therefore appear to be of intrinsic importance to the postulated functional rotation within the MdtBC transporter. Cross-linking studies within the MdtB or MdtC subunits might be needed to indicate whether conformational flexibility within the periplasmic domain is necessary for the cycling function of the trimer.

The interdependence of the protomers within the trimeric RND pump is recently also further substantiated by a fascinating observation: For the AcrAB-TolC system, very strong cooperative kinetics with Hill coefficients of 4–5 were measured by the Nikaido lab (Lim and Nikaido 2010; Nagano and Nikaido 2009) for cephalosporins (with low apparent affinity to AcrB, but not for nitrocefin, a cephalosporin with high affinity) extrusion, which might indicate bi-site activation as postulated (Pos 2009).

This observation immediately raises many interesting questions, e.g., the physiological role for cooperativity of this pump system but also on the determinants for cooperativity within the AcrAB-TolC complex. Cooperative effects can be anticipated on different levels within the complex: e.g., each protomer appears to be able to bind multiple drugs, and binding of one drug might be directly [by providing direct interaction interfaces (homotropic or heterotropic)] or indirectly (by allosteric effects within the protomer) influencing subsequent drug binding. Further cooperative effects can be anticipated by protomers within the trimer, where binding of drug(s) to one monomer influences binding/activity of the neighboring protomer as seems to be the case for the $MdtB_2C$ system. Another speculation concerns the role of the large (lipid-filled) cavity formed by the TM domains of the protomers. Activity might be controlled via the concentration of membrane-soluble drugs residing in the cavity area, by, e.g., changing lateral pressure compared to exclusively lipid-filled conditions and therefore stimulating the turnover rates of the trimeric complex. Another regulatory role might be given for the lipid composition, where specific lipids might bind preferably to (a) specific conformational state(s) of the AcrB trimer. BetP, e.g., is also trimeric, but no cooperative transport behavior can be shown for the transport for its substrate betaine. But interdependence has been shown by heterotropic allostery depending on osmotic stress (e.g., internal K^+ concentration) (Ressl et al. 2009), as well as a periplasmic ionic network involved in crosstalk between the monomers affecting turnover rates (Gärtner et al. 2011). The concerted conformational changes in case of trimeric AcrB resembles more catalytic cooperativity as shown for the F_1F_o-ATP-synthase (Boyer 1997) rather than the regulatory cooperativity seen for, e.g., the ACTase (Lipscomb and Kantrowitz 2011). Further cooperativity might involve the membrane adaptor component AcrA. Although its role in drug binding is not yet completely clear, shared (allosteric) binding sites with AcrB might exist. Its homolog CusB from the CusCBA Ag^+ and Cu^+ transport system has been shown to bind monovalent cations independent from the RND component (Bagai et al. 2007). AcrA might furthermore be involved in conformational energy conservation by, e.g., conserving energy in conformational strain delivered by the proton motive force via the RND AcrB and being released at a subsequent step, e.g., the O to L transition. The interaction between each AcrB monomer and its affiliated AcrA molecule has been investigated very recently (Symmons et al. 2009) and might suggest that during functional rotation of AcrB, conformational changes are transmitted towards AcrA. Rather than just being a rigid adapter, AcrA might actively transduce energy generated by AcrB and induce subtle peristaltic motions in the TolC channel. Similar to the tunnels observed in a single AcrB protomer during functional rotation (Murakami et al. 2006; Seeger et al. 2006; Sennhauser et al. 2007), TolC is possibly not a rigid hollow cylinder but changes its diameter by conformational change of each monomer, thereby pushing substrate unidirectionally from the closing aperture into the media (Pos 2009; Vaccaro et al. 2006, 2008).

9.5 The Role of a Switch Loop in Multiple Drug Binding in the Access and Deep Binding Pockets

Recent AcrB X-ray structures by two groups (Eicher et al. 2012; Nakashima et al. 2011) have substantiated multiple drug binding within the AcrB trimer as was postulated by the functional rotation hypothesis. The PC1/PC2 cleft in the L protomer of AcrB (Fig. 9.4) was proposed to function as a putative initial binding site for substrates (Murakami et al. 2006; Pos 2009; Seeger et al. 2006). A 2.25 Å resolution AcrB/doxorubicin co-crystal structure (pdb entry 4DX7) solved in our lab (Eicher et al. 2012) indicated simultaneous binding of a doxorubicin molecule in the deep binding pocket of the T protomer, a twofold-symmetric sandwich of stacked doxorubicin molecules (Fig. 9.7), and is likely to represent a structural indication for an access site for drugs into AcrB. This site was designated as the "access pocket."

The two doxorubicin molecules are tightly interacting with each other by stacking interactions of their anthracyclic rings. The stacked dimer itself is predominantly coordinated by hydrophobic interactions in the access pocket. In vivo substrate cross-linking experiments support the crystallographic observation that substrates bind to the access pocket (Fig. 9.7). Cysteine substitutions of those residues interacting with the doxorubicin dimer were also strongly cross-linked by the maleimide-labeled AcrB-substrate Bodipy-FL (Husain and Nikaido 2010). Strong cross-linking efficiencies indicate that these residues are part of the substrate pathway and accessible for substrate molecules under in vivo conditions. Moreover, AcrB variants F617A, F666A, and R717A conferred significantly decreased resistances against various compounds (Bohnert et al. 2008; Yu et al. 2005) (Fig. 9.7) and are therefore indicative to play an important role in AcrB-mediated transport of substrates.

Most strikingly, high molecular mass drugs rifampicin and erythromycin were identified in X-ray structures solved in the Yamaguchi lab in 2011 (Nakashima et al. 2011) within the L protomers of asymmetric AcrB co-crystal structures [pdb entries 3AOB (resolution = 3.35 Å), 3AOC (resolution = 3.35 Å) and 3AOD (resolution = 3.3 Å)] (Fig. 9.8). One of the structures showed the remarkable

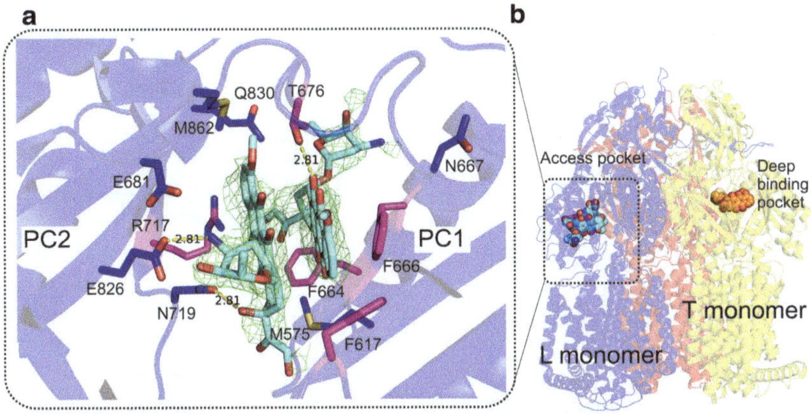

Fig. 9.7 Binding of doxorubicin to the access pocket of the L protomer. (**a**) Close-up view of the access pocket with two doxorubicin molecules in *stick representation* (*cyan*: carbon, *red*: oxygen, *blue*: nitrogen) and $2F_o$–F_c electron density (2.25 Å resolution) displayed as *green* mesh after refinement and contoured at 1σ. Side chain residues within 3.5 Å distance from the doxorubicin molecules are shown in *stick representation* (*blue*: carbon, *red*: oxygen, *light blue*: nitrogen, *yellow*: sulfur). Residues identified as part of the drug export pathway by Bodypy-FL maleimide labeling (Husain and Nikaido 2010) are depicted as *magenta sticks* (*magenta*: carbon, *red*: oxygen, *blue*: nitrogen) as well as alanine-substituted residues (F617A, F666A, R717A), causing reduced *E. coli* resistance towards drugs. (**b**) Side view of the asymmetric AcrB trimer in *cartoon representation* (L in *blue*, T in *yellow*, O in *red*) with two doxorubicin molecules bound to the access or one doxorubicin molecule bound to the deep binding pocket in *sphere representation* (*cyan* or *orange*, respectively: carbon, *red*: oxygen, *blue*: nitrogen)

co-crystallization of rifampicin and minocycline in the L and T protomers, respectively, within the same AcrB trimer, indicating true "multiple drug" binding. Superimpositions of these structures with the 2.25 Å resolution doxorubicin co-crystal structure [pdb entry 4DX7, also containing drugs in both access and deep binding pockets (Eicher et al. 2012)] indicated that the position of rifampicin was largely congruent with the position of the doxorubicin dimer in 4DX7 (Fig. 9.8, D), whereas erythromycin was positioned more towards the deep binding pocket relative to the location of the doxorubicin dimer. The structures might indicate the intermediate phases of drug transport through the AcrB porter domain (Fig. 9.8). Of essential importance for the binding of substrates is a large loop (AcrB amino acids 613–623) of the PC1 subdomain protruding into the core of the porter domain of each AcrB protomer and is noticeably positioned between access and deep binding pocket (Fig. 9.9). This loop was designated as the "switch loop" and is partially defining the deep binding pocket as well as the access binding pocket. The switch loop adopts a specific conformation in the L protomer of the AcrB/doxorubicin co-crystal structure (pdb entry 4DX7) and interacts with the doxorubicin dimer in the access pocket.

A single amino acid exchange of G616N within the switch loop of AcrB was able to inhibit the efflux activity mainly for rather large substrates, like macrolide antibiotics (Eicher et al. 2012). Nevertheless, the transport of another high

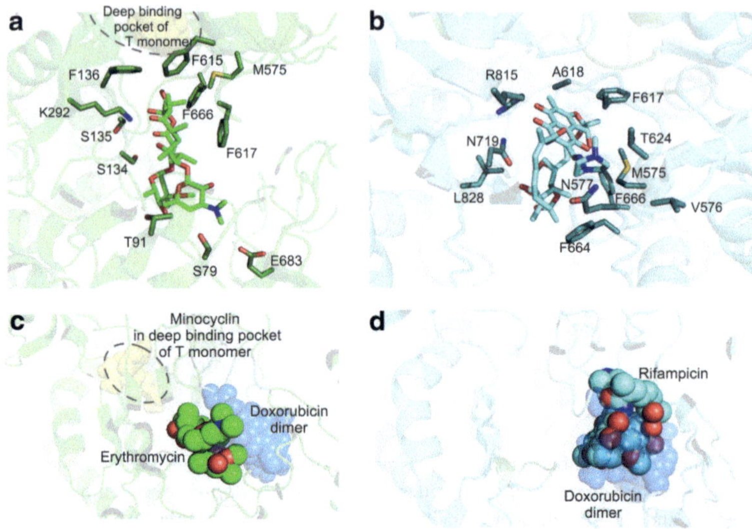

Fig. 9.8 Detailed view on the local structure of the porter domain of L protomers represented as *transparent cartoons* of the AcrB/substrate co-crystal structures (**a**) pdb entry 3AOC (erythromycin bound) and (**b**) pdb entry 3AOD (rifampicin and minocycline bound) solved by Nakashima et al. (2011). (**a**) Erythromycin is represented in *green sticks*. Residues interacting with erythromycin are shown in *dark green sticks*. (**b**) Rifampicin is displayed in *cyan sticks*. All residues involved in binding of rifampicn are shown in *dark cyan sticks*. (**c**) Porter domain of pdb entry 3AOC represented as *transparent green cartoon*. Erythromycin (*green spheres*) within pdb entry 3AOC is substantially displaced from the superimposed doxorubicin dimer (*transparent blue spheres*) identified in pdb entry 4DX7 (Eicher et al. 2012) and is nearer to the theoretical position of the deep binding pocket (presented by *yellow spheres*). (**d**) Porter domain of pdb entry 3AOD represented as *transparent cyan cartoon*. Rifampicin (*cyan spheres*) within pdb entry 3AOD is almost similarly positioned within the access pocket as the superimposed doxorubicin dimer (*transparent blue spheres*) identified in pdb entry 4DX7

molecular mass drug, novobiocin, was not affected by the G616 substitution. Moreover, resistance against small molecular mass drugs ciprofloxacin, linezolid, and TPP^+ was affected as well by the G616N substitution. Clearly, other physicochemical parameters of drug molecules, besides the molecular mass, are causing the rather puzzling phenotype. The crystal structure of AcrB G616N revealed an alternative switch-loop conformation in the L protomer of AcrB compared to the wild-type conformation. This altered switch-loop conformation would prevent or inhibit binding of a doxorubicin dimer in the access pocket by sterical clash. The multiple distinct L-switch loop conformations, observed in various crystal structures, indicate that the access pocket might be a highly flexible binding pocket, where the switch loop acts as an adaptor module. The putatively flexible adaptor module like the switch loop appears to be used to exclusively alter the shape of the binding pocket without altering the global protein conformation. This "adaptor-mediated binding mechanism" might be a general feature of multidrug recognition

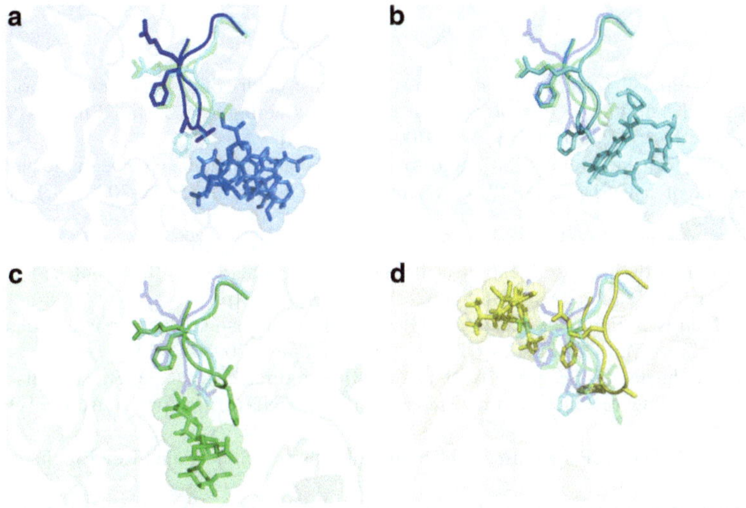

Fig. 9.9 Detailed view on the switch loop in cartoon representation of AcrB L monomers with a doxorubicin dimer from pdb entry 4DX7 (**a**, *blue sticks*), rifampicin from pdb entry 3AOD (**b**, *cyan sticks*) and erythromycin from 3AOC (**c**, *green sticks*) bound to the access pocket. The observed switch loop conformations correlate with the different substrates used for co-crystallization and might indicate that the switch loop conformation is substrate induced. (**d**) All switch loop conformations of the L monomers differ significantly from the switch loop conformation seen in the T monomer of pdb entry 4DX5 (AcrB with minocycline bound to only the deep binding pocket). In all pictures the switch loops of the respective other structures are displayed as *transparent cartoons* to demonstrate the reorientation of switch loop. Side chain residues are displayed in *stick* representation and substrate molecules as *stick* and *transparent spheres* to demonstrate the van der Waals volumes

in membrane transporters and can either function according to an induced fit or conformational selection mechanism.

Combining the results obtained from the structural and functional data described above presents the tripartite drug efflux proteins as highly dynamic entities with allosteric and cooperative properties. Gaining deeper insight into the underlying mechanisms of substrate specificity of AcrB and the initial hypothesis of peristaltic motion driving drug efflux (Seeger et al. 2006) is in need of in vitro (reconstitution) assays (for AcrAB-TolC or any of the homologue tripartite systems) and time-resolved structural analysis. Moreover, the assignment of the physiological function of many of these tripartite pumps will help to understand why these machineries are highly cooperative and in need of allosteric regulation.

Acknowledgments The work of the Pos lab presented in this chapter was supported by the Swiss National Foundation, the German Research Foundation (SFB 807, Transport and Communication across Biological Membranes), the DFG-EXC115 (Cluster of Excellence Macromolecular Complexes at the Goethe-University Frankfurt), the Innovative Medicine Initiative (IMI), Project TRANSLOCATION (http://www.imi.europa.eu/content/translocation) and by grants from Europe Aspire and Human Frontier Science Program.

References

Akama H, Matsuura T, Kashiwagi S et al (2004) Crystal structure of the membrane fusion protein, MexA, of the multidrug transporter in Pseudomonas aeruginosa. J Biol Chem 279:25939–25942, doi: 10.1074/jbc.C400164200 C400164200 [pii]

Bagai I, Liu W, Rensing C et al (2007) Substrate-linked conformational change in the periplasmic component of a Cu(I)/Ag(I) efflux system. J Biol Chem 282:35695–702. doi:10.1074/jbc.M703937200

Balakrishnan L, Hughes C, Koronakis V (2001) Substrate-triggered recruitment of the TolC channel-tunnel during type I export of hemolysin by Escherichia coli. J Mol Biol 313:501–510, doi: 10.1006/jmbi.2001.5038 S0022-2836(01)95038-7 [pii]

Bavro VN, Pietras Z, Furnham N et al (2008) Assembly and channel opening in a bacterial drug efflux machine. Mol Cell 30:114–21. doi:10.1016/j.molcel.2008.02.015

Bohnert JA, Schuster S, Seeger MA et al (2008) Site-directed mutagenesis reveals putative substrate binding residues in the Escherichia coli RND efflux pump AcrB. J Bacteriol 190:8225–8229

Boyer PD (1997) The ATP synthase–a splendid molecular machine. Annu Rev Biochem 66:717–749

Brandstatter L, Sokolova L, Eicher T et al (2011) Analysis of AcrB and AcrB/DARPin ligand complexes by LILBID MS. Biochim Biophys Acta 1808:2189–96. doi:10.1016/j.bbamem.2011.05.009

Bush K, Courvalin P, Dantas G et al (2011) Tackling antibiotic resistance. Nat Rev Microbiol 9:894–6. doi:10.1038/nrmicro2693

Davin-Regli A, Bolla J-M, James CE et al (2008) Membrane permeability and regulation of drug "influx and efflux" in enterobacterial pathogens. Curr Drug Targets 9:750–9

de Cristóbal RE, Vincent PA, Salomón RA (2006) Multidrug resistance pump AcrAB-TolC is required for high-level, Tet(A)-mediated tetracycline resistance in Escherichia coli. J Antimicrob Chemother 58:31–6. doi:10.1093/jac/dkl172

Eicher T, Cha HJ, Seeger MA et al (2012) Transport of drugs by the multidrug transporter AcrB involves an access and a deep binding pocket that are separated by a switch-loop. Proc Natl Acad Sci U S A 109:5687–5692, doi: 1114944109 [pii] 10.1073/pnas.1114944109

Elkins CA, Nikaido H (2002) Substrate specificity of the RND-type multidrug efflux pumps AcrB and AcrD of Escherichia coli is determined predominantly by two large periplasmic loops. J Bacteriol 184:6490–6498. doi:10.1128/JB.184.23.6490

Fischer N, Kandt C (2011) Three ways in, one way out: water dynamics in the trans-membrane domains of the inner membrane translocase AcrB. Proteins 79:2871–2885. doi:10.1002/prot.23122

Fujihira E, Tamura N, Yamaguchi A (2002) Membrane topology of a multidrug efflux transporter, AcrB, in Escherichia coli. J Biochem 131:145–51

Gärtner RM, Perez C, Koshy C, Ziegler C (2011) Role of bundle helices in a regulatory crosstalk in the trimeric betaine transporter BetP. J Mol Biol 414:327–36. doi:10.1016/j.jmb.2011.10.013

Higgins MK, Bokma E, Koronakis E et al (2004) Structure of the periplasmic component of a bacterial drug efflux pump. Proc Natl Acad Sci U S A 101:9994–9999, doi: 10.1073/pnas.0400375101

Husain F, Nikaido H (2010) Substrate path in the AcrB multidrug efflux pump of Escherichia coli. Mol Microbiol 78:320–330. doi:10.1111/j.1365-2958.2010.07330.x

Kawabe T, Fujihira E, Yamaguchi A (2000) Molecular construction of a multidrug exporter system, AcrAB: molecular interaction between AcrA and AcrB, and cleavage of the N-terminal signal sequence of AcrA. J Biochem 128:195–200

Kim H-S, Nikaido H (2012) Different functions of MdtB and MdtC subunits in the heterotrimeric efflux transporter MdtB(2)C complex of Escherichia coli. Biochemistry 51:4188–97. doi:10.1021/bi300379y

Kim HS, Nagore D, Nikaido H (2010) Multidrug efflux pump MdtBC of Escherichia coli is active only as a B2C heterotrimer. J Bacteriol 192:1377–1386, doi: JB.01448-09 [pii] 10.1128/JB.01448-09

Koronakis V, Sharff A, Koronakis E et al (2000) Crystal structure of the bacterial membrane protein TolC central to multidrug efflux and protein export. Nature 405:914–919. doi:10.1038/35016007

Lee A, Mao W, Warren M (2000) Interplay between efflux pumps may provide either additive or multiplicative effects on drug resistance. J Bacteriol 182(11):3142–3150, doi: 10.1128/JB.182.11.3142-3150.2000

Lim SP, Nikaido H (2010) Kinetic parameters of efflux of penicillins by the multidrug efflux transporter AcrAB-TolC of Escherichia coli. Antimicrob Agents Chemother 54:1800–1806, doi: AAC.01714-09 [pii] 10.1128/AAC.01714-09

Lipscomb W, Kantrowitz E (2011) Structure and mechanisms of Escherichia coli aspartate transcarbamoylase. Acc Chem Res 45:444–53. doi:10.1021/ar200166p

Lu S, Zgurskaya HI (2012) Role of ATP binding and hydrolysis in assembly of MacAB-TolC macrolide transporter. Mol Microbiol 86:1132–43. doi:10.1111/mmi.12046

Ma D, Cook DN, Alberti M et al (1995) Genes acrA and acrB encode a stress-induced efflux system of Escherichia coli. Mol Microbiol 16:45–55

Mao W, Warren MS, Black DS et al (2002) On the mechanism of substrate specificity by resistance nodulation division (RND)-type multidrug resistance pumps: the large periplasmic loops of MexD from Pseudomonas aeruginosa are involved in substrate recognition. Mol Microbiol 46:889–901

Mokhonov VVV, Mokhonova EIEI, Akama H, Nakae T (2004) Role of the membrane fusion protein in the assembly of resistance-nodulation-cell division multidrug efflux pump in Pseudomonas aeruginosa. Biochem Biophys Res Commun 322:483–489. doi:10.1016/j.bbrc.2004.07.140

Murakami S, Nakashima R, Yamashita E, Yamaguchi A (2002) Crystal structure of bacterial multidrug efflux transporter AcrB. Nature 419:587–93. doi:10.1038/nature01050

Murakami S, Nakashima R, Yamashita E et al (2006) Crystal structures of a multidrug transporter reveal a functionally rotating mechanism. Nature 443:173–9. doi:10.1038/nature05076

Nagano K, Nikaido H (2009) Kinetic behavior of the major multidrug efflux pump AcrB of Escherichia coli. Proc Natl Acad Sci U S A 106:5854–8. doi:10.1073/pnas.0901695106

Nakamura H, Hachiya N, Tojo T (1978) Second acriflavine sensitivity mutation, acrB, in Escherichia coli K-12. J Bacteriol 134(3):1184–1187

Nakashima R, Sakurai K, Yamasaki S et al (2011) Structures of the multidrug exporter AcrB reveal a proximal multisite drug-binding pocket. Nature 480:565–569. doi:10.1038/nature10641

Nikaido H (1998) Antibiotic resistance caused by gram-negative multidrug efflux pumps. Clin Infect Dis 27(Suppl 1):S32–41

Nikaido H, Pagès J-M (2012) Broad-specificity efflux pumps and their role in multidrug resistance of Gram-negative bacteria. FEMS Microbiol Rev 36:340–63. doi:10.1111/j.1574-6976.2011.00290.x

Nikaido H, Takatsuka Y (2009) Mechanisms of RND multidrug efflux pumps. Biochim Biophys Acta 1794:769–781, doi: S1570-9639(08)00334-8 [pii]10.1016/j.bbapap.2008.10.004

Nishino K, Yamaguchi A (2001) Analysis of a complete library of putative drug transporter genes in Escherichia coli. J Bacteriol 183:5803–5812. doi:10.1128/JB.183.20.5803

Pei X, Hinchliffe P (2011) Structures of sequential open states in a symmetrical opening transition of the TolC exit duct. Proc Natl Acad Sci U S A 108(5):2112–2117, doi: 10.1073/pnas.1012588108/-/DCSupplemental.www.pnas.org/cgi/doi/10.1073/pnas.1012588108

Perez C, Khafizov K, Forrest LR et al (2011) The role of trimerization in the osmoregulated betaine transporter BetP. EMBO Rep 12:804–10. doi:10.1038/embor.2011.102

Petrek M, Kosinova P, Koca J, Otyepka M (2007) MOLE: a Voronoi diagram-based explorer of molecular channels, pores, and tunnels. Structure 15:1357–1363

Piddock LJ (2006) Clinically relevant chromosomally encoded multidrug resistance efflux pumps in bacteria. Clin Microbiol Rev 19:382–402. doi:10.1128/CMR.19.2.382

Pos KM (2009) Drug transport mechanism of the AcrB efflux pump. Biochim Biophys Acta 1794:782–93. doi:10.1016/j.bbapap.2008.12.015

Pos KM, Diederichs K (2002) Purification, crystallization and preliminary diffraction studies of AcrB, an inner-membrane multi-drug efflux protein. Acta Crystallogr D Biol Crystallogr 58:1865–1867

Ressl S, Terwisscha van Scheltinga AC, Vonrhein C et al (2009) Molecular basis of transport and regulation in the Na(+)/betaine symporter BetP. Nature 458:47–52. doi:10.1038/nature07819

Saier MH, Tran CV, Barabote RD (2006) TCDB: the Transporter Classification Database for membrane transport protein analyses and information. Nucleic Acids Res 34:D181–6. doi:10.1093/nar/gkj001

Schulz R, Vargiu AV, Collu F et al (2010) Functional rotation of the transporter AcrB: insights into drug extrusion from simulations. PLoS Comput Biol 6:e1000806. doi:10.1371/journal.pcbi.1000806

Schulz R, Vargiu AV, Ruggerone P et al (2011) Role of water during the extrusion of substrates by the efflux transporter AcrB. J Phys Chem B 115:8278–8287. doi:10.1021/jp200996x

Seeger MA, Schiefner A, Eicher T et al (2006) Structural asymmetry of AcrB trimer suggests a peristaltic pump mechanism. Science 313:1295–1298. doi:10.1126/science.1131542

Seeger MA, von Ballmoos C, Eicher T et al (2008) Engineered disulfide bonds support the functional rotation mechanism of multidrug efflux pump AcrB. Nat Struct Mol Biol 15:199–205

Seeger MA, von Ballmoos C, Verrey F, Pos KM (2009) Crucial role of Asp408 in the proton translocation pathway of multidrug transporter AcrB: evidence from site-directed mutagenesis and carbodiimide labeling. Biochemistry 48:5801–5812. doi:10.1021/bi900446j

Sennhauser G, Amstutz P, Briand C et al (2007) Drug export pathway of multidrug exporter AcrB revealed by DARPin inhibitors. PLoS Biol 5:e7, doi: 06-PLBI-RA-1517R2 [pii] 10.1371/journal.pbio.0050007

Sennhauser G, Bukowska MA, Briand C et al (2009) Crystal structure of the multidrug exporter MexB from Pseudomonas aeruginosa. J Mol Biol 389:134–145, doi: S0022-2836(09)00401-X [pii] 10.1016/j.jmb.2009.04.001

Su CC, Li M, Gu R et al (2006) Conformation of the AcrB multidrug efflux pump in mutants of the putative proton relay pathway. J Bacteriol 188:7290–7296, doi: 188/20/7290 [pii] 10.1128/JB.00684-06

Su CC, Yang F, Long F et al (2009) Crystal structure of the membrane fusion protein CusB from Escherichia coli. J Mol Biol 393:342–355, doi: S0022-2836(09)01025-0 [pii] 10.1016/j.jmb.2009.08.029

Su C-C, Long F, Zimmermann MT et al (2011a) Crystal structure of the CusBA heavy-metal efflux complex of Escherichia coli. Nature 470:558–62. doi:10.1038/nature09743

Su CC, Long F, Zimmermann MT et al (2011b) Crystal structure of the CusBA heavy-metal efflux complex of Escherichia coli. Nature 470:558–562, doi: nature09743 [pii] 10.1038/nature09743

Sulavik MC, Houseweart C, Cramer C et al (2001) Antibiotic susceptibility profiles of Escherichia coli strains lacking multidrug efflux pump genes. Antimicrob Agents Chemother 45:1126–1136

Symmons MF, Bokma E, Koronakis E et al (2009) The assembled structure of a complete tripartite bacterial multidrug efflux pump. Proc Natl Acad Sci U S A 106:7173–7178, doi: 0900693106 [pii] 10.1073/pnas.0900693106

Takatsuka Y, Nikaido H (2006) Threonine-978 in the transmembrane segment of the multidrug efflux pump AcrB of Escherichia coli is crucial for drug transport as a probable component of the proton relay network. J Bacteriol 188:7284–7289, doi: 188/20/7284 [pii] 10.1128/JB.00683-06

Takatsuka Y, Nikaido H (2007) Site-directed disulfide cross-linking shows that cleft flexibility in the periplasmic domain is needed for the multidrug efflux pump AcrB of Escherichia coli. J Bacteriol 189:8677–8684, JB.01127-07 [pii] 10.1128/JB.01127-07

Takatsuka Y, Nikaido H (2009) Covalently linked trimer of the AcrB multidrug efflux pump provides support for the functional rotating mechanism. J Bacteriol 191:1729–1737, JB.01441-08 [pii] 10.1128/JB.01441-08

Tal N, Schuldiner S (2009) A coordinated network of transporters with overlapping specificities provides a robust survival strategy. Proc Natl Acad Sci U S A 106:9051–9056, doi: 0902400106 [pii] 10.1073/pnas.0902400106

Tamura N, Murakami S, Oyama Y et al (2005) Direct interaction of multidrug efflux transporter AcrB and outer membrane channel TolC detected via site-directed disulfide cross-linking. Biochemistry 44:11115–11121

Tanabe M, Szakonyi G, Brown KA et al (2009) The multidrug resistance efflux complex, EmrAB from Escherichia coli forms a dimer in vitro. Biochem Biophys Res Commun 380:338–42. doi:10.1016/j.bbrc.2009.01.081

Tikhonova EB, Zgurskaya HI (2004) AcrA, AcrB, and TolC of Escherichia coli form a stable intermembrane multidrug efflux complex. J Biol Chem 279:32116–32124. doi:10.1074/jbc.M402230200

Tikhonova EB, Yamada Y, Zgurskaya HI (2011) Sequential mechanism of assembly of multidrug efflux pump AcrAB-TolC. Chem Biol 18:454–463, doi: S1074-5521(11)00090-1 [pii] 10.1016/j.chembiol.2011.02.011

Touzé T, Eswaran J, Bokma E et al (2004) Interactions underlying assembly of the Escherichia coli AcrAB-TolC multidrug efflux system. Mol Microbiol 53:697–706. doi:10.1111/j.1365-2958.2004.04158.x

Tsukazaki T, Mori H, Echizen Y et al (2011) Structure and function of a membrane component SecDF that enhances protein export. Nature 474:235–8. doi:10.1038/nature09980

Vaccaro L, Koronakis V, Sansom MS (2006) Flexibility in a drug transport accessory protein: molecular dynamics simulations of MexA. Biophys J 91:558–564

Vaccaro L, Scott KA, Sansom MS (2008) Gating at both ends and breathing in the middle: conformational dynamics of TolC. Biophys J 95:5681–5691, doi: S0006-3495(08)81985-6 [pii] 10.1529/biophysj.108.136028

Vargiu AV, Collu F, Schulz R et al (2011) Effect of the F610A mutation on substrate extrusion in the AcrB transporter: explanation and rationale by molecular dynamics simulations. J Am Chem Soc 133:10704–10707. doi:10.1021/ja202666x

Vila J, Martínez JL (2008) Clinical impact of the over-expression of efflux pump in nonfermentative Gram-negative bacilli, development of efflux pump inhibitors. Curr Drug Targets 9:797–807

Walsh C (2000) Molecular mechanisms that confer antibacterial drug resistance. Nature 406:775–81. doi:10.1038/35021219

Weeks JW, Celaya-Kolb T, Pecora S, Misra R (2010) AcrA suppressor alterations reverse the drug hypersensitivity phenotype of a TolC mutant by inducing TolC aperture opening. Mol Microbiol 75:1468–83. doi:10.1111/j.1365-2958.2010.07068.x

Xu Y, Lee M, Moeller A et al (2011) Funnel-like hexameric assembly of the periplasmic adapter protein in the tripartite multidrug efflux pump in gram-negative bacteria. J Biol Chem 286:17910–20. doi:10.1074/jbc.M111.238535

Yao X-Q, Kimura N, Murakami S, Takada S (2013) Drug uptake pathways of multidrug transporter AcrB studied by molecular simulations and site-directed mutagenesis experiments. J Am Chem Soc. doi:10.1021/ja310548h

Yu EW, Aires JR, Mcdermott G, Nikaido H (2005) A periplasmic drug-binding site of the AcrB multidrug efflux pump: a crystallographic and site-directed mutagenesis study. J Bacteriol 187:6804–6815, doi:187/19/6804 [pii]10.1128/JB.187.19.6804-6815.2005

Yu L, Lu W, Wei Y (2011) AcrB trimer stability and efflux activity, insight from mutagenesis studies. PLoS One 6:e28390. doi:10.1371/journal.pone.0028390

Zgurskaya HI, Nikaido H (1999) AcrA is a highly asymmetric protein capable of spanning the periplasm. J Mol Biol 285:409–420, doi: 10.1006/jmbi.1998.2313

Zgurskaya HI, Nikaido H (2000) Cross-linked complex between oligomeric periplasmic lipoprotein AcrA and the inner-membrane-associated multidrug efflux pump AcrB from Escherichia coli. J Bacteriol 182:4264–4267

Zgurskaya HI, Yamada Y, Tikhonova EB et al (2009) Structural and functional diversity of bacterial membrane fusion proteins. Biochim Biophys Acta 1794:794–807, doi: S1570-9639(08)00335-X [pii]10.1016/j.bbapap.2008.10.010

Chapter 10
What Can a Living Fossil Tell Us About Evolution and Mechanism of Ion-Coupled Transporters: The Story of Small Multidrug Transporters

Shimon Schuldiner

Abstract Small Multidrug Resistance (SMR) transporters are small homo- or hetero-dimers that confer resistance to multiple toxic compounds by exchanging substrate with protons. They reside in the inner membrane of bacteria and in halophilic and methanogenic archaea and because many of their substrates are routinely used as antibiotics and antiseptics, they have been associated with the phenomenon of multidrug resistance. EmrE, the most studied SMR member, has presented biochemists with unusual surprises regarding its topology and raised an interesting controversy since structural information was in an apparent conflict with biochemical data. One of the reasons for the controversy was the assumption that, to ensure proper function, membrane proteins must be inserted by a mechanism that warrants a unique topology. As it turns out, EmrE and other SMR transporters display a remarkable plasticity regarding topology in the membrane, interaction between subunits, and interaction with substrates. This plasticity implies a high evolvability of these proteins and, as a consequence, a lack of commitment that facilitates acquisition of new functions and topologies. Because of this high evolvability, we suggest that SMRs are living fossils at an evolutionary junction. Study of their properties provides a wonderful glimpse at the evolution and mechanism of ion-coupled transporters.

Keywords Antiporter • Multidrug transporters • Parallel and antiparallel topology • Proton release • Transient kinetics • Tryptophan fluorescence • Uncoupled mutants

S. Schuldiner (✉)
Department of Biological Chemistry, Alexander Silberman Institute of Life Sciences, Hebrew University of Jerusalem, 91904 Jerusalem, Israel
e-mail: shimon.schuldiner@huji.ac.il

R. Krämer and C. Ziegler (eds.), *Membrane Transport Mechanism*, Springer Series in Biophysics 17, DOI 10.1007/978-3-642-53839-1_10,

10.1 Living Fossils and Modern Mechanisms

Living fossil is an informal term for any living species or organism which appears to be the same as a species otherwise only known from fossils and which has no close living relatives. These species have all survived major extinction events and generally retain low taxonomic diversities. I suggest that EmrE and SMR proteins resemble, in a way, living fossils that are widespread only in bacteria and in very few groups of archaea (Schuldiner 2009; Kolbusz et al. 2013; Bay and Turner 2009). Because of their high evolvability we suggest that SMRs are at an evolutionary junction. Evolvability can be seen also as lack of commitment: EmrE and other SMRs are not fully committed to a certain substrate and can easily change specificity (Brill et al. 2012) and even mode of coupling (Brill et al. 2012; Adam et al. 2007); they are not committed to a certain topology (Nasie et al. 2010) and not even committed to a certain partner to form the oligomer (Nasie et al. 2010; Schuldiner 2012).

In this chapter, I will concentrate on what we have learned from the study of EmrE and I will compare the mechanism proposed for EmrE to that in other H^+-coupled antiporters. EmrE is a small (110 residues) SMR transporter from *Escherichia coli* that functions as a dimer and extrudes positively charged drugs in exchange for two protons, thus rendering bacteria resistant to a variety of toxic compounds. Study of this small, 110-residue multidrug transporter from *E. coli* has provided valuable information for the understanding of the coupling mechanism of the ion-coupled transporter family (Yerushalmi and Schuldiner 2000a, c; Soskine et al. 2004; Schuldiner 2007, 2009; Adam et al. 2007). The simplicity of the coupling mechanism in EmrE allows for the remarkable flexibility in the structure of the protein that will be discussed below.

10.1.1 Competition as a Way of Life for H^+-Coupled Antiporters

H^+-coupled antiporters are ubiquitous proteins that utilize proton electrochemical gradients generated by primary pumps. As a result, these proteins remove their substrates from the cytoplasm into acidic intracellular compartments or out to the medium. Well-known examples of this type of transporters are the multidrug transporters, vesicular neurotransmitter transporters, organic anion, cation transporters, and Na^+/H^+ antiporters.

H^+-coupled antiporters have been proposed to couple transport by utilizing a sequential binding and translocation mechanism, through which the substrate must be released prior to binding and translocation of the counter-transported ion (Fig. 10.1). Such a mechanism anticipates two major conformations of the transporter, facing alternatively each side of the membrane (C_o and C_i), which can interconvert only when one of the substrates is bound. Furthermore, binding of

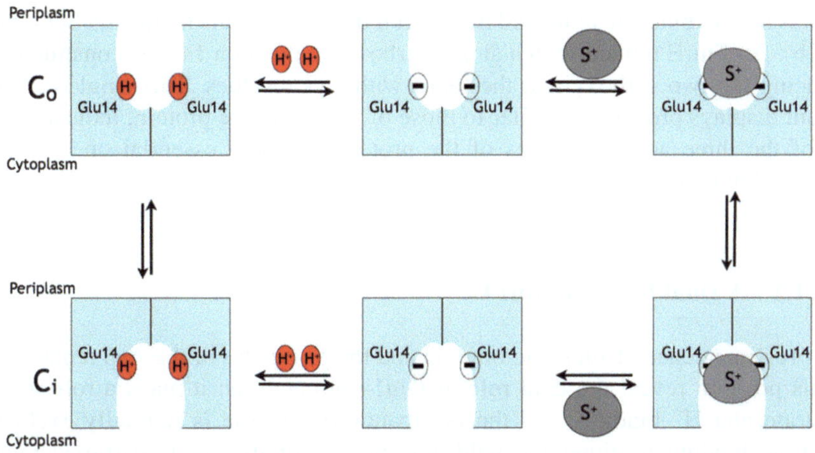

Fig. 10.1 *Alternating access model of EmrE transport*. Each state in the transport cycle is only open to one side of the membrane, and the two states only interconvert when either two protons (*small red circles*) or one substrate molecule (*grey circle*) are bound. Glu14 is the only membrane embedded carboxyl (*small circles* with *negative sign*) in each one of the protomers in the functional dimer

the counter substrates together is prohibited. This implies a competition between the two substrates that maybe direct as in the case of EmrE or indirect, in other examples that are discussed in detail below.

10.1.1.1 The Unbearable Simplicity of the Coupling Mechanism in EmrE

EmrE provides a unique experimental paradigm not only because of its size and stability but because, under proper conditions, the detergent solubilized protein binds substrate and releases protons in a mode that reflects with high fidelity its catalytic activity in the membrane. This property has enabled a detailed study of the molecular basis of substrate recognition and the coupling between protons and substrate (Yerushalmi and Schuldiner 2000a, b, c; Yerushalmi et al. 2001; Soskine et al. 2004; Gutman et al. 2003; Weinglass et al. 2005). EmrE contains eight charged residues, seven of them located in the hydrophilic loops and only one membrane-embedded charged residue, Glu14, which is also conserved in more than 200 homologous proteins in bacteria and archaea (Bay et al. 2008; Schuldiner 2009). Replacement of Glu14 in EmrE or Smr from *S. aureus* (equivalent Glu13) with Cys, Gln, His, Tyr, or Asp had a profound effect on the phenotype (Grinius and Goldberg 1994; Yerushalmi et al. 1995; Muth and Schuldiner 2000; Yerushalmi and Schuldiner 2000b). Further characterization of the mutants showed that the E14C mutation yielded a protein completely devoid of activity, while the E14D mutation was impaired in its ability to couple substrate fluxes to the proton gradient but was able to bind substrate and transport it downhill (Yerushalmi and Schuldiner 2000b). An EmrE mutant with a single carboxyl at position 14 was constructed by replacing the two carboxyls in the loops with Cys residues. This single carboxyl mutant displays properties similar to those of the wild-type protein, indicating that out of the three acidic residues of the protein, the only essential one is Glu14 (Yerushalmi et al. 2001).

10.1.1.2 A Dual Role for Glu14

The pH dependence of substrate binding and the properties of the Asp replacement at this position reveal the dual role of Glu14 in catalysis: it plays a role in both substrate and H^+ binding, and the occupancy of the site is mutually exclusive. Substrate binding to either the wild-type protein or the single carboxyl mutant increases dramatically with increasing pH suggesting that deprotonation of Glu14 is required [Fig 10.2 and Muth and Schuldiner (2000), Yerushalmi and Schuldiner (2000a), Yerushalmi and Schuldiner (2000c)]. Conversely, substrate induces proton release, and both reactions (substrate binding and proton release) have been observed directly in the detergent solubilized preparation of EmrE (Soskine et al. 2004).

In addition to the steady state measurements, the use of transient kinetics allowed us to break down this multistep process to its individual steps (Adam et al. 2007). A powerful tool was provided by the study of the fluorescent properties of EmrE. The fluorescence of Trp63 in EmrE reflects the occupancy of the binding site in the protein: the highest fluorescence is observed when the protein is fully deprotonated; the protonated protein displays a lower fluorescence, and the substrate bound protein displays the lowest [Fig. 1 in Adam et al. (2007)]. Interestingly, the environment to which Trp63 is exposed, as reflected by the wavelength of its fluorescence peak, does not change significantly at the three different states suggesting that Trp63 does not go through large conformational changes that expose it to different environments. A possible reason for the different fluorescence levels could be due to quenching or energy transfer to the substrate or to other amino acids not yet identified.

These findings led us to study the transient kinetics of substrate binding and proton release using spectroscopic techniques: the fluorescence of either the wild-type protein or one with a single Trp at position 63 was used to measure binding rates (Adam et al. 2007). The maximal rates of substrate binding of about $2 \times 10^7\ M^{-1}\ s^{-1}$ are relatively close to typical values in diffusion limited processes, implying that binding does not necessitate major conformational changes of the protein. The k_{on} values for binding to the single Trp63 mutant in the pH range measured are essentially identical to the k_{on} values for binding to the wild-type protein. These findings further validate the use of the mutant to study the mechanism of catalysis. Moreover, the results suggest that Trp63 is the one responsible for most of the fluorescence changes in the wild-type protein. The changes in

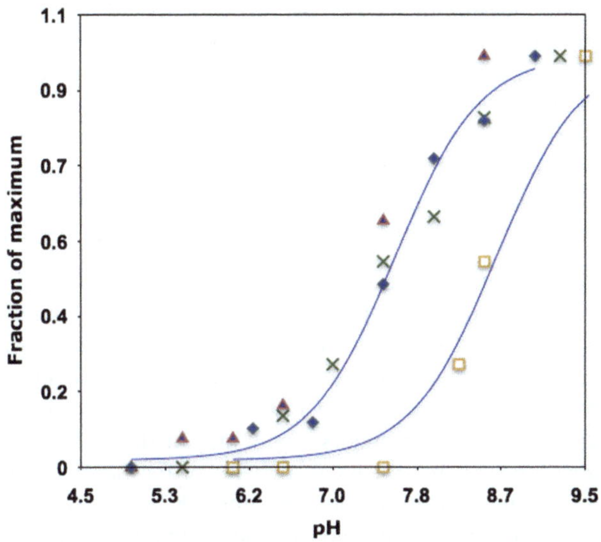

Fig. 10.2 *pH dependence of substrate binding and transport in EmrE.* Rates (*multiplication symbol*) and steady state (*filled diamond*) levels of TPP^+ binding to detergent solubilized EmrE increase with pH in a manner practically identical to the increase of downhill rates of efflux of methyl viologen (*filled triangles*) from proteoliposomes reconstituted with EmrE. A pH gradient of 2 units was imposed to drive methyl viologen accumulation into proteoliposomes. In these experiments the rate increases with increasing pH, but there is a shift of approximately 1 pH unit to the alkaline side (squares). Data from references Adam et al. (2007), Yerushalmi and Schuldiner (2000b) and Muth and Schuldiner (2000)

fluorescence of Trp63 are extremely large, and under optimal conditions most of the fluorescence is quenched either by protons or by substrate.

10.1.1.3 The Proposed Mechanism

Taken as a whole, the results support a binding mechanism where binding of either protons or substrate requires dissociation of the other. The suggested reaction path implies that first $EH_2 + S$ dissociate to $E + S + 2H^+$, and it rapidly leads to the formation of $ES + 2H^+$. The model assumes, as a first approximation, no binding to either the singly or doubly protonated protein and that both protons have a single or very similar pK_a [summarized in Adam et al. (2007)]. The experimental support for this model is based on the following findings:

1. Stoichiometry of substrate-induced proton release as determined from steady state measurements with detergent solubilized purified EmrE is $2H^+$/functional unit (dimer) (Soskine et al. 2004). This finding is supported also by stoichiometry of the whole transport process as measured in the intact *E. coli* cell: $2H^+$/substrate (Rotem and Schuldiner 2004).

2. In steady state measurements the pH dependence of the substrate-induced release of protons supports the contention that the pK_A must be well above 7 (Soskine et al. 2004). This contention is supported also by the pH dependence of the Trp63 fluorescence (Adam et al. 2007).
3. Although EmrE releases two protons upon substrate binding, the substrate induced proton release is a first order reaction, supporting the contention that the two protons display identical or very similar pK_A values.
4. Substrate binding shows a time course fit with a single exponential.

When results from the pre-steady state measurements are fitted to this simple model, the pre-equilibrium pK_A is predicted to be ~7.3 for the wild-type carboxyl at position 14 (Glu14) and ~5.8 for the aspartate replacement (Asp14). The pK_A values from this fit are lower than the ones estimated from steady state measurements (8.3–8.5 and 6.7, respectively) (Soskine et al. 2004). The reason for this apparent discrepancy is not fully understood. It may be due to technical limitations: a lower accuracy of the steady state measurements or for the need to obtain a model that better describes the reaction and predicts other pK_As. It may also reflect the fact that in the described experiments we are measuring two different parameters: substrate-induced initial rates of proton release in one case and quantity of protons released in the steady state measurements.

10.1.1.4 Uncoupled Mutants Raise Interesting Questions

The mechanism suggested above necessitates the fine-tuning of the pK_A of the carboxyl or the carboxyls participating in the protonation and deprotonation reactions. This contention was experimentally demonstrated in EmrE where replacement of Glu14 with Asp results in a decrease of the pK_A of the carboxyl. The mutant protein with a lower pK_A is mostly deprotonated at physiological pH, so it binds substrate but cannot couple the substrate flux to the proton gradient (Yerushalmi and Schuldiner 2000b; Soskine et al. 2004).

The behavior of this mutant raises a general question regarding antiporter mechanisms. In a classical scheme of an antiporter cycle, it is assumed that the transition of the "unloaded" transporter between the conformations facing the opposite sides of a membrane is not allowed (Fig. 10.1). Should such a transition be permitted, the coupling between the two fluxes would be compromised. A single conservative mutation of Glu14 to Asp transforms an antiporter into a uniporter suggesting that, in the mutant, this transition is permitted. It is possible that the mutation has an effect on the transporter that modifies the likelihood of this step to occur. Another more interesting possibility is that the transition is always allowed, even in the wild-type EmrE. The reason that it is not detectable is that the concentration of the deprotonated form of the wild-type transporter is, under most conditions, extremely low. In the E14D mutant that displays a lower pK_A, a large fraction of the transporter is deprotonated and can then interconvert freely between C_o and C_i conformation without the protonation reaction.

10.1.1.5 Validation of the Binding Measurements

Although they provide invaluable information, the measurements of the binding and release reaction with a detergent solubilized protein need validation with transport experiments performed with the protein in the membrane. Downhill efflux and exchange experiments were performed with proteoliposomes reconstituted with purified EmrE (Yerushalmi and Schuldiner 2000b). Downhill transport of substrate involves all the steps described in Fig. 10.1 but is driven by the substrate electrochemical gradient and does not necessitate a gradient of H^+ ions, i.e., the pH inside and outside the proteoliposome are the same. When the rate of Methyl Viologen efflux was measured at various pH values, it was found that the increase in the rate with an increase in pH was identical to that of the binding reaction measured in detergent [Fig. 10.2 and see also Yerushalmi and Schuldiner (2000b)]. In exchange experiments, a saturating concentration of unlabeled substrate is included in the outside solution and the transporter thus interconverts between the inside and the outside conformations in the "substrate-loaded" mode. Also in these experiments the pH dependence of the reactions was similar to that of the binding reaction (Yerushalmi and Schuldiner 2000b). When active uptake was measured, a pH gradient of 2 units was imposed to drive the accumulation. In these experiments the rate increases with increasing pH, but there is a shift of approximately 1 pH unit to the alkaline side suggesting that Glu14 is now "sensing" the average between the internal and external pH of the proteoliposome [Fig. 10.2 and see also Yerushalmi and Schuldiner (2000b)]. This "averaging" may be the result of the alternate exposure of the carboxyl to the acidic and alkaline face of the liposome. When similar experiments are performed with the "uncoupled" E14D mutant, the rates of efflux and exchange are practically independent of pH in the range 5–9 as expected from a protein with the essential carboxyl deprotonated at these pH values (Yerushalmi and Schuldiner 2000b).

These results validate the experiments performed with the detergent solubilized transporter and support the dual role proposed for Glu14.

10.1.1.6 Direct and Indirect Competition in Other Antiporters

As described in detail above, in the case of EmrE, the coupling mechanism is based on a direct competition of the substrate and the H^+ for the same binding site. Is there evidence for a similar mechanism for the larger modern antiporters? The answer is most likely a complex one.

NhaA provides another well-documented case of a direct competition between the two substrates (Mager et al. 2011). NhaA is a Na^+/H^+ antiporter from *E. coli* that belongs to the CPA family of monovalent cation/proton antiporters (Padan et al. 2009). It plays a central role in sodium and proton homeostasis by exchanging two H^+ with one Na^+ ion (Taglicht et al. 1993). Electrophysiological measurement of the activity of NhaA was tested over a wide pH range from pH 5.0 to 9.5.

Forward and reverse transport directions were investigated at zero membrane potential using preparations with inside-out and right side-out oriented transporters with Na^+ or H^+ gradients as the driving force. Under symmetrical pH conditions with a Na^+ gradient for activation, transport exhibited highly symmetrical activity with a bell-shaped pH dependence. The pH dependence was associated with a systematic increase of the *Km* for Na^+ at acidic pH. Under symmetrical Na^+ concentration with a pH gradient for NhaA activation, an unexpected novel characteristic of the antiporter was revealed; rather than being downregulated, it remained active even at a pH as low as 5 (Mager et al. 2011). This classical competition seems to be at a common binding site, as suggested by biochemical and structural evidence that support a mechanism whereby two Asp residues, Asp163 and Asp164, provide the site for both H^+ and Na^+ (Arkin et al. 2007; Padan et al. 2009).

On the other hand, at least in two cases, the counter-transported substrates seem to bind at different sites without compromising the competitive nature of their binding. Thus, in the case of MdfA, a proton/drug antiporter from *E. coli* that belongs to the MFS family, genetic and biochemical evidence support the contention that two distinct binding sites may exist for substrates and protons (Fluman et al. 2012). PfMATE, a H^+/drug antiporter from *Pyrococcus furiosus* that belongs to the MATE family, was recently crystallized (Tanaka et al. 2013). Structural and biochemical evidence support the contention that the binding sites for H^+ and substrates do not overlap.

An extreme example of "long-distance" competition is provided by the AcrAB-TolC complex from *E. coli*, a large tripartite complex that belongs to the RND family and provides the major intrinsic resistance of these cells. It actively removes a large variety of drugs from the periplasm by a mechanism that utilizes the proton electrochemical gradient across the cytoplasmic membrane. In the case of the AcrAB-TolC complex, there is a complete spatial separation of the sites: the proton binding site is in the membrane domain while the substrate binding site is in the periplasmic portion of this large complex (Seeger et al. 2009). The conformational changes occurring upon protonation and deprotonation of carboxylic residues in the membrane domain are coupled to conformational changes in distant domains where the substrate transport takes place (Seeger et al. 2009).

A notable, and maybe not unique, exception to the mechanism depicted in Fig. 10.1 is presented by the ClC Cl^-/H^+ antiporter (Miller and Nguitragool 2009). Antiporters bind their substrates alternately in the two different states and shuttle them across the membrane with ping-pong like conformational changes according to the alternating access model. In contrast, ClC is able to bind both its substrates simultaneously, and, importantly, proton movement along its pathway seems to be possible only when the central anion binding site is occupied by Cl^- (Miller and Nguitragool 2009).

Noteworthy, in three of the cases described above, NhaA, MdfA, and AcrAB-TolC complex, the pK_A of the important residues has been estimated to be at around 7.5 (Seeger et al. 2009; Fluman et al. 2012; Arkin et al. 2007; Mager et al. 2011), a value well within the range of the intracellular pH of *E. coli* cells (Padan et al. 1981). Thus, it seems that regardless of their specific structures or mechanisms, the transporters have evolved so that they are exquisitely tuned to function at the very constant cytoplasmic pH maintained by *E. coli* cells.

10.1.2 Mechanism and Structure

Much has been said about the structure and topology of EmrE [reviewed in Schuldiner (2012), see also Chap. 11]. The available models of the structure are derived from low-resolution 2D and 3D-crystals (Fleishman et al. 2006; Chen et al. 2007). To explain the quasi-symmetry observed in the 3D structure of EmrE acquired by electron cryo-microscopy (cryo-EM), Ubarretxena-Belandia et al suggested the possibility of an antiparallel topology of the protomers in the EmrE homodimer (Fig. 10.3) (Ubarretxena-Belandia et al. 2003). Very soon thereafter, a model derived from a low-resolution 3D-crystal structure supported this suggestion (Pornillos et al. 2005). Even though the structure was withdrawn (Chang et al. 2006), it stimulated investigators in the field to pursue this unique avenue. Fleishman et al. used the symmetry relationship mentioned above, combined with sequence conservation data, to assign the transmembrane segments in EmrE to the densities seen in the cryo-EM structure. The C-α model of the transmembrane region was constructed so that the helices of one protomer have the topology opposite to the ones of the other protomer (Fig. 10.3, antiparallel) (Fleishman et al. 2006). A recalculated model based on the 3D-crystals is similar to the model suggested from the electron microscopy data (Chen et al. 2007). The crystals used to derive the C-α model X-ray structure were obtained with protein solubilized with detergents that inhibit activity by disrupting the oligomer (Chen et al. 2007; Soskine et al. 2006). The protomers thus obtained arrange in the crystal in a conformation that minimizes the energy needed for crystal formation and do not necessarily reflect the topology in the membrane.

In addition, genetic experiments were designed to support the claim for an antiparallel topology. EmrE was fused to the "topology-reporter proteins" alkaline phosphatase and green fluorescent protein, and the results showed that the topology of the EmrE fusion proteins in the membrane is sensitive to the distribution of positive charges in the protein (Rapp et al. 2006). Manipulation of the positive charges generated a set of mutants, some with N_oC_o, and others with N_iC_i apparent topology (Rapp et al. 2006, 2007). Since neither mutant conferred resistance to ethidium, the authors concluded that this was due to the modified topology. Co-expression of the inactive mutants restored the ethidium resistance phenotype to the same level as seen with wild-type EmrE (Rapp et al. 2007). The suggested interpretation of this finding was that co-expression results in the generation of a functional, antiparallel heterodimer. However, this conclusion is based solely on the contention that the N_iC_i and N_oC_o mutants are inactive. This conclusion necessitates further experimental support, because the mutants that were designed to insert into the membrane with the N_iC_i and N_oC_o topology are both functional

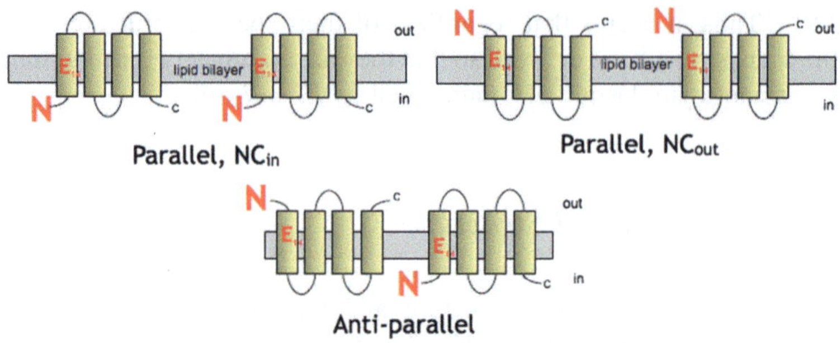

Fig. 10.3 *Three topoforms of EmrE*. The topogenic determinants in the wild type EmrE are nonexistent and therefore parallel dimers insert in random topologies: about half N_iC_i and half N_oC_o. After insertion, the interaction of the protomers will result in a parallel or antiparallel mode depending only on their relative affinities. Adapted from reference Schuldiner (2012)

and capable of removing substrate from the cell when the phenotype is assayed in cells with the proper genetic background (Nasie et al. 2010) or following growth continuously in liquid medium (McHaourab et al. 2008). The lower activity of the single mutants compared with the co-expressed ones is most likely due to impaired dimerization and not necessarily due to a different topology (Nasie et al. 2010).

10.1.2.1 Evidence Supporting Parallel Topology

To validate the above-mentioned structural models, we tested one of their most basic predictions: the existence of antiparallel dimers. All our experiments were consistent only with parallel dimers. Unique cysteines were engineered in the hydrophilic loops or at the termini, and crosslinkers that could react with residues 9–11 Å apart quantitatively cross-linked the protomers (Soskine et al. 2002). This finding is inconsistent with an antiparallel topology, since in such a case the Cys would be at least 35–40 Å apart. Since cross-linking could be amenable to artifacts, we also purified two proteins cross-linked in different positions and showed that they are fully active supporting the contention that the crosslinking experiments reflect the situation in functional dimers (Soskine et al. 2006). Generation of genetic fusions (tandems) where the C-terminus of one protomer was fused to the N-terminus of the second protomer provided a paradigm to study the topology both in vivo and in vitro. The linkers were designed such that they are very short or very hydrophilic. This ensured that both termini are on the same side of the membrane and force the dimer into a parallel topology (Steiner-Mordoch et al. 2008). All the tandems built as described were functional in vivo and after purification. Control experiments were performed to ensure that the functional unit is the dimer and not a result of interaction of dimers.

10.1.2.2 Multiple Topologies of SMRs

Any single dimeric membrane protein may, in theory, adopt three different topologies (denoted here as "topoforms") schematically shown in Fig. 10.3. Relative to each other, the two protomers can theoretically adopt either a parallel (N and C-termini of both protomers on the same side of the membrane) or anti-parallel orientation (N and C-termini of each protomer on opposite sides of the membrane). Additionally, relative to the lipid bilayer, the ensemble of parallel dimers can allegedly adopt a single topology or a dual topology where dimers are either all in a N_iC_i (or N_oC_o) orientation or a mixture of both. However, while all the potential topologies could occur what topology makes biological sense? The answer to that question is open since there are constraints such as the way the protein is inserted into the membrane during synthesis and the requirement for the dimer to be catalytically active.

One of the reasons for the controversy about the topology of EmrE was the assumption that to ensure proper function membrane proteins must be inserted in a unique topology. Indeed, it is obvious that proteins such as receptors must face the environment they are probing, channels that sense the electrical field across membranes depend on the polarity of the field and cannot be inserted randomly, pumps that derive their energy from molecules such as ATP, available only on one side of the membrane, need to have the ATP utilization machinery in the right location. However, is a unique topology a necessary prerequisite for uniporters or ion-coupled transporters?

H^+-coupled transporters couple the free energy released from downhill translocation of H^+ in response to an H^+ electrochemical gradient to drive the energetically uphill stoichiometric flux (in or out) of their substrates. As such, they function as thermodynamic nano-devices capable of transporting in either direction, the direction dictated solely by the driving gradients. There is no need to assume differences in affinities from both sides of the membrane for such a mechanism to hold [see for example, Guan and Kaback (2004)]. Therefore, H^+-coupled transporters should be essentially symmetrical in their function unless there is a specific need for rectification or regulation of their function.

To account for the observation that both parallel and antiparallel dimers are functional, we propose that this is possible because the mechanism of coupling ion and substrate transport in EmrE is so simple: two carboxyls with a properly tuned pK_A located approximately in the middle of the membrane provide the core of the coupling mechanism. This simplicity provides the robustness necessary to tolerate such a unique and unprecedented ambiguity in the interaction of the subunits and in the dimer topology relative to the membrane.

10.1.3 SMRs and Evolution of Transporters

The behavior observed with EmrE and other SMRs may represent a stage in the evolution of the topology of membrane proteins (Schuldiner 2007, 2009, 2012). The existence of homodimers and heterodimers in this family of small H^+-coupled multidrug transporters provides an exceptional paradigm to study the evolution of heterodimers and of the larger transport proteins. The heterodimers have apparently originated from gene duplication of the more ancient homodimers. After gene duplication, a relatively small number of mutations may be sufficient to convert a homodimer into a heterodimer. The large transporters may then have arisen from a series of gene fusions. The small SMRs did not survive as such during evolution and genes coding for them are found mainly in the bacterial world with a few representatives in the Archaea but none, thus far, in the eukaryotes.

We hypothesize that the appearance of heterodimers may have provided an evolutionary advantage. Thus, for example, a single mutation in an essential residue of a functional homodimer would lead to complete loss of activity. Partial activities have been observed in in vitro-produced heterodimers even when one of the protomers carries a mutation in an essential residue (Ninio et al. 2001; Sharoni et al. 2005; Rotem et al. 2001; Steiner-Mordoch et al. 2008). Thus, the above handicap is partially overcome in the heterodimers that have evolved from the primitive homodimer after gene duplication.

Another advantage for heterodimers is the potentially higher diversity that can be achieved by combinations of mutations, a diversity that may have been critical in proteins that should recognize a wide range of substrates. Yet, an additional advantage for such functional dimers is also hinted at by the ongoing controversy regarding their topology. In many of the putative heterodimers, the topological determinants are apparent and predict an antiparallel topology. In some, as judged from the same determinants, the interaction may be parallel. On the other hand, in homodimers such as EmrE, the need for topological determinants in such a small protein may be too much of a burden, a burden that can be alleviated in larger proteins where single signals may be better tolerated. When these determinants are engineered by mutagenesis, they impair activity (Rapp et al. 2007), most likely because of the diminished interaction between monomers dictated by the high density of positive charges in the loops (Nasie et al. 2010).

The next step in the evolution of transporters has been the generation of the larger ones that usually function as monomers. A structural approach has revealed that many ion-coupled transporters are found as oligomers in crystals and most likely also in the membrane. However, the functional significance of the finding is not yet evident since in many cases it has been shown that the substrate translocation pathway and the coupling gear are in the monomer (Geertsma et al. 2005; Yernool et al. 2004; Veenhoff et al. 2002; Lanyi 2004; Bamber et al. 2007; Williams et al. 1999; Fujiyoshi et al. 2002). Even though the monomers are the functional unit, in some cases, their interaction appears necessary for catalysis [e.g., in the case of the RND transporters (Eicher et al. 2009) and lactose transport

(Veenhoff et al. 2002)], stability (Herz et al. 2009), or for regulation (Perez et al. 2011).

Why did the small proteins not survive the process of evolution? Maybe it simply was an evolutionary dead end. However, it could be that the necessity to add soluble loops for regulation, targeting to specific membranes and interaction with other components of the cell, selected in favor of the larger transporters where these additions are less of a burden.

Inverted repeats have been identified in many large modern transporters and have been suggested to play important roles in the transport mechanism (Forrest and Rudnick 2009). We propose that the plasticity described here could provide an example of how inverted repeats in modern large polytopic transporters may have developed. The interaction interfaces must be appropriate for generation of a stable oligomer, and a substrate binding cavity must supply interaction points at given and fixed locations. We suggest that the promiscuity among interacting protomers reported here is tolerated also functionally because in EmrE, as in other proteins interacting with multiple substrates, the size of the binding pocket must be large enough to allow different molecules to reside in it in different orientations and to establish interactions with different sets of residues on the pocket walls (Seeger et al. 2006; Murakami et al. 2006; Godsey et al. 2002). Therefore, the minimal structure where we find the necessary elements for substrate recognition and coupling may be provided by a dimer with the protomers in either one of the possible topologies.

Acknowledgments Work in our laboratory is supported by National Institutes of Health Grant NS16708 and Grant 97/12 from the Israel Science Foundation. I thank all the colleagues and friends that critically read this manuscript and suggested improvements.

References

Adam Y, Tayer N, Rotem D, Schreiber G, Schuldiner S (2007) The fast release of sticky protons: Kinetics of substrate binding and proton release in a multidrug transporter. Proc Natl Acad Sci U S A 104(46):17989–17994. doi:10.1073/pnas.0704425104

Arkin IT, Xu H, Jensen MO, Arbely E, Bennett ER, Bowers KJ, Chow E, Dror RO, Eastwood MP, Flitman-Tene R, Gregersen BA, Klepeis JL, Kolossvary I, Shan Y, Shaw DE (2007) Mechanism of Na+/H + antiporting. Science 317(5839):799–803. doi:10.1126/science.1142824

Bamber L, Harding M, Monne M, Slotboom DJ, Kunji ER (2007) The yeast mitochondrial ADP/ATP carrier functions as a monomer in mitochondrial membranes. Proc Natl Acad Sci U S A 104(26):10830–10834, doi:0703969104 [pii]10.1073/pnas.0703969104

Bay DC, Turner RJ (2009) Diversity and evolution of the small multidrug resistance protein family. BMC Evol Biol 9:140, doi:1471-2148-9-140 [pii] 10.1186/1471-2148-9-140

Bay DC, Rommens KL, Turner RJ (2008) Small multidrug resistance proteins: A multidrug transporter family that continues to grow. Biochim Biophys Acta 1778:1814–1838

Brill S, Falk OS, Schuldiner S (2012) Transforming a drug/H + antiporter into a polyamine importer by a single mutation. Proc Natl Acad Sci U S A 109:16894–16899. doi:10.1073/pnas.1211831109

Chang G, Roth CB, Reyes CL, Pornillos O, Chen Y-J, Chen AP (2006) Retraction. Science 314 (5807):1875. doi:10.1126/science. 314.5807.1875b

Chen YJ, Pornillos O, Lieu S, Ma C, Chen AP, Chang G (2007) X-ray structure of EmrE supports dual topology model. Proc Natl Acad Sci U S A 104(48):18999–19004

Eicher T, Brandstatter L, Pos KM (2009) Structural and functional aspects of the multidrug efflux pump AcrB. Biol Chem 390(8):693–699. doi:10.1515/BC.2009.090

Fleishman SJ, Harrington SE, Enosh A, Halperin D, Tate CG, Ben-Tal N (2006) Quasi-symmetry in the cryo-EM structure of EmrE provides the key to modeling its transmembrane domain. J Mol Biol 364(1):54–67

Fluman N, Ryan CM, Whitelegge JP, Bibi E (2012) Dissection of mechanistic principles of a secondary multidrug efflux protein. Mol Cell 47(5):777–787. doi:10.1016/j.molcel.2012.06.018

Forrest LR, Rudnick G (2009) The rocking bundle: a mechanism for ion-coupled solute flux by symmetrical transporters. Physiology (Bethesda) 24:377–386, doi:24/6/377 [pii] 10.1152/physiol.00030.2009

Fujiyoshi Y, Mitsuoka K, de Groot BL, Philippsen A, Grubmüller H, Agre P, Engel A (2002) Structure and function of water channels. Curr Opin Struct Biol 12(4):509–515

Geertsma ER, Duurkens RH, Poolman B (2005) Functional Interactions between the subunits of the lactose transporter from Streptococcus thermophilus. J Mol Biol 350(1):102–111

Godsey MH, Zheleznova Heldwein EE, Brennan RG (2002) Structural biology of bacterial multidrug resistance gene regulators. J Biol Chem 277(43):40169–40172

Grinius L, Goldberg E (1994) Bacterial multidrug resistance is due to a single membrane protein which functions as a drug pump. J Biol Chem 269:29998–30004

Guan L, Kaback HR (2004) Binding affinity of lactose permease is not altered by the H + electrochemical gradient. Proc Natl Acad Sci 101(33):12148–12152. doi:10.1073/pnas.0404936101

Gutman N, Steiner-Mordoch S, Schuldiner S (2003) An amino acid cluster around the essential Glu-14 is part of the substrate and proton binding domain of EmrE, a multidrug transporter from Escherichia coli. J Biol Chem 278:16082–16087

Herz K, Rimon A, Jeschke G, Padan E (2009) Beta-sheet-dependent dimerization is essential for the stability of NhaA Na+/H + antiporter. J Biol Chem 284(10):6337–6347. doi:10.1074/jbc.M807720200

Kolbusz MA, Slotboom DJ, Lolkema JS (2013) Genomic distribution of the small multidrug resistance protein EmrE over 29 Escherichia coli strains reveals two forms of the protein. FEBS J 280(1):244–255. doi:10.1111/febs.12065

Lanyi JK (2004) Bacteriorhodopsin. Annu Rev Physiol 66(1):665–688, doi:10.1146/annurev.physiol.66.032102.150049

Mager T, Rimon A, Padan E, Fendler K (2011) Transport mechanism and pH regulation of the Na+/H + antiporter NhaA from Escherichia coli: An electrophysiological study. J Biol Chem 286:23570–23581, doi:M111.230235 [pii]10.1074/jbc.M111.230235

McHaourab HS, Mishra S, Koteiche HA, Amadi SH (2008) Role of sequence bias in the topology of the multidrug transporter EmrE. Biochemistry 47(31):7980–7982. doi:10.1021/bi800628d

Miller C, Nguitragool W (2009) A provisional transport mechanism for a chloride channel-type Cl-/H + exchanger. Philos Trans R Soc Lond B Biol Sci 364(1514):175–180, doi:Q311319160335168 [pii] 10.1098/rstb.2008.0138

Murakami S, Nakashima R, Yamashita E, Matsumoto T, Yamaguchi A (2006) Crystal structures of a multidrug transporter reveal a functionally rotating mechanism. Nature 443(7108):173–179

Muth TR, Schuldiner S (2000) A membrane-embedded glutamate is required for ligand binding to the multidrug transporter EmrE. EMBO J 19(2):234–240

Nasie I, Steiner-Mordoch S, Gold A, Schuldiner S (2010) Topologically random insertion of EmrE supports a pathway for evolution of inverted repeats in ion-coupled transporters. J Biol Chem 285(20):15234–15244, doi:M110.108746 [pii]10.1074/jbc.M110.108746

Ninio S, Rotem D, Schuldiner S (2001) Functional analysis of novel multidrug transporters from human pathogens. J Biol Chem 276(51):48250–48256
Padan E, Zilberstein D, Schuldiner S (1981) pH Homeostasis in Bacteria. Biochim Biophys Acta 650:151–166
Padan E, Kozachkov L, Herz K, Rimon A (2009) NhaA crystal structure: functional-structural insights. J Exp Biol 212(Pt 11):1593–1603. doi:10.1242/jeb.026708
Perez C, Khafizov K, Forrest LR, Kramer R, Ziegler C (2011) The role of trimerization in the osmoregulated betaine transporter BetP. EMBO Rep 12(8):804–810. doi:10.1038/embor.2011.102
Pornillos O, Chen YJ, Chen AP, Chang G (2005) X-ray structure of the EmrE multidrug transporter in complex with a substrate. Science 310(5756):1950–1953
Rapp M, Granseth E, Seppala S, von Heijne G (2006) Identification and evolution of dual-topology membrane proteins. Nat Struct Mol Biol 13(2):112–116
Rapp M, Seppala S, Granseth E, von Heijne G (2007) Emulating membrane protein evolution by rational design. Science 315(5816):1282–1284
Rotem D, Schuldiner S (2004) EmrE, a multidrug transporter from Escherichia coli, transports monovalent and divalent substrates with the same stoichiometry. J Biol Chem 279: 48787–48793
Rotem D, Sal-man N, Schuldiner S (2001) In vitro monomer swapping in EmrE, a multidrug transporter from Escherichia coli, reveals that the oligomer is the functional unit. J Biol Chem 276(51):48243–48249
Schuldiner S (2007) When biochemistry meets structural biology: the cautionary tale of EmrE. Trends Biochem Sci 32(6):252–258
Schuldiner S (2009) EmrE, a model for studying evolution and mechanism of ion-coupled transporters. Biochim Biophys Acta 1794:748–762, doi:S1570-9639(08)00398-1 [pii] 10.1016/j.bbapap.2008.12.018
Schuldiner S (2012) Undecided membrane proteins insert in random topologies. Up, down and sideways: it does not really matter. Trends Biochem Sci 37(6):215–219. doi:10.1016/j.tibs.2012.02.006
Seeger MA, Schiefner A, Eicher T, Verrey F, Diederichs K, Pos KM (2006) Structural asymmetry of AcrB trimer suggests a peristaltic pump mechanism. Science 313(5791):1295–1298
Seeger MA, von Ballmoos C, Verrey F, Pos KM (2009) Crucial role of Asp408 in the proton translocation pathway of multidrug transporter AcrB: evidence from site-directed mutagenesis and carbodiimide labeling. Biochemistry 48(25):5801–5812. doi:10.1021/bi900446j
Sharoni M, Steiner-Mordoch S, Schuldiner S (2005) Exploring the binding domain of EmrE, the smallest multidrug transporter. J Biol Chem 280(38):32849–32855
Soskine M, Steiner-Mordoch S, Schuldiner S (2002) Crosslinking of membrane-embedded cysteines reveals contact points in the EmrE oligomer. Proc Natl Acad Sci U S A 99(19): 12043–12048
Soskine M, Adam Y, Schuldiner S (2004) Direct evidence for substrate induced proton release in detergent solubilized EmrE, a multidrug transporter. J Biol Chem 279:9951–9955
Soskine M, Mark S, Tayer N, Mizrachi R, Schuldiner S (2006) On parallel and antiparallel topology of an homodimeric multidrug transporter. J Biol Chem 281(47):36205–36212
Steiner-Mordoch S, Soskine M, Solomon D, Rotem D, Gold A, Yechieli M, Adam Y, Schuldiner S (2008) Parallel topology of genetically fused EmrE homodimers. EMBO J 27(1):17–26
Taglicht D, Padan E, Schuldiner S (1993) Proton-sodium stoichiometry of NhaA, an electrogenic antiporter from Escherichia coli. J Biol Chem 268:5382–5387
Tanaka Y, Hipolito CJ, Maturana AD, Ito K, Kuroda T, Higuchi T, Katoh T, Kato HE, Hattori M, Kumazaki K, Tsukazaki T, Ishitani R, Suga H, Nureki O (2013) Structural basis for the drug extrusion mechanism by a MATE multidrug transporter. Nature 496(7444):247–251. doi:10.1038/nature12014

Ubarretxena-Belandia I, Baldwin JM, Schuldiner S, Tate CG (2003) Three-dimensional structure of the bacterial multidrug transporter EmrE shows it is an asymmetric homodimer. EMBO J 22(23):6175–6181

Veenhoff LM, Heuberger EHML, Poolman B (2002) Quaternary structure and function of transport proteins. Trends Biochem Sci 27(5):242–249

Weinglass AB, Soskine M, Vazquez-Ibar JL, Whitelegge JP, Faull KF, Kaback HR, Schuldiner S (2005) Exploring the role of a unique carboxyl residue in EmrE by mass spectrometry. J Biol Chem 280(9):7487–7492

Williams KA, Geldmacher-Kaufer U, Padan E, Schuldiner S, Kuhlbrandt W (1999) Projection structure of NhaA, a secondary transporter from Escherichia coli, at 4.0 A resolution. EMBO J 18(13):3558–3563

Yernool D, Boudker O, Jin Y, Gouaux E (2004) Structure of a glutamate transporter homologue from Pyrococcus horikoshii. Nature 431(7010):811–818

Yerushalmi H, Schuldiner S (2000a) A common binding site for substrates and protons in EmrE, an ion-coupled multidrug transporter. FEBS Lett 476:93–97

Yerushalmi H, Schuldiner S (2000b) An Essential Glutamyl Residue in EmrE, a Multidrug Antiporter from Escherichia coli. J Biol Chem 275:5264–5269

Yerushalmi H, Schuldiner S (2000c) A model for coupling of H(+) and substrate fluxes based on "time-sharing" of a common binding site. Biochemistry 39(48):14711–14719

Yerushalmi H, Lebendiker M, Schuldiner S (1995) EmrE, an Escherichia coli 12-kDa multidrug transporter, exchanges toxic cations and H + and is soluble in organic solvents. J Biol Chem 270(12):6856–6863

Yerushalmi H, Mordoch SS, Schuldiner S (2001) A single carboxyl mutant of the multidrug transporter EmrE is fully functional. J Biol Chem 276(16):12744–12748

Chapter 11
Symmetrically Asymmetric: EmrE Seen from the NMR Perspective

Johanna Becker-Baldus and Clemens Glaubitz

Abstract Secondary active transporters couple substrate with ion translocation across the membrane. The actual transport mechanism must involve a number of distinct steps with alternatingly accessible substrate/ion binding sites and potentially occurring occluded states. To understand their functional mechanism, it is necessary to link transport kinetics to conformational dynamics. Furthermore, various biophysical studies indicate high conformational flexibility in secondary transporters, which raises the question whether transport requires switching between well-defined conformational states or is merely based on shifted conformational equilibriums. Addressing this question requires biophysical approaches sensitive to different timescales and length scales such as solid-state NMR, liquid-state NMR, EPR, optical- and infrared spectroscopy in close combination with mutagenesis-based functional studies. In this chapter, we demonstrate and discuss for the case of EmrE, how liquid- and solid-state NMR spectroscopy can contribute to resolving the mechanism of secondary transporters. EmrE, a homo-dimeric multidrug antiporter from the SMR family, has been suggested to offer a convenient paradigm for secondary transporters due to its small size. It has attracted almost all available NMR methods making it a good case for demonstrating advanced methodology by which fundamental properties of secondary transporters can be obtained.

Keywords EmrE • Solution-state NMR • Solid-state NMR • DNP

J. Becker-Baldus • C. Glaubitz (✉)
Institute for Biophysical Chemistry and Centre for Biomolecular Magnetic Resonance, Goethe Universität Frankfurt, Frankfurt am Main, Germany
e-mail: glaubitz@em.uni-frankfurt.de

R. Krämer and C. Ziegler (eds.), *Membrane Transport Mechanism*, Springer Series in Biophysics 17, DOI 10.1007/978-3-642-53839-1_11,

Abbreviations

CCCP	Carbonyl cyanide 3-chlorophenylhydrazone
CODEX	Centerband only detection of exchange
CP	Cross polarisation
CSA	Chemical shift anisotropy
DMPC	1,2-dimyristoyl-sn-glycero-3-phosphocholine
DHPC	1,2-dihexanoyl-sn-glycero-3-phosphocholine
DNP	Dynamic nuclear polarisation
EM	Electron microscopy
EPR	Electron paramagnetic resonance
FRET	Förster resonance energy transfer
GPCR	G-protein coupled receptor
MAS	Magic angle spinning
MTP	Methyltriphenylphosphonium
NMR	Nuclear magnetic resonance
NOE	Nuclear Overhauser effect
O-SSNMR	Oriented solid-state NMR
PISA wheel	Polarity index slant angle
PISEMA	Polarisation inversion spin exchange at magic angle
SMR protein	Small multidrug resistance protein
TMH	Transmembrane helix
TPP	Tetraphenylphosphonium
TROSY	Transverse relaxation optimised spectroscopy

11.1 Introduction

The discovery of EmrE, the prototype member of the small multidrug resistance (SMR) protein family by Schuldiner and coworkers, had been greeted with great enthusiasm, as this protein appeared to be an ideal paradigm for resolving the structure–function relationship in an ion-coupled transporter (Schuldiner et al. 1997): With just four transmembrane helices and 110 residues, it is the smallest transporter reported to date and its functional mechanism appeared to be associated with only a few highly conserved residues. It resides in the inner *E. coli* membrane and provides efflux of multiple, positively charged, toxic compounds such as TPP^+, ethidium bromide, methyl viologen and others in exchange for protons. Despite its apparent simplicity, it posed a significant challenge to resolve and to understand its topology, its oligomeric state and to determine its structure.

Early studies on the oligomerisation state of EmrE hinted towards a homo-trimer (Yerushalmi et al. 1996), which was later corrected based on accumulating evidence towards a homo-dimer as the main functional unit (Yerushalmi et al. 1996; Chen et al. 2007; Muth and Schuldiner 2000; Rotem et al. 2001; Ubarretxena-Belandia et al. 2003). Freeze-fracture electron microscopy on SMR proteins also indicated that larger complexes could form depending on the lipid environment (Mors et al. 2013).

So far, no high-resolution structure of EmrE is available, but 3D X-ray crystallography (Chen et al. 2007) and cryo-EM data (Ubarretxena-Belandia et al. 2003) obtained from 2D crystals converge into a consensus model, which is also in line with the EPR- and NMR data (Amadi et al. 2010; Gayen et al. 2013; Morrison et al. 2012). The data show unambiguously that EmrE forms an antiparallel homodimer with a high degree of structural asymmetry between the two protomers. This is visualised in Fig. 11.1a. Cryo-EM data revealed a quasi-symmetry of parts of the EmrE structure with respect to the membrane plane resulting in a model for conformational changes with inward and outward facing binding sites (Fig. 11.2a) (Fleishman et al. 2006). Further evidence for an antiparallel dimer topology comes from numerous other studies including single-molecule FRET (Morrison et al. 2012), EPR (Amadi et al. 2010) or cross-linking (Gayen et al. 2013). For a membrane protein, such a topology is unusual and has sparked a debate in the light of apparently contradicting in vivo and cross-linking data (Rapp et al. 2006; Schuldiner 2007; Lloris-Garcera et al. 2012) as highlighted in Chap. 10.

The current model for the transport EmrE mechanism, as proposed by Schuldiner and co-workers, is shown in Fig. 11.1b. The residue E14 in TMH1, highly conserved within the SMR family, has been shown to be essential for both substrate and proton binding, and the occupancy of the binding site with protons or substrates is mutually exclusive (Muth and Schuldiner 2000; Rotem et al. 2001; Rotem and Schuldiner 2004; Steiner-Mordoch et al. 2008; Yerushalmi et al. 2001). It was shown that both E14s in the EmrE dimer release their protons upon substrate binding. This must be followed by a conformational change during which EmrE switches from an inward to an outward facing conformation followed by substrate release and proton uptake. The pKa of E14 is relatively high and was found to be between 7 and 8.5, depending on the experiment. Substrate binding and transport was also found to depend sensitively on the lipid- or membrane mimicking environment chosen for EmrE (Miller et al. 2009). The current ideas about the functional mechanism of EmrE are discussed in great detail in Chap. 10.

During the transport cycle, an intermediate state is formed during which the bound substrate is at least partially excluded from water (Fig. 11.1b). Evidence comes from real-time fluorescence spectroscopy on EmrE and on its homologue TBsmr (Basting et al. 2008; Lehner et al. 2008; Lorch et al. 2005). Briefly, EmrE is co-reconstituted with inside-out-oriented bacteriorhodopsin. The later protein generates a proton gradient, which triggers transport by EmrE. The occluded state is revealed by the fact that the quantum yield of the substrate ethidium is significantly increased when it is partially or completely excluded from water. This experiment is also a very rare example for a spectroscopy study on an ion-coupled transporter while it is undergoing active substrate transport. Usually, an equilibrium state with the same pH on both sides of the protein solubilised in detergent micelles is studied, while this approach could potentially allow gaining a closer understanding of transport cycle kinetics if, e.g. combined with single-molecule fluorescence spectroscopy.

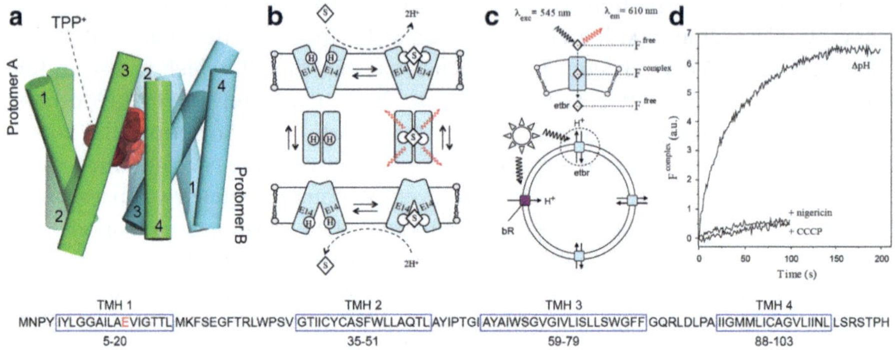

Fig. 11.1 (**a**) EmrE is a 110-residue multidrug antiporter. It forms an antiparallel homo-dimer with four transmembrane helices in each protomer, as shown in the structural consensus model (Chen et al. 2007; Fleishman et al. 2006). The four helices stretching from residues 5–20, 35–51, 59–79 and 88–103 are *highlighted* in the protein sequence. (**b**) EmrE translocates one substrate in exchange with two protons (Adam et al. 2007). The highly conserved residue E14 in TMH1 is essential for both substrate and proton binding. EmrE undergoes a cycle of alternatingly inward and outward facing binding sites. The transport cycle includes the formation of an intermediate state, during which substrate is partially excluded from water. (**c**) Evidence comes from real-time fluorescence spectroscopy (Basting et al. 2008; Lehner et al. 2008; Lorch et al. 2005). In this experiment, the fluorescence intensity of the substrate ethidium bromide has been monitored during the transport process. A stable pH gradient can be generated by co-reconstitution with inside-out-oriented bacteriorhodopsin, which is activated by light (545 nm). Protons are pumped inside and are subsequently exchanged by EmrE for ethidium. The same light source used for bacteriorhodopsin also excites ethidium fluorescence. Detecting its fluorescence emission (610 nm) in a time-resolved manner in the presence of ΔpH revealed a significant increase in fluorescence intensity of ethidium bound to EmrE (**d**). The addition of ionophores, nigericin and CCCP, which abolish ΔpH and ΔpH + $\Delta\Psi$, respectively, causes a collapse of this fluorescence signal. This observation is explained by the formation of a transport intermediate state during which ethidium is partially excluded from water explaining its enhanced quantum yield

In addition to the need of obtaining highly resolved structures, in-depth data for understanding the link between molecular dynamics, functional mechanism and structural properties are in demand and could be provided especially by NMR spectroscopy. One of the key requirements for liquid-state NMR is the choice of a suitable membrane-mimicking environment (Fig. 11.2a): The protein/detergent complex has to be small enough to ensure fast isotropic tumbling while preserving at the same time functional integrity and providing good spectral resolution. One major disadvantage of detergent micelles is their high curvature providing usually a rather unnatural environment for membrane proteins. The situation is improved by the use of bicelles (Fig. 11.2a), which are mixtures of short- and long-chain lipids. Their ratio determines whether they align themselves in the magnetic field (anisotropic bicelle) or whether they undergo fast isotropic tumbling (isotropic bicelle). Even more bilayer-like are nanodiscs (Fig. 11.2a), which are larger than bicelles, but still soluble. They are lipid bilayer patches, surrounded by a scaffold protein. All of these three systems have been used successfully for liquid-state NMR on membrane proteins (Raschle et al. 2010) and recently determined structures

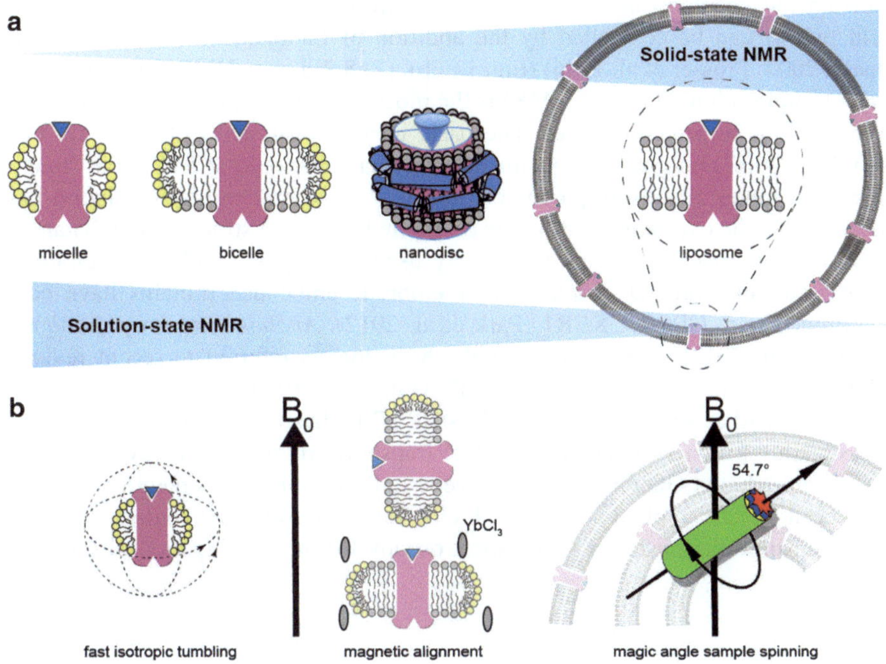

Fig. 11.2 (**a**) Membrane mimicking environments for membrane proteins, which are compatible with NMR spectroscopy, include micelles, bicelles, nanodiscs and liposomes. Liquid-state NMR relies on fast isotropic tumbling. The larger the protein/lipid/detergent complexes, the less suitable they are. For probing the protein directly within the lipid bilayer, solid-state NMR is the method of choice. Reconstituting membrane proteins in anisotropic bicelles allows aligning them within the magnetic field for oriented solid-state NMR (O-SSNMR) through which helix orientations can be determined. (**b**) Bicelles align spontaneously with their normal perpendicular to the magnetic field, but the addition of lanthanides flips them to a parallel orientation (Prosser et al. 1996), which is required for O-SSNMR. For magic angle sample spinning (MAS-NMR), proteoliposomes can be used directly. Fast sample rotation about the magic angle emulates fast, molecular isotropic tumbling resulting in well-resolved spectra

include, for example, human VDAC-1 (Bayrhuber et al. 2008), sensory rhodopsin II (Gautier et al. 2010) and proteorhodopsin (Reckel et al. 2011). To be able to work under even more native-like conditions using proteoliposomes, solid-state NMR has to be applied (Fig. 11.2a). The term refers to the fact that fast isotropic tumbling of the molecule of interest is not required, hence they appear "solid" on the NMR timescale and all anisotropic interactions such as dipole couplings, quadrupole couplings or chemical shift anisotropies are preserved and not averaged to zero as in solution. For membrane proteins, two major approaches have been established: (1) Oriented solid-state NMR (O-SSNMR) relies on macroscopically ordered samples. NMR resonances of membrane embedded proteins in such preparations are orientation dependent and structural details can be extracted directly. Oriented samples are either prepared on glass plates as solid support or more elegantly by utilising the spontaneous alignment of anisotropic bicelles within the magnetic field, which can be controlled by the addition of Lanthanides (De Angelis and Opella 2007; Prosser et al. 1996) (Fig. 11.2b). (2) Solid-state NMR based on magic angle sample spinning (MAS-NMR) is the most universal approach as it is directly compatible with proteolipsomes. The basic idea is illustrated in Fig. 11.2b. Fast sample rotation about an axis inclined to the magnetic field at 54.7° (the "magic angle") averages all anisotropic interactions leading to well-resolved, solution-like spectra. Specific experiments can now be applied to extract structure and dynamics data. This approach has been extensively used for hypothesis-driven membrane protein studies, and 3D structures of membrane embedded proteins have been determined [e.g. GPCR CXCR1 (Park et al. 2012), *Anabaena* sensory rhodopsin (Wang et al. 2013)]. For further details the reader is referred to recent reviews (Hong et al. 2012; Lakatos et al. 2012; Renault et al. 2010).

So far, neither solid-state nor liquid-state NMR has been used extensively for investigating primary or secondary active transport proteins (for an overview see, e.g. (Hellmich and Glaubitz 2009). However, all of the methods mentioned above have been already applied to EmrE, which offers therefore a convenient showcase to demonstrate the power of NMR spectroscopy for the investigation of transport proteins.

11.2 Solution-State NMR on EmrE in Isotropic Bicelles

Soon after its discovery, EmrE appeared rather attractive to solution-state NMR due to its apparently small size. First solution-state NMR experiments made use of the stability of EmrE dissolved in chloroform/methanol/water mixtures (Schwaiger et al. 1998). An extensive resonance assignment together with some distance constraints based on NOE data enabled a first secondary structure analysis in organic solvent. However, to keep the membrane protein in a more functional and correctly folded state, a better membrane mimicking environment provided by micelles, bicelles, nanodiscs or lipsomes is required (Fig. 11.2a). For solution-state NMR on EmrE, additional constraints apply: the protein/detergent/lipid complex has to be small enough (approx. <70 kDa) so that fast isotropic tumbling can take place, the EmrE dimer should remain intact and as functional as possible and the particular preparation conditions should result in high quality NMR spectra. Finding such conditions is not straightforward and very much protein dependent requiring often careful and extensive screening. Especially finding the suitable detergent micelles can be difficult. They are highly dynamic assemblies, which can destabilise membrane proteins and their high curvature is very different from lipid bilayers, which can also affect protein integrity (Raschle et al. 2010; Tate 2010). Liposomes would be ideal from the functional point of view, but require solid-state NMR (see below). Nanodiscs are an attractive option but the overall complex involving lipids, membrane protein and scaffold protein tends to be large. In case of EmrE and its homologues, isotropic bicelles were shown to offer a very attractive membrane environment fulfilling all the before mentioned criteria (Morrison and Henzler-Wildman 2012; Poget et al. 2007). Isotropic or better fast tumbling bicelles were first described by Vold and co-workers (Vold et al. 1997). Bicelles in general consist of a long-chain phospholipid such as DMPC and a short-chain one such as DHPC, respectively. Their size and shape are controlled by the ratio of DMPC:DHPC (q). At $q > 2.5$, disc-shaped structures with 30–100 nm diameter and 4–5 nm in thickness are obtained, which align in the magnetic field and can be used as alignment media in solution state NMR to obtain residual dipole couplings (Tjandra and Bax 1997) or for oriented solid-state NMR studies (see below). In contrast, fast tumbling bicelles are obtained for $q = 0.1–0.5$. They are slightly larger than micelles and are assumed to be also disc-shaped with long-chain lipids inside surrounded by a rim of DHPC (Glover et al. 2001). They offer a very attractive alternative to micelles as the lipid composition can be easily varied to better mimic biological membranes, and incorporated membrane proteins experience a lateral pressure profile comparable to liposomes. Henzler-Wildman and co-workers have described an optimised reconstitution procedure based on an extensive screen, by which EmrE is incorporated into isotropic bicelles in correctly folded and fully functional form (Morrison and Henzler-Wildman 2012). The resulting bicelle system was used for structure and dynamic studies by solution-state NMR on the TPP^+-bound form of EmrE (Morrison et al. 2012). The authors observed doubling of almost all peaks in $^1H–^{15}N$ TROSY HSQC spectra, which represent a "finger print" of the proteins backbone fold. Each peak in the resulting spectra corresponds to the amide nitrogen/proton spin pair in the protein backbone of each residue. In general, peak multiplets could be either caused by a structural asymmetry in the antiparallel EmrE dimer, by slow exchange between two structurally different conformations of a symmetric, parallel EmrE dimer or by slow exchange between different conformations of an asymmetric, antiparallel EmrE dimer. A parallel EmrE topology has been ruled out by the authors based on single molecule FRET analysis in agreement with a multitude of other studies (see above). Furthermore, the authors used the pseudo-twofold symmetry of EmrE around an axis parallel to the membrane plane found by EM and X-ray analysis (Fleishman et al. 2006) to exclude that the observed doubling stems directly from conformational exchange: Slow exchange between an inward and an outward-facing open state would result in two equal populations with identical conformation. Although their orientations will be different, their spectra will be undistinguishable (Fig. 11.3b, c). Therefore, the observed peak doubling is caused by the structural asymmetry between monomers A and B in the antiparallel EmrE dimer. Whether an additional slow interconversion between an inward- and outward-facing form takes place could not be decided based on these data. To test this hypothesis, special exchange experiments were carried out, by which slow conformational changes between defined states can be monitored (Li and Palmer 2009). In this experiment, magnetisation exchange between nuclei in monomer A and B will occur during a variable mixing time if conformational exchange takes place. As a result, additional cross peaks are observed (Fig. 11.3c) for many sites in EmrE indicating a global exchange process. Analysing cross peak volumes as a function of the mixing time

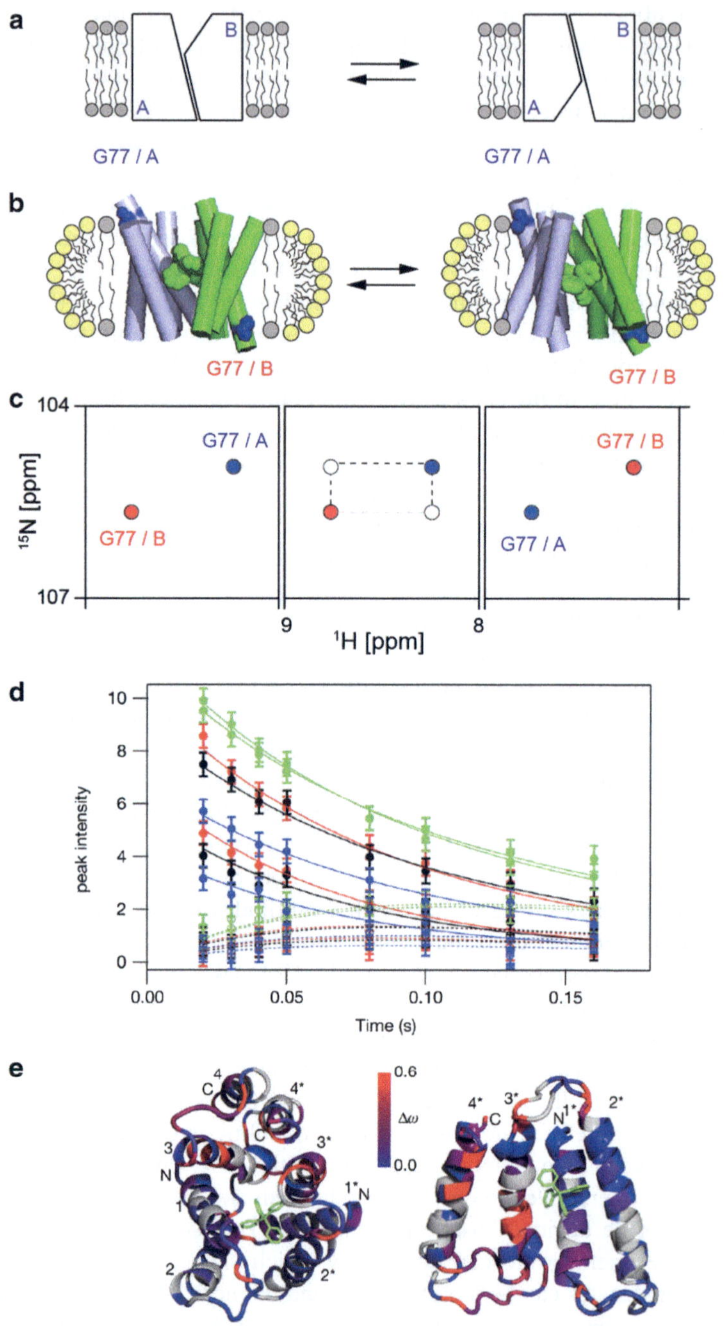

Fig. 11.3 (**a**) Monomers A and B in the antiparallel EmrE dimer are structurally asymmetric and swap conformation between and inward and the outward-facing state. (**b**) First experimental evidence for conformational exchange between an inward and outward facing conformation was provided by solution-state NMR (Morrison et al. 2012): Monomers A and B in the antiparallel EmrE dimer are structurally asymmetric giving rise to peak doublets as illustrated here for the G77

enables determining exchange rates. The data obtained show that the rates for many individual sites are rather similar resulting in a global rate of 4.8 s^{-1}. This means that the whole protein undergoes conformational exchange at this timescale, which can be explained by rigid body movements of its helices.

As indicated above, both monomers swap conformation during exchange (Fig. 11.3a). If helices move as rigid bodies, secondary structure changes should mainly occur at the end of the helices, which is indeed in agreement with larger chemical shift changes in this region (Fig. 11.3e) besides of a kink in TMH3.

The presented data strongly support a model in which the asymmetric and antiparallel EmrE dimer undergoes a symmetric equilibrium exchange between inward and outward facing states involving structure swapping between both monomers (Fig. 11.3a). The exchange between two forms as observed here in the present of ligand are in line with an alternating access model. For understanding conformational dynamics during transport, pH and substrate gradients would have to be applied, which is not compatible with the sample geometry required for NMR experiments, but could be probed potentially, e.g. by fluorescence spectroscopy on unilamellar vesicles (Basting et al. 2008).

11.3 O-SSNMR on EmrE in Magnetically Aligned Bicelles

Solid-state NMR on macroscopically ordered membrane protein samples (O-SSNMR) offers under certain circumstances an attractive alternative to MAS-NMR. A high degree of order can be achieved either by mechanical alignment through solid-supported bilayers usually prepared on thin glass plates or more elegantly by utilising magnetic alignment of bicelles in strong magnetic fields (Fig. 11.2). A high degree of macroscopic order causes narrow lines in the spectrum. In contrast to MAS-NMR, the observed NMR signals, often the amide ^{15}N-chemical shift and the ^{15}N-^{1}H dipole coupling between the nitrogen and its amide proton, remain anisotropic and depend on their orientation with respect to the

Fig. 11.3 (continued) amide in TMH3. (**c**) Slow interconversion between an inward and outward facing equally populated conformation would result in indistinguishable spectra for both states [*left* and *right* in (**c**)], because the conformation of monomer A converts into that of B and vice versa along a symmetry axis in plane with the membrane (Fleishman et al. 2006). (**d**) Evidence of actual conformational interconversion was provided by exchange spectroscopy by detecting cross peaks between resonances belonging to the same residue in two different conformations [(**c**), *middle*]. Analysing peak intensities as a function of mixing time shows a built-up of cross peaks (*open symbols*) and a decay for auto-peaks (*filled symbols*) from which a global exchange rate of 4.9 ± 0.5 s^{-1} was determined (**d**). The chemical shift differences observed for all residues are plotted in the antiparallel dimer model [(**e**), *left*] and in an overlay of both monomers [(**e**), *right*]. Figure (**b**) was made based on the model by Fleishman et al. (Fleishman et al. 2006). Figure (**c**) was adopted and (**d**) and (**e**) were taken from Fig. 2 in Morrison et al. (2012). Copyright for (**d**) and (**e**) Nature Publishing Group

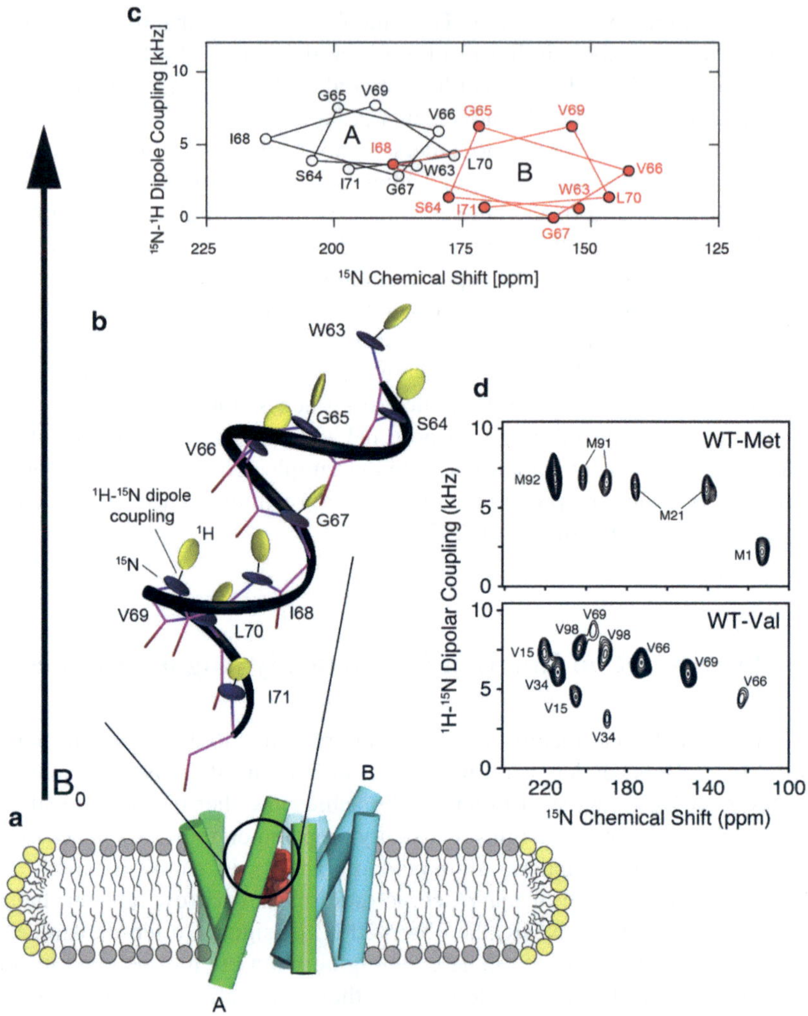

Fig. 11.4 Basic experimental concept of oriented solid-state NMR (O-SSNMR) on EmrE as used by Traaseth and co-workers (Gayen et al. 2013). The protein dimer is incorporated into DMPC/DHPC bicelles aligned with their membrane normal parallel to the applied magnetic field (**a**). Such an alignment can be achieved by the addition of lanthanides (Prosser et al. 1996). NMR interactions utilised in O-SSNMR are illustrated here for residues 63–71 in TMH3 of EmrE monomer A. The ^{15}N chemical shielding tensor (*blue*) and the ^{1}H–^{15}N dipole coupling cause orientation-dependent NMR signals, from which conclusions about helix orientation and structure can be drawn. The model was calculated from PDB-2I68 (Fleishman et al. 2006) using SIMMOL (Bak et al. 2002) (**b**). Typical PISEMA spectrum simulated for residues 63–71 in TM3 of monomer A and B. The significant differences in both spectral patterns illustrate structural differences. The spectra were calculated using SIMPSON (Bak et al. 2000). Experimental PISEMA spectra of ^{15}N-Val and ^{15}N-Met labelled EmrE show peak doublings in agreement with a structural asymmetry (Gayen et al. 2013). Data in (**d**) were kindly provided by the authors. Copyright for (4d) John Wiley and Sons

applied external magnetic field. This allows directly determining helix orientations with respect to each other. The typical pattern of ideal alpha-helices observed in PISEMA spectra (Wu et al. 1994), which correlate ^{15}N chemical shifts with ^{1}H–^{15}N dipole couplings, have been called PISA wheels and represent a direct spectroscopic representation of a helical wheel projected onto the plane of the 2D spectrum (Marassi and Opella 2000; Wang et al. 2000). The basic concept is illustrated and explained in Fig. 11.4.

O-SSNMR has been utilised by Traaseth and co-workers for EmrE (Gayen et al. 2013). The protein was reconstituted into anisotropic DMPC/DHPC bicelles ($q = 3.2$), and it was carefully verified that EmrE forms antiparallel dimers. Bicelles align spontaneously in the magnetic field with their bilayer normal perpendicular to the field direction. However, PISEMA spectroscopy works best with a parallel arrangement, which can be achieved by adding Lanthanides such as $YbCl_3$ (De Angelis and Opella 2007). EmrE contains four methionines and five valines, which were ^{15}N labelled and their signals were assigned through site-directed mutagenesis. Typical PISEMA spectra of apo-state EmrE are shown in Fig. 11.4d. EmrE contains four methionines and five valines. The spectra however reveal a peak doubling for all valines (15, 34, 66, 69, 98) as well as for M91 and M121. Since the location of each peak in the spectrum depends on the secondary structure and on the orientation of the peptide plane with respect to the membrane normal, the observed peak doubling strongly supports a structural asymmetry in the antiparallel quaternary arrangement of ligand-free EmrE. Such an asymmetry is already present in the model by Fleischman et al. (Fleishman et al. 2006) as illustrated here: PISEMA spectra were simulated for a stretch of nine residues (W63-I71) in TMH3 (Fig. 11.4b, c). A spectroscopic "helical wheel representation", so-called PISA wheels, is obtained by connecting cross peaks from neighbouring residues (Fig. 11.4c). The simulation reveals striking differences between TMH3 in monomers A and B, which are mainly due to differences in the helix tilt. Large differences were also predicted for TMH2, while TMH1 and TMH4 were more similar (Gayen et al. 2013). Upon saturating EmrE with high affinity substrate TPP^+, the observed asymmetry remains but chemical shift changes indicative of small changes in structure and helical tilt angles are observed for each transmembrane helix. Especially, the differential behaviour of resonances V66 and V69 has been pointed out: Only V66 in monomer A and V69 in monomer B change, which is compatible with a predicted kink in TM3 between both residues (Fleishman et al. 2006) and also agrees with EPR studies (Amadi et al. 2010). A comparison between predicted PISEMA spectra based on the calculated EmrE model with the experimental data obtained by the authors also highlights important differences, especially with respect to the relative helix orientation around their long axis.

11.4 MAS-NMR on EmrE in Liposomes

Membrane proteins reconstituted into lipid bilayers are preparations in which the proteins are even closer to their native state compared to micelles, bicelles or nanodiscs. Liposome samples can be studied directly using MAS-NMR (see Sect. 11.1 and Fig. 11.2b). After reconstitution into the liposomes of choice, the sample is pelleted and transferred to an MAS NMR rotor. In the following, three different types of MAS-NMR applications to EmrE will be presented and discussed. For a broader overview and discussion of practical aspects of solid-state NMR on transporters, the reader is referred to Lakatos et al. (2012).

11.4.1 Detecting Ligands Bound to EmrE by MAS-NMR

Bound ligands in membrane protein complexes can be detected by MAS-NMR if they carry NMR active nuclei. This is for example the case for ^{31}P in the high affinity EmrE substrate TPP$^+$. Alternatively, isotopes such as ^{13}C, ^{15}N or ^{2}H have to be incorporated by synthetic means. Substrate detection relies on differences in the molecular dynamics between the bound and the non-bound population. These differences are utilised by the cross-polarisation experiment: ^{1}H magnetisation is transferred to nuclei such as ^{31}P, ^{13}C or ^{15}N via dipole couplings in order to enhance their signals. However, dipole couplings will be scaled down or even averaged out completely by fast molecular motions if the non-bound substrate tumbles isotropically in solution. Therefore, this substrate population becomes invisible. In contrast, bound substrates can be cross-polarised and therefore detected since their motions are restricted while residing in the binding pocket. This straightforward but powerful approach has been first applied to probe ^{13}C-labelled sugars bound to MFS-type sugar transporter GalP in *E. coli* directly within its native *E. coli* membrane (Spooner et al. 1994). An example for EmrE is shown in Fig. 11.5 (Ong et al. 2013). The high affinity substrate TPP$^+$ has been titrated to EmrE/DMPC proteoliposomes. ^{1}H–^{31}P-CP MAS NMR enabled detecting the TPP$^+$ population bound to EmrE. The observed signal intensity saturates at a molar TPP$^+$:EmrE ratio of 1:2 as expected for the dimer. Once the high affinity-binding site is saturated, a second bound population is observed, which can be removed by one additional washing step. These experiments show that TPP$^+$ specifically bound to the EmrE dimer within the lipid bilayer can be detected. Based on this approach, it has been shown that both substrates TPP$^+$ and MTP$^+$ share the same binding pocket but show a certain degree of structural heterogeneity (Ong et al. 2013). This could be consistent with a large binding cavity, which is able to accommodate all the different substrates transported by EmrE. The data also show that both TPP$^+$ and MTP$^+$ are significantly immobilised within the binding pocket.

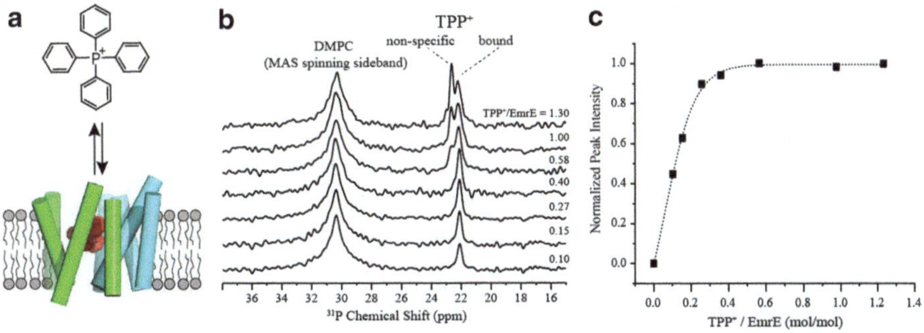

Fig. 11.5 Observing substrates bound to EmrE within the lipid bilayer by ^{31}P MAS-NMR under cross polarisation. The differences in their mobility are used by the cross polarisation experiment to suppress signals from fast tumbling non-bound substrate, while the signals of the bound population are enhanced (**a**). Titrating TPP^+ to EmrE/DMPC proteoliposomes reveals a ^{31}P resonance for specifically bound TPP^+ (**b**). Under excess of TPP^+, a second non-specifically bound population occurs, which can be removed by a simple washing step. The signal intensity of specifically bound TPP^+ saturates above a TPP^+:EmrE ratio of 1:2 as expected for a dimer (**c**). Data were taken from Ong et al. (2013)

11.4.2 MAS-NMR on Labelled EmrE

So far, three studies have been reported in which MAS-NMR was applied to isotope labelled EmrE within lipid bilayers (Lehner et al. 2008; Agarwal et al. 2007; Banigan et al. 2013). As in liquid-state NMR, a crucial step for obtaining side resolved information is signal assignment. This can routinely be done in solid-sate NMR for small systems with narrow resonance lines as observed for crystalline proteins (Castellani et al. 2002; Franks et al. 2008; Zhang et al. 2010). In favourable cases extended assignments have also been obtained for large membrane proteins in the lipid bilayer. Examples include a proton pump (Shi et al. 2009) and a toxin bound to an ion channel (Lange et al. 2006). However, up to now, no transport protein could be assigned by solid-state NMR. Their spectral resolution tends to be not as good as obtained for other types of membrane proteins, which is probably caused by unfavourable motions causing line-broadening effects. Due to these difficulties, all MAS-NMR studies reported on EmrE up to date have in common that they employ advanced MAS-NMR experiments often in combination with sophisticated and selective labelling schemes to reduce spectral complexity.

In a study reported by Reif and co-workers, uniformly ^{13}C, ^{15}N-labelled EmrE was used (Agarwal et al. 2007). Despite the strong signal overlap in their ^{13}C–^{13}C correlation spectra, more than the five expected proline signals could be observed, which would agree with the asymmetry seen in other experiments as discussed above. Furthermore, the carbon Cδ resonance of the essential residue E14 could be identified using spectral editing via a REDOR filter. Its chemical shift indicates a protonated apo-state EmrE under the conditions used in line with the current model (Fig. 11.1b). Upon addition of TPP^+, the resonance could no longer be detected,

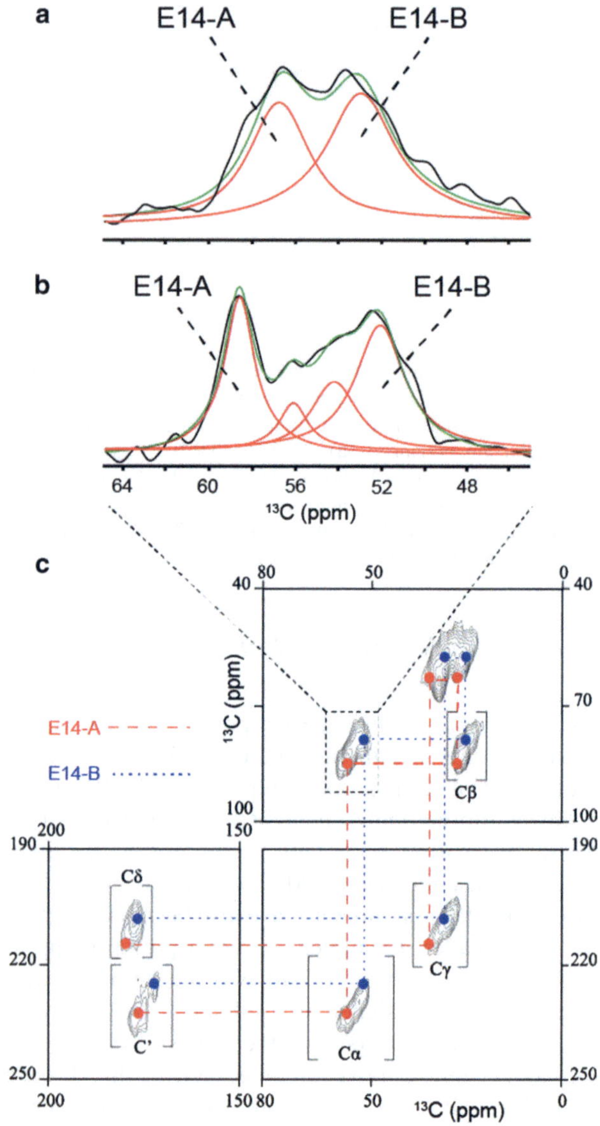

Fig. 11.6 ^{13}C MAS-NMR spectra of EmrE reconstituted into *E. coli* lipids. Residue E14 was selectively ^{13}C labelled. In the apo-state, the Cα resonance splits into two equal populations (**a**), which both shift upon saturation with ethidium (**b**). Two-dimensional double-quantum-single quantum spectra reveal peak doublings for all carbons in E14, which agrees with the structural asymmetry of the EmrE dimer as discussed above. Data were taken from Lehner et al. (2008). For this study, EmrE was produced using cell free expression, and the mutation E25A was introduced to avoid signal overlap from the second glutamate

probably due to a chemical shift change and/or line broadening. The latter was observed for many other resonances, which could be caused by intermediate exchange between apo-state and substrate-bound EmrE or increased heterogeneity upon substrate binding.

The residue E14 was also within the focus of a study in which EmrE was produced using cell free synthesis (Lehner et al. 2008). Lehner et al. overcame the assignment problem using mutagenesis and residue selective labelling. The primary sequence of EmrE contains the essential glutamic acid E14 and the non-essential E25. Therefore, the mutant E25A was used. Labelling of E14 was possible, since isotope scrambling, which is severe for glutamic acid in vivo, was suppressed in vitro making an assignment straightforward. Carbon spectra of apo-state EmrE and of EmrE saturated with ethidium were recorded using a double quantum filtering technique, which suppresses the natural abundance ^{13}C background (Fig. 11.6). The observed signal pattern agrees well with the signals expected for a glutamic acid residue, but in the Cα region for both samples a peak splitting is observed. Using double-quantum single-quantum correlation spectroscopy, two complete glutamic acid spin systems could be identified for each sample. This supports the view that also the binding site of EmrE is asymmetric. Upon binding of ethidium, both E14 signal sets shift indicating that each of them is in contact with the ligand. The observed Cδ chemical shift of E14 upon substrate binding is in agreement with a deprotonated side chain. The apo-state chemical shift differs from the one reported before (Agarwal et al. 2007), which could be due to different pH, temperature, lipid and buffer conditions resulting in E14 protonation state differences. However, both studies are strongly consistent with an asymmetric dimer.

In a study by Traaseth and co-workers, further efforts were made to obtain more assignments of EmrE based on a combination of extensive-selective labelling and novel NMR experiments (Banigan et al. 2013). EmrE was uniformly ^{13}C and ^{15}N labelled, while labelling of all isoleucines was suppressed and leucines were only ^{15}N labelled. Special filtering techniques were employed on these samples, which resulted in highly selective spectra. In particular, a so-called after-glow NCOCX spectrum and a frequency-selective-dipolar-recoupling-NCACX spectrum were recorded. These spectra showed promising resolution and much reduced spectral crowding and led to a few assignments. However, to obtain more complete assignments shedding light onto the extent of the asymmetry of the EmrE dimer, more selectively labelled samples are needed.

In summary, using MAS-NMR, structure and dynamic data on EmrE in a close-to-native environment could be obtained. Especially, the asymmetry in the binding site of the dimer is revealed. The studies indicate that a brute force approach combining selective isotope labelling with state-of-the-art MAS-NMR might lead to an assignment of the majority of the residues of the protein, which will give more insight into the structure, asymmetry and function of EmrE within the lipid bilayer.

11.4.3 Dynamic Nuclear Polarisation Enhanced MAS-NMR on EmrE

As highlighted above, both liquid- and solid-state NMR offer unique possibility to obtain dynamic data, which are highly complementary to the structural insight provided by crystallography. But as with most quantitative biophysical methods, NMR sensitivity is an important concern, as the observable nuclear magnetisation is generally small. This is especially true for membrane protein applications in which limited sample availability and low protein concentration or low protein-to-lipid ratios add additional constraints to the detectable signal intensity. The most fundamental solution relies on the fact that the electron magnetic moment is about 650-fold stronger than that of protons and 2600-fold stronger than for carbons. Doping the protein sample of interest with stable radicals acting as polarising agents, a magnetisation transfer can be achieved under special experimental conditions leading to orders of magnitude higher nuclear polarisation, which results in a much higher NMR signal intensity. The different mechanisms leading to this sensitivity enhancement are summarised under the term "dynamic nuclear polarisation" (DNP) [for a comprehensive technical review see, e.g. (Maly et al. 2008)]. So far, signal enhancement in proteins by DNP works best for solid-state NMR at low temperatures (<120 K). Especially, the so-called cross effect has been shown to work most efficiently, for which bi-radicals with strongly dipolar coupled paramagnetic centres such as TOTAPOL or AMUPol are used (Sauvee et al. 2013; Song et al. 2006). Saturating the electron transitions with a strong and continuous microwave radiation alters the energy level populations of adjacent proton spins coupled to these electrons resulting in an enhanced non-Boltzmann nuclear magnetisation. In the solid state, all protons are strongly coupled to each other, which causes an efficient distribution of magnetisation throughout the whole sample via spin diffusion. Therefore, the magnetisation of all protons, and not only of those in close proximity to the polarising agents, is enhanced. Transfer to ^{13}C or ^{15}N is done via cross polarisation. The basic experimental scheme is shown in Fig. 11.7a: A high power continuous microwave source, usually a gyrotron, is coupled to an NMR spectrometer. The microwave frequency matches the electron Larmor frequency in the NMR magnet (e.g. 263 GHz vs. 400 MHz at 9.7 T). For MAS NMR, microwaves are guided via a waveguide directly to the sample rotating at a temperate of usually 80–110 K.

EmrE is the only secondary transporter analyzed so far by DNP (Ong et al. 2013). The aim of this study was to provide direct evidence for close proximity or interaction between substrate and the side chain of highly conserved residue E14. As outlined above, E14 is essential for both proton and substrate binding (Muth and Schuldiner 2000; Rotem et al. 2001; Rotem and Schuldiner 2004; Steiner-Mordoch et al. 2008; Yerushalmi et al. 2001), but close proximity to bound substrate has not been shown yet. The polarising agent TOTAPOL was therefore added to proteoliposome samples containing ^{13}C-labelled EmrE in complex with ^{13}C-labelled substrate TPP^+ (Fig. 11.7b). The application of DNP caused a 20-fold signal enhancement resulting in a 400-fold faster data acquisition (Fig. 11.7c). In order to visualise potential contacts between substrate and E14, suitable labelling strategies have to be considered. In this study, the sparse isotope-labelling pattern produced when using 2-^{13}C-glycerol as a carbon source resulting in 100 % isotope enrichment of the carboxylic carbon Cδ of glutamate (Castellani et al. 2002) was utilised. The spectra were further simplified by suppressing unwanted signals from aromatic residues by supplementing the minimal growth medium with unlabelled amino acids.

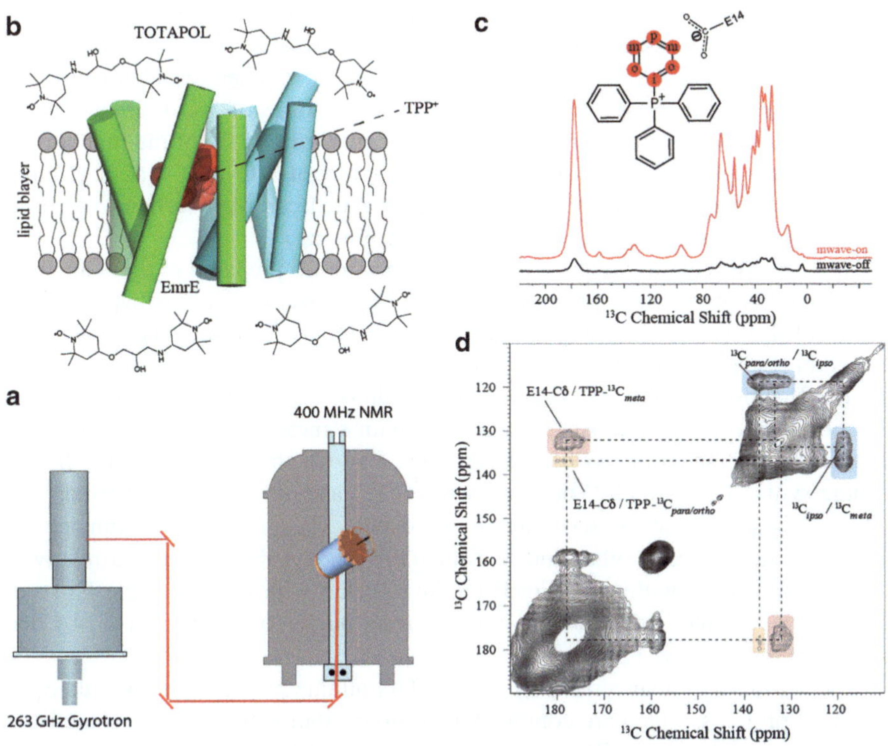

Fig. 11.7 DNP enables a significant sensitivity enhancement of solid-state NMR experiments. Schematic set-up for DNP-enhanced MAS-NMR: A high power continuous wave microwave source (gyrotron) is coupled via a waveguide to the sample in the NMR spectrometer [for a review see Maly et al. (2008)] (**a**). Samples have to be doped with polarising agents such as the biradical TOTAPOL and are usually frozen in a glycerol/water matrix (Song et al. 2006) (**b**). For EmrE reconstituted into DMPC lipid bilayers, a 20-fold signal enhancement was achieved (20 mM TOTAPOL, 100 K) (**c**). Under DNP conditions, a ^{13}C–^{13}C DARR spectrum (300 ms mixing time) reveals specific through-space correlations between ^{13}C-TPP$^+$ and the side chain of E14 in EmrE (Ong et al. 2013)

Contacts between the substrate TPP$^+$ to ^{13}C nuclei in EmrE could be either measured directly using its phosphorous nuclei or by introducing ^{13}C labels in its phenyl rings via chemical synthesis. Since the phosphorus in TPP$^+$ is sterically shielded by the four phenyl rings, the distance to a ^{13}C nuclei in an adjacent amino acid will be relatively long (>8 Å) resulting in weak ^{31}P–^{13}C dipole couplings, which could be hard to observe. In contrast, a closer proximity and stronger ^{13}C–^{13}C dipole couplings are expected for ^{13}C-labelled phenyl rings in TPP$^+$. One drawback often encountered with phenyl rings is that they fluctuate on timescales on which NMR has a blind spot making their detection difficult. However, these motions can be reduced at lower temperatures, which are required for DNP anyway. Under these conditions, a ^{13}C–^{13}C DARR spectrum was recorded, an experiment monitoring ^{13}C–^{13}C through-space correlations. The spectrum in

Fig. 11.7d shows intramolecular cross peaks within ^{13}C-TPP$^+$ as well as intermolecular cross peaks between ^{13}C-TPP$^+$ and E14-Cδ. A number of controls using complementary labelling schemes excluded that other residues cause this contact. One can therefore conclude that the phenyl rings of ^{13}C-TPP$^+$ within the EmrE binding pocket are found in close spatial proximity to Cδ of the functionally essential residue E14. The main cross peak between ^{13}C-TPP$^+$ and ^{13}C-EmrE reveals a chemical shift expected for a deprotonated carboxyl carbon and shows that E14 becomes deprotonated upon substrate binding.

TPP$^+$ was titrated to reconstituted apo-state samples prepared at pH 8 at which E14 should be protonated due to its unusually high pKa of 7–8.5 (Chap. 10). The chemical shift is the same range as observed previously for EmrE in complex with ethidium (Lehner et al. 2008). The heterogeneous line shape of the cross peaks in the EmrE-substrate complex observed here could be caused by the low temperature conditions but are also consistent with a certain structural asymmetry (Morrison et al. 2012; Lehner et al. 2008). Interestingly, E14-Cδ shows a strong and distinct contact with $^{13}C_{meta}$ of TPP$^+$ but none with $^{13}C_{ipso}$ and only weak correlations with $^{13}C_{para}/^{13}C_{ortho}$. This could mean that E14-Cδ and $^{13}C_{meta}$ are found within approximately 6 Å from each other and that the TPP$^+$ phenyl ring assumes a relatively defined orientation within the EmrE binding pocket with respect to the E14 side chain. The distance between the phosphorous nuclei of TPP$^+$ and E14-Cδ is estimated to be below 12 Å.

The sensitivity enhancement by DNP offers unique possibilities for studies on transport proteins, which are complementary to crystallography and to established liquid-state/solid-state NMR methods on non-frozen samples. Especially, the use of low temperatures to freeze out unfavourable motions or to freeze-trap functionally relevant states could provide important mechanistic insight.

11.4.4 Perspectives for Probing Slow Motions Within Membrane Proteins in Liposomes?

The data on EmrE in isotropic bicelles presented and discussed in Sect. 11.2 show already that relatively slow domain movements associated with an alternating access model occur. One can assume that domain movements with correlation times in the ms–s regime are a common feature of many membrane proteins. However, such motions are often difficult to observe and even more difficult to quantify. In general, NMR exchange spectroscopy makes these timescales accessible, but a number of boundary conditions complicate its use for domain movements in membrane proteins: Exchange spectroscopy is best applied if chemical shift changes are associated with the two or more exchanging states. In such a case, exchange rates can be extracted from the built-ups of cross peaks occurring between the exchanging states. However, the exchange rate has to be smaller than the frequency difference between both resonances. Furthermore, selective isotope labelling is required in order to avoid cross peaks due to spin diffusion processes as observed in uniformly labelled samples. Another problem is that exchange processes, e.g. between an inward and outward facing conformation, as in the case of EmrE, only involve small changes on the secondary structure level. This means that spectra of these states will be almost identical and exchange cross peaks are more difficult to detect. In addition, a series of well-resolved 2D spectra is necessary to obtain the desired exchange rates. Thus, it is often very challenging using MAS-NMR.

For reorientation processes without changes in the isotropic chemical shift, CODEX (Centerband Only Detection of EXchange) spectroscopy offers promising opportunities (deAzevedo et al. 2000). In principle, this technique measures exchange processes between reorienting CSA tensors directly from peak intensities in 1D spectra. CSA tensors are fixed within the molecular frame and change their orientation with a moving transmembrane helix during opening and closing processes (Fig. 11.8a). This results in intensity changes in the CODEX experiment from which motional correlation times can be derived and by which motion models can be tested. Proof of concept data for EmrE is shown in Fig. 11.8b. The experiment was performed on ^{15}N-labelled EmrE within DMPC lipid bilayers. The signal intensity was averaged over all amide ^{15}N signals and plotted as a function of the mixing time. The exchange curve is best described by a bi-exponential function with time constants of 15 ms and 1 s, which correspond to correlation times of different motional processes. The longer correlation time is difficult to interpret because it is dominated by spin diffusion. The faster process corresponds most likely to slow domain movements and is on the same order as reported for EmrE in isotropic bicelles (Sect. 11.2). Integrating these experiments with multidimensional spectra will allow obtaining a detailed and site-resolved view of domain motions within membrane proteins such as EmrE.

11.5 Conclusions: What Can NMR Provide for a Mechanistic Understanding of Transporters

The most advanced liquid-state and solid-state NMR methods have been extensively used to gain a better view at the structure–function relationship of EmrE. Although based on quite different approaches, the data converge into a surprisingly consistent picture.

Data from liquid-state NMR on EmrE in isotropic bicelles, O-SSNMR on EmrE in anisotropic bicelles and MAS-NMR on EmrE in liposomes clearly show a high level of asymmetry in an antiparallel dimer. An antiparallel topology has been confirmed by cryo-EM, X-ray, SM-FRET and cross-linking studies. MAS-NMR data also show that this asymmetry extends into the EmrE binding pocket, which can accommodate many different substrates. The essential residue E14 in TMH1 gets deprotonated upon substrate binding as expected from biochemical assays, but

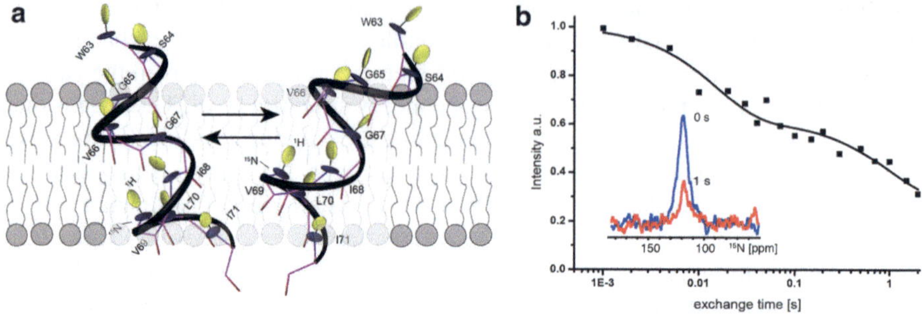

Fig. 11.8 *Probing slow motions within membrane proteins by MAS-NMR within the lipid bilayer.* Transport by EmrE based on an alternating access model requires helix reorientations. Such motions could be measured by exchange spectra in which, e.g. the reorientations of ^{15}N chemical shift anisotropy tensors are observed (**a**). The *blue* ellipsoids in (**a**) correspond to ^{15}N CSA tensors in TMH3 of EmrE. Exchange rates can be measured by CODEX experiments (deAzevedo et al. 2000). A ^{15}N-CODEX exchange curve obtained from ^{15}N-EmrE in DMPC liposomes in complex with ethidium reveals a double exponential behaviour. The fast decay has a time constant of 15 ms. The slow decay was here fitted with a time constant of 1 s but also contains contributions from spin diffusion and has to be interpreted with great care (**b**). Data were acquired at a Bruker 850 WB spectrometer at 270 K and 10 kHz sample rotation rate (Ong, Becker-Baldus, personal communication). CODEX spectra with 1 ms and 2 s mixing time are shown in the *inset*

only the use of DNP-enhanced MAS-NMR enabled a direct proof for close proximity between bound substrate and E14 within the EmrE dimer. Understanding transport mechanisms also requires access to conformational dynamics of these systems. Different types of exchange spectroscopy based on liquid- and solid-state in isotropic bicelles and liposomes, respectively, show that domain movement in the upper ms timescale take place. Such movements are required to make the substrate-binding site alternatingly accessible from both sides.

EmrE is structurally asymmetric but functionally symmetric as both in and out conformations are identical. The asymmetry adds additional complexity for structure determination but at the same time, functional studies are simplified as directional reconstitution becomes unnecessary. EmrE has been termed a “living fossil at an evolutionary junction” in Chap. 10 and understanding its structure, mechanism and conformational dynamics offers therefore a fascinating view at the development and general mechanism of ion-coupled transporters.

The case of EmrE presented here demonstrates that transport proteins are still challenging targets for NMR spectroscopy, but the use of advanced methodology with carefully selected sample preparation conditions shows that unique data can be obtained. Especially, the possibility to analyse dynamic processes such as slow domain movements, the direct use of lipids instead of detergents and the high sensitivity of DNP will establish NMR spectroscopy in the field of transport proteins as a highly complementary method to crystallography-based structural biology.

Acknowledgements We are grateful to Andrea Lakatos and Yean-Sin Ong for helpful discussions and comments to the manuscript.

References

Adam Y, Tayer N, Rotem D, Schreiber G, Schuldiner S (2007) The fast release of sticky protons: kinetics of substrate binding and proton release in a multidrug transporter. Proc Natl Acad Sci U S A 104:17989–17994

Agarwal V, Fink U, Schuldiner S, Reif B (2007) MAS solid-state NMR studies on the multidrug transporter EmrE. Biochim Biophys Acta Biomembr 1768:3036–3043

Amadi ST, Koteiche HA, Mishra S, Mchaourab HS (2010) Structure, dynamics, and substrate-induced conformational changes of the multidrug transporter EmrE in liposomes. J Biol Chem 285:26710–26718

Bak M, Rasmussen JT, Nielsen NC (2000) SIMPSON: a general simulation program for solid-state NMR spectroscopy. J Magn Reson 147:296–330

Bak M, Schultz R, Vosegaard T, Nielsen NC (2002) Specification and visualization of anisotropic interaction tensors in polypeptides and numerical simulations in biological solid-state NMR. J Magn Reson 154:28–45

Banigan JR, Gayen A, Traaseth NJ (2013) Combination of N-15 reverse labeling and afterglow spectroscopy for assigning membrane protein spectra by magic-angle-spinning solid-state NMR: application to the multidrug resistance protein EmrE. J Biomol NMR 55:391–399

Basting D, Lorch M, Lehner I, Glaubitz C (2008) Transport cycle intermediate in small multidrug resistance protein is revealed by substrate fluorescence. FASEB J 22:365–373

Bayrhuber M, Meins T, Habeck M, Becker S, Giller K, Villinger S, Vonrhein C, Griesinger C, Zweckstetter M, Zeth K (2008) Structure of the human voltage-dependent anion channel. Proc Natl Acad Sci U S A 105:15370–15375

Castellani F, van Rossum B, Diehl A, Schubert M, Rehbein K, Oschkinat H (2002) Structure of a protein determined by solid-state magic-angle-spinning NMR spectroscopy. Nature 420: 98–102

Chen YJ, Pornillos O, Lieu S, Ma C, Chen AP, Chang G (2007) X-ray structure of EmrE supports dual topology model. Proc Natl Acad Sci U S A 104:18999–19004

De Angelis AA, Opella SJ (2007) Bicelle samples for solid-state NMR of membrane proteins. Nat Protoc 2:2332–2338

deAzevedo ER, Hu WG, Bonagamba TJ, Schmidt-Rohr K (2000) Principles of centerband-only detection of exchange in solid-state nuclear magnetic resonance, and extension to four-time centerband-only detection of exchange. J Chem Phys 112:8988–9001

Fleishman SJ, Harrington SE, Enosh A, Halperin D, Tate CG, Ben-Tal N (2006) Quasi-symmetry in the cryo-EM structure of EmrE provides the key to modeling its transmembrane domain. J Mol Biol 364:54–67

Franks WT, Wylie BJ, Schmidt HLF, Nieuwkoop AJ, Mayrhofer R-M, Shah GJ, Graesser DT, Rienstra CM (2008) Dipole tensor-based atomic-resolution structure determination of a nanocrystalline protein by solid-state NMR. Proc Natl Acad Sci U S A 105:4621–4626

Gautier A, Mott HR, Bostock MJ, Kirkpatrick JP, Nietlispach D (2010) Structure determination of the seven-helix transmembrane receptor sensory rhodopsin II by solution NMR spectroscopy. Nat Struct Mol Biol 17:768–U147

Gayen A, Banigan JR, Traaseth NJ (2013) Ligand-induced conformational changes of the multidrug resistance transporter EmrE probed by oriented solid-state NMR spectroscopy. Angew Chem Int Ed Engl 52(39):10321–10324

Glover KJ, Whiles JA, Wu GH, Yu NJ, Deems R, Struppe JO, Stark RE, Komives EA, Vold RR (2001) Structural evaluation of phospholipid bicelles for solution-state studies of membrane-associated biomolecules. Biophys J 81:2163–2171

Hellmich UA, Glaubitz C (2009) NMR and EPR studies of membrane transporters. Biol Chem 390:815–834

Hong M, Zhang Y, Hu F (2012) Membrane protein structure and dynamics from NMR spectroscopy. In: Johnson MA, Martinez TJ (eds) Annual review of physical chemistry, vol 63. Annual Reviews, Palo Alto, CA, pp 1–24

Lakatos A, Mors K, Glaubitz C (2012) How to investigate interactions between membrane proteins and ligands by solid-state NMR. Meth Mol Biol 914:65–86

Lange A, Giller K, Hornig S, Martin-Eauclaire MF, Pongs O, Becker S, Baldus M (2006) Toxin-induced conformational changes in a potassium channel revealed by solid-state NMR. Nature 440:959–962

Lehner I, Basting D, Meyer B, Haase W, Manolikas T, Kaiser C, Karas M, Glaubitz C (2008) The key residue for substrate transport (Glu(14)) in the EmrE dimer is asymmetric. J Biol Chem 283:3281–3288

Li Y, Palmer AG III (2009) TROSY-selected ZZ-exchange experiment for characterizing slow chemical exchange in large proteins. J Biomol NMR 45:357–360

Lloris-Garcera P, Bianchi F, Slusky JS, Seppala S, Daley DO, von Heijne G (2012) Antiparallel dimers of the small multidrug resistance protein EmrE are more stable than parallel dimers. J Biol Chem 287:26052–26059

Lorch M, Lehner I, Siarheyeva A, Basting D, Pfleger N, Manolikas T, Glaubitz C (2005) NMR and fluorescence spectroscopy approaches to secondary and primary active multidrug efflux pumps. Biochem Soc Trans 33:873–877

Maly T, Debelouchina GT, Bajaj VS, Hu KN, Joo CG, Mak-Jurkauskas ML, Sirigiri JR, van der Wel PCA, Herzfeld J, Temkin RJ, Griffin RG (2008) Dynamic nuclear polarization at high magnetic fields. J Chem Phys 128:052211–052219

Marassi FM, Opella SJ (2000) A solid-state NMR index of helical membrane protein structure and topology. J Magn Reson 144:150–155

Miller D, Charalambous K, Rotem D, Schuldiner S, Curnow P, Booth PJ (2009) In vitro unfolding and refolding of the small multidrug transporter EmrE. J Mol Biol 393:815–832

Morrison EA, Henzler-Wildman KA (2012) Reconstitution of integral membrane proteins into isotropic bicelles with improved sample stability and expanded lipid composition profile. Biochim Biophys Acta Biomembr 1818:814–820

Morrison EA, DeKoster GT, Dutta S, Vafabakhsh R, Clarkson MW, Bahl A, Kern D, Ha T, Henzler-Wildman KA (2012) Antiparallel EmrE exports drugs by exchanging between asymmetric structures. Nature 481:45–U50

Mors K, Hellmich UA, Basting D, Marchand P, Wurm JP, Haase W, Glaubitz C (2013) A lipid-dependent link between activity and oligomerization state of the M. tuberculosis SMR protein TBsmr. Biochim Biophys Acta 1828:561–567

Muth TR, Schuldiner S (2000) A membrane-embedded glutamate is required for ligand binding to the multidrug transporter EmrE. EMBO J 19:234–240

Ong YS, Lakatos A, Becker-Baldus J, Pos KM, Glaubitz C (2013) Detecting substrates bound to the secondary multidrug efflux pump EmrE by DNP-enhanced solid-state NMR. J Am Chem Soc 135:15754–15762

Park SH, Das BB, Casagrande F, Tian Y, Nothnagel HJ, Chu M, Kiefer H, Maier K, De Angelis AA, Marassi FM, Opella SJ (2012) Structure of the chemokine receptor CXCR1 in phospholipid bilayers. Nature 491:779–783

Poget SF, Cahill SM, Girvin ME (2007) Isotropic bicelles stabilize the functional form of a small multidrug-resistance pump for NMR structural studies. J Am Chem Soc 129:2432–2433

Prosser RS, Hunt SA, DiNatale JA, Vold RR (1996) Magnetically aligned membrane model systems with positive order parameter: Switching the sign of S-zz with paramagnetic ions. J Am Chem Soc 118:269–270

Rapp M, Granseth E, Seppala S, von Heijne G (2006) Identification and evolution of dual-topology membrane proteins. Nat Struct Mol Biol 13:112–116

Raschle T, Hiller S, Etzkorn M, Wagner G (2010) Nonmicellar systems for solution NMR spectroscopy of membrane proteins. Curr Opin Struct Biol 20:471–479

Reckel S, Gottstein D, Stehle J, Lohr F, Verhoefen MK, Takeda M, Silvers R, Kainosho M, Glaubitz C, Wachtveitl J, Bernhard F, Schwalbe H, Guntert P, Dotsch V (2011) Solution NMR structure of proteorhodopsin. Angew Chem Int Ed Engl 50:11942–11946

Renault M, Cukkemane A, Baldus M (2010) Solid-state NMR spectroscopy on complex biomolecules. Angew Chem Int Ed 49:8346–8357

Rotem D, Schuldiner S (2004) EmrE, a multidrug transporter from Escherichia coli, transports monovalent and divalent substrates with the same stoichiometry. J Biol Chem 279: 48787–48793

Rotem D, Sal-man N, Schuldiner S (2001) In vitro monomer swapping in EmrE, a multidrug transporter from Escherichia coli, reveals that the oligomer is the functional unit. J Biol Chem 276:48243–48249

Sauvee C, Rosay M, Casano G, Aussenac F, Weber RT, Ouari O, Tordo P (2013) Highly efficient, water-soluble polarizing agents for dynamic nuclear polarization at high frequency. Angew Chem Int Ed Engl 125(41):11058–11061

Schuldiner S (2007) When biochemistry meets structural biology: the cautionary tale of EmrE (vol 32, pg 252, 2007). Trends Biochem Sci 32:300

Schuldiner S, Lebendiker M, Yerushalmi H (1997) EmrE, the smallest ion-coupled transporter, provides a unique paradigm for structure-function studies. J Exp Biol 200:335–341

Schwaiger M, Lebendiker M, Yerushalmi H, Coles M, Groger A, Schwarz C, Schuldiner S, Kessler H (1998) NMR investigation of the multidrug transporter EmrE, an integral membrane protein. Eur J Biochem 254:610–619

Shi L, Ahmed MA, Zhang W, Whited G, Brown LS, Ladizhansky V (2009) Three-dimensional solid-state NMR study of a seven-helical integral membrane proton pump–structural insights. J Mol Biol 386:1078–1093

Song CS, Hu KN, Joo CG, Swager TM, Griffin RG (2006) TOTAPOL: A biradical polarizing agent for dynamic nuclear polarization experiments in aqueous media. J Am Chem Soc 128: 11385–11390

Spooner PJR, Rutherford NG, Watts A, Henderson PJF (1994) NMR observation of substrate in the binding-site of an active sugar-h + symport protein in native membranes. Proc Natl Acad Sci U S A 91:3877–3881

Steiner-Mordoch S, Soskine M, Solomon D, Rotem D, Gold A, Yechieli M, Adam Y, Schuldiner S (2008) Parallel topology of genetically fused EmrE homodimers. EMBO J 27:17–26

Tate CG (2010) Practical considerations of membrane protein instability during purification and crystallisation. Meth Mol Biol 601:187–203

Tjandra N, Bax A (1997) Direct measurement of distances and angles in biomolecules by NMR in a dilute liquid crystalline medium. Science 278:1111–1114

Ubarretxena-Belandia I, Baldwin JM, Schuldiner S, Tate CG (2003) Three-dimensional structure of the bacterial multidrug transporter EmrE shows it is an asymmetric homodimer. EMBO J 22:6175–6181

Vold RR, Prosser RS, Deese AJ (1997) Isotropic solutions of phospholipid bicelles: A new membrane mimetic for high-resolution NMR studies of polypeptides. J Biomol NMR 9: 329–335

Wang J, Denny J, Tian C, Kim S, Mo Y, Kovacs F, Song Z, Nishimura K, Gan Z, Fu R, Quine JR, Cross TA (2000) Imaging membrane protein helical wheels. J Magn Reson 144:162–167

Wang S, Munro RA, Shi L, Kawamura I, Okitsu T, Wada A, Kim SY, Jung KH, Brown LS, Ladizhansky V (2013) Solid-state NMR spectroscopy structure determination of a lipid-embedded heptahelical membrane protein. Nat Methods 10(10):1007–1012

Wu CH, Ramamoorthy A, Opella SJ (1994) High-resolution heteronuclear dipolar solid-state NMR-spectroscopy. J Magn Reson A 109:270–272

Yerushalmi H, Lebendiker M, Schuldiner S (1996) Negative dominance studies demonstrate the oligomeric structure of EmrE, a multidrug antiporter from Escherichia coli. J Biol Chem 271:31044–31048

Yerushalmi H, Mordoch SS, Schuldiner S (2001) A single carboxyl mutant of the multidrug transporter EmrE is fully functional. J Biol Chem 276:12744–12748

Zhang Y, Doherty T, Li J, Lu W, Barinka C, Lubkowski J, Hong M (2010) Resonance assignment and three-dimensional structure determination of a human alpha-Defensin, HNP-1, by solid-state NMR. J Mol Biol 397:408–422

MIX
Papier aus verantwortungsvollen Quellen
Paper from responsible sources
FSC® C105338

If you have any concerns about our products,
you can contact us on
ProductSafety@springernature.com

In case Publisher is established outside the EU,
the EU authorized representative is:
Springer Nature Customer Service Center GmbH
Europaplatz 3, 69115 Heidelberg, Germany

Printed by Libri Plureos GmbH
in Hamburg, Germany